The Worlds of Oronce Fine
Mathematics, Instruments and Print
in Renaissance France

THE WORLDS OF ORONCE FINE MATHEMATICS, INSTRUMENTS AND PRINT IN RENAISSANCE FRANCE

EDITED BY
ALEXANDER MARR

SHAUN TYAS
DONINGTON
2009

Published in 2009 by
SHAUN TYAS
(an imprint of 'Paul Watkins')
c/o 1 High Street
Donington
Lincolnshire
PE11 4TA

ISBN
978-1-900289-96-2

Typeset and designed from
the discs of the editor
by Shaun Tyas

Printed in Great Britain by the MPG Books Group,
Bodmin and King's Lynn

CONTENTS

NOTES ON CONTRIBUTORS

Angela Axworthy is a doctoral student in Philosophy at the CESR, Tours. She specializes in the history of Early Modern Italian and French philosophy of mathematics, having worked between 2002 and 2005 on the conception of mathematics of Niccolò Tartaglia, Christophorus Clavius and Federico Commandino. She is currently working on Oronce Fine's definition of the status of mathematics.

Jean-Marc Besse is Director of Research at the CNRS, Paris. He has published widely on the history of geography and is the author of *Les grandeurs de la Terre: Aspects du savoir geographique a la Renaissance* (2003).

Jean-Jacques Brioist has a PhD in Materials Science. He is the author of 'Entre astrologie et cartographie, la genèse des logarithmes' (*Journal de la Renaissance*, 2004) and 'L'ingénierie cartésienne de Renau d'Eliçagaray' (*Documents pour l'Histoire des Techniques*, forthcoming).

Pascal Brioist is a member of the CESR, Tours. An expert on mathematics and military culture in the Renaissance, he is the author (with Hervé Drévillon and Pierre Serna) of *Croiser le fer. Violence et culture de l'epee dans la France moderne (XVIe–XVIIIe siecle)* (2002).

Giovanna Cifoletti is a member of the EHESS, Paris. She has published on mathematics (especially algebra) in the sixteenth and seventeenth centuries, and is currently preparing critical editions of works by Guillaume Gosselin and François Viète.

Stephen Clucas is Reader in Intellectual History at Birkbeck. His most recent publication on Early Modern learned culture is *John Dee: Interdisciplinary Studies in Renaissance Thought* (2006).

Sven Dupré is a research fellow of the Flemish Research Foundation in Belgium and a founding member of the Centre for History of Science at Ghent University. He has published widely on the history of optics and the material culture of Early Modern science.

Catherine Eagleton is a Curator at the British Museum, London, and an Affiliated Research Scholar at the Department of History and Philosophy of Science, Cambridge. She is particularly interested in the relationship between books, instruments, and diagrams, and in the history of collecting.

Henrique Leitão is a member of the Centro de História das Ciências, University of Lisbon. He has published extensively on Portuguese science in the Renaissance and has edited several works by Pedro Nunes.

Alexander Marr is a Lecturer in Art History at the University of St Andrews, specializing in Early Modern art and science. He is the editor (with R. J. W. Evans) of *Curiosity and Wonder from the Renaissance to the Enlightenment* (2006) and the author of *Between Raphael and Galileo: Mutio Oddi and the Mathematical Culture of Late Renaissance Italy* (forthcoming from University of Chicago Press).

Adam Mosley is a Senior Lecturer in History at Swansea University, and the author of *Bearing the Heavens: Tycho Brahe and the Astronomical Community of the Late-Sixteenth Century* (2007). Cosmography is the current focus of his ongoing interests in the mathematical sciences, scientific instruments, and the history of the book and of reading.

Isabelle Pantin is Professor of Renaissance Literature at the Ecole Normale Supérieure, and she participates in the research program of the Observatoire de Paris (CNRS, SYRTE), in the section dedicated to the history of astronomy and related fields.

Anthony Turner is an independent historian, antiquarian bookseller and consultant. He is currently working on a bio-bibliographical dictionary of French scientific instrument-makers and on a study of practical mathematics and natural philosophy in the provincial culture of Early Modern France and Britain.

LIST OF ILLUSTRATIONS

5. Printing Practical Mathematics: Oronce Fine's *De speculo ustorio* between Paper and Craft

5.1: G. Tanstetter and P. Apian (eds.), *Vitellonis Mathematici Doctissimi Peri Optikes, id est de Natura, Ratione et Projectione Radiorum Visus, Luminum, Colorum Atque Formarum Quam Vulgo Perspectivam Vocant, Libri X...* (Nuremberg, 1535), frontispiece. The British Library Board. All rights reserved. Shelfmark 536.I.22.

5.2: F. Risner (ed.), *Opticae Thesaurus Alhazeni Arabis Libri Septem, Nuncprimùm Editi. Eiusdem Liber de Crepusculis et Nubium Ascensionibus. Item Vitellionis Thuringopoloni Libri X* (Basel, 1572), title-page. Photographic Archive: Institute and Museum of the History of Science, Florence.

5.3: Detail from: *Cl. Ptolemaei Pelusiensis Mathematici operis quadriparti ... De sectione conica, orthogona, quae parabola dicitur: De Speculo ustorio, libelli duo, hactenus desiderati: restituti ab Antonio Gogava Graviensi* (Louvain, 1548). The British Library Board. All rights reserved. Shelfmark 531.g.2(4).

5.4: O. Fine, *De speculo ustorio...* (Paris, 1551), p. 17^{v}. The British Library Board. All rights reserved. Shelfmark 529.g.7(1).

5.5: O. Fine, *De speculo ustorio...* (Paris, 1551), p. 18^{v}. The British Library Board. All rights reserved. Shelfmark 529.g.7(1).

5.6: Detail from G. B. Della Porta, *Magiae naturalis libri viginti* (Frankfurt, 1591), Ghent University Library, Ma 1232a., p. 602.

6. Oronce Fine's Sundials: the Sources and Influences of *De solaribus horologiis*

6.1a and b: Ship-shaped dial in the Museo Poldi Pezzoli, Milan (inv. no. 4277). Reproduced by kind permission of the Museo Poldi Pezzoli.

6.2a and b: Fine's printed version of the ship-shaped dial, from O. Fine, *De solaribus horologiis...* (Paris, 1560), pp. 184 and 187. Reproduced by kind permission of the Whipple Library, University of Cambridge.

6.3: Diagram of the *organum ptolomei*, from Österreichische Nationalbibliothek, Vienna, MS Lat 5418, fol. 182^{r}.

6.4: Fine's printed version of a universal rectilinear dial, now often known as the Regiomontanus dial. From O. Fine, *De solaribus horologiis...* (Paris, 1560), p. 179. Reproduced by kind permission of the Whipple Library, University of Cambridge.

6.5a and b: The Whipple ship-shaped dial (inv. no. 731). Reproduced by kind permission of the Whipple Museum of the History of Science.

6.6a, b and c: Diagrams showing (a) the construction of the ship-shaped dial, and (b and c) showing the two parts from which the dial can be made, by fixing them together at the centre and adding a jointed arm. From J. Bullant, *Petit*

7. Cosmography and Geography in the Sixteenth Century, between Mathematics and History: the Position of Oronce Fine

9. Oronce Fine and Cartographical Methods

10. Pedro Nunes against Oronce Fine: Content and Context of a Refutation

10.1: P. Nunes, *De erratis Orontii Finaei* (Coimbra, 1546), title-page. By permission of the University of Lisbon.10.2: P. Nunes, *De erratis Orontii Finaei* (Coimbra, 1571), title-page. By permission of the University of Lisbon.

11. Oronce Fine's Legacy in the French Algebraic Tradition: Peletier, Ramus and Gosselin

11.1: O. Fine, *Arithmetica practica...* (Paris, 1544), title-page. Reproduced courtesy of St Andrews University Library.

12. Dropped Out of Sight: Oronce Fine and the Water-clock in the Sixteenth and Seventeenth Centuries

12.1: Oronce Fine's water-clock, from O. Fine, *De solaribus horologiis...* (Paris, 1560). Reproduced courtesy of St Andrews University Library.

12.2a and b: Section and arrangement of the drum for a compartmented cylindrical clepsydra from 'Traité des horloges Elementaires, ou de la manière de faire des Horloges avec l'Eau, la Terre, l'Air & Feu', translation of D. Martinelli, *Horologi elementari divisi in quattro parti...* (Venice, 1669) in J. Ozanam, *Récreations mathématiques et physiques* (Paris, 1694). Reproduced courtesy of the editor.

12.3: Water-clock by William Winde (1582), from *Gulielmi Windaei Bergannsis ad Zoman Cursus Mathematicus Tomus Secundus*. The British Library, Kings MS 267.

13. Epilogue

13.1: Polyhedral dial, from O. Fine, *De solaribus horologiis...* (Paris, 1560). Reproduced courtesy of St Andrews University Library.

13.2: Woodcut diagram with decorative flourish, from O. Fine, *De solaribus horologiis...* (Paris, 1560). Reproduced courtesy of St Andrews University Library.

LIST OF WORKS BY ORONCE FINE CITED IN THE TEXT

Fine, 1526: *Aequatorium planetorum, unico instrumentum comprehensum...* (Paris: N. Savetier, 1526), in-4°.

Fine, 1526: *Descriptio partum et succincta eucidatio quadrantis cuiusdam universalis...* (Paris: N. Savetier, 1527), in-8°.

Fine, 1528: *La théorique des cielz, mouvemens, et termes practiques des sept planètes...* (Paris: J. Pierre, 1528), in-fol.

Fine, 1532a: *Epistre exhortative, touchant la perfection et commodité des ars libéraulx mathématiques...* (Paris: P. Leber, 1532), in-8°.

Fine, 1532*b*: *Protomathesis ... [De arithmetica practica libri IIII; De geometria libri duo; De cosmographia sive mundi sphaera libri V; De solaribus horologiis et quadrantibus libri IIII]...* (Paris: G. Morhii, 1532), in-fol.

Fine, 1536: *In sex priores libros geometricorum elementorum Euclidis...* (Paris: S. de Colines, 1526), in-fol.

Fine, 1542: *De mundi sphaera, sive cosmographia...* (Paris: S. de Colines, 1542), in-fol.

Fine, 1544*a*: *Arithmetica practica...* (Paris: S. de Colines, 1544), in-8°.

Fine, 1544*b*: *Liber de geometria practica...* (Strasbourg: J. Knobloch, 1544), in-4°.

Fine, 1544*c*: *Quadratura circuli...* (Paris: Simon de Colines, 1544), in-fol.

Fine, 1550: *De universali quadrante, sinuumve organo...* (Paris: R. and C. Chaudière, 1550), in-4°.

Fine, 1551*a*: *In sex priores libros geometricorum elementorum Euclidis...* (Paris: R. Chaudière, 1551), in-4°.

Fine, 1551*b*: *Sphaera mundi, sive cosmographia...* (Paris: M. de Vascosan, 1551), in-4°.

Fine, 1551*c*: *La sphére du monde, proprement dite cosmographie...* (Paris: M. de Vascosan, 1551), in-4°.

1551*d*: *De speculo ustorio...* (Paris: M. de Vascosan, 1551), in-4°.

Fine, 1552: *La sphére du monde, proprement dite cosmographie...* (Paris: M. de Vascosan, 1552), in-4°.

Fine, 1553: *De duodecim caeli domicilis et horis inequalibus libellus...* (Paris: M. de Vascosan, 1553), in-4°.

Fine, 1555: *De mundi sphaera, sive Cosmographia...* (Paris: M. de Vascosan, 1555), in-4°.

Fine, 1556*a*: *La composition et usage du quarré geometrique...* (Paris: G. Gourbin, 1556), in-4°.

Fine, 1556*b*: *De rebus mathematicis...* (Paris: M. de Vascosan, 1556), in-fol.

Fine, 1557: *Compendaria tractatio de fabrica et usu annuli astronomici...* (Paris: G. Cavellat, 1557), in-8°.

Fine, 1560: *De soloraibus horologiis...* (Paris: G. Cavellat, 1560), in-4°.

Fine, trans. Bartoli, 1587: *Opere di Orontio Fineo ... divise in cinque parti: Aritmetica, geometria, cosmografia, e oriuoli, tradotte da Cosimo Bartoli...* (Venice: F. Franceschi, 1587), in-4°.

Borrhaus, ed. Fine, 1551: Martin Borrhaus, *Elementale cosmographicum...* (Paris: G. Cavellat, 1551), in-8°.

Martinez, ed. Fine, 1519: Juan Martinez, *Arithmetica Joannes Martini Silicei in theoricen et praxim scissa...* (Paris: H. Stefans, 1519), in-8°.

Peurbach, ed. Fine, 1515: Georg Peurbach, *Theoricarum novarum textus...* (Paris: J. Petit, 1515), in-fol.

Reisch, ed. Fine, 1535: Gregor Reisch, *Margarita philosophica ... nuper autem ab Orontio Fineo...* (Basel: H. Petri, 1535), in-4°.

Ricius, ed. Fine, 1521: Augustinus Ricus, *De motu octavae sphaerae...* (Paris: S. de Colines, 1521), in-4°.

Sacrobosco, ed. Fine, 1516: Johannes de Sacrobosco, *Mundialis sphaere opusculum...* (Paris: R. Chaudière, 1516), in-4°.

Sacrobosco, ed. Fine, 1538: Johannes de Sacrobosco, *Sphaera mundialis...* (Paris: R. Chaudière, 1538), in-4°.

ACKNOWLEDGMENTS

The essays collected here were first presented at a conference, 'The Worlds of Oronce Fine: Mathematics, Instruments and Print in Renaissance France', held in the School of Art History, University of St Andrews, 12–14 May 2006. The conference was generously supported by The British Academy, The French Vernacular Book Project, and the British Society for the History of Science. I am grateful to all these bodies, and indeed to the participants, for helping to make the conference such a success. In particular, I should like to thank J. V. Field, who delivered a highly stimulating keynote lecture and made important contributions to our discussions. The administrative staff of the School of Art History, Dawn Waddell and Margaret Hall, ensured that the conference ran smoothly, while the staff of Special Collections in the University Library enriched the event by mounting an exhibition of Fine's printed works. In the course of preparing these essays for publication I have benefited greatly from discussions with Timothy Chesters, Suzanne Karr Schmidt, and Jason and Alice König. As ever, I owe an enormous debt of gratitude to my wife, Christie, for her enduring support and enthusiasm.

A.J.M., St Andrews, 2009

Introduction

ALEXANDER MARR

History has not been terribly kind to Oronce Fine (1494–1555, Fig. 1.1).[1] Despite his having been one of the most famous mathematicians of Renaissance France, he is represented in modern scholarship by only a handful of articles and just one monograph – Gallois's 1890 Latin dissertation on his geographical work.[2] Indeed, while his activities as a cartographer and book illustrator have attracted sporadic attention from students of visual culture and book history, his contribution to the mathematical arts has been largely neglected.[3] The reason for this is essentially

1 There is some dispute over whether or not the last letter of Fine's surname should be accented. I follow Poulle (who cites the arguments of scholars from the Dauphiné, from whence Fine hailed) in omitting the accent. For the sake of consistency, I have requested this usage in all the contributions to the present volume. For the arguments for and against (evidence from Fine's contemporaries is inconclusive) see E. Poulle, 'Fine, Oronce', in C. C. Gillispie (ed.), *Dictionary of Scientific Biography*, 18 vols. (New York, 1970–1990), vol. 15 (Supp.), pp. 153–157, at p. 156, n. 1; R. P. Ross, 'Studies on Oronce Finé (1494–1555)', unpublished PhD. diss. (Columbia University, 1971), pp. 8–9, n. 2.

2 L. Gallois, *De Orontio Finaeo gallico geographo* (Paris, 1890). The most significant study of Fine since Gallois is Ross's important (but now out of date) doctoral dissertation, cited above, which spawned a handful of articles, cited in the contributions to the present volume. See also D. Hillard and E. Poulle, 'Oronce Fine et l'horloge planétaire de la Bibliothèque Sainte-Geneviève', *Bibliothèque d'Humanisme et Renaissance*, vol. 33 (1971), pp. 311–351; E. Poulle, 'Les mécanisations de l'astronomie des épicycles: l'horloge d'Oronce Fine', *Comptes rendus des séances, Académie des Inscriptions et Belles-Lettres* (Paris, 1974), pp. 59–79; L. Maierù, 'Il meraviglioso problema in Oronce Finé, Girolamo Cardano e Jacques Peletier', *Bolletino di storia delle scienze matematiche*, vol. 4, no. 1 (1984), pp. 141–170.

3 A notable exception, in addition to the studies by Ross, Hillard and Poulle, is Natalie Zemon Davis's pioneering work on sixteenth-century French mathematics: 'Sixteenth-Century French Arithmetics on the Business Life', *Journal of the History of Ideas*, vol. 21, no. 1 (1960), pp. 188–222. For Fine's work in cartography see J.-J. Brioist, in this volume. Fine is mentioned in passing in L. Febvre and H.-J. Martin's classic study of French printing, *The Coming of the Book: The Impact of Printing 1450–1800*, trans. D. Gerard, ed. G. Nowell-Smith and D. Wootton (London, 1976 [first ed., 1958]), p. 98. For his work as book illustrator and in Parisian printing houses, see R. Brun, 'Un illustrateur

threefold. First, the sheer diversity of Fine's activities in the sciences and arts renders him a challenging figure to study; second, the enormous weight of his published output is daunting; third, despite his prodigious production, Fine has generally been regarded as a populariser rather than an original thinker. Yet it is these very qualities that make him such a valuable figure through whom to study mathematical culture of the Renaissance.

Fine was a quintessential polymath, equally at home in the workshop as in the study. Not only did he make and sell his own instruments, he also designed maps, prepared the illustrations for his many books, and worked as an editor and overseer in Paris's bustling printing houses.[4] One of his paper instruments, an equatorium that he designed and published in 1526, perfectly illustrates the way in which he combined mathematics, instruments, and print for commercial gain, while demonstrating his superb skills as a draftsman (Fig. 1.2a and b).[5] Indeed, Fine – entrepreneur, technician and professor – is thoroughly indicative of the fruitful intersections between scholarship and craft, business and the academy that played such a crucial role in scientific endeavours of the sixteenth century.[6]

Furthermore, the range of his published works allows us to examine more or less the entire spectrum of Renaissance mathematical arts. The bibliographies

---

méconnu: Oronce Finé', *Arts et métiers graphiques*, vol. 41 (1934), pp. 51–57; A. F. Johnson, 'Oronce Finé as an Illustrator of Books', *Gutenberg-Jahrbuch*, vol. 3 (1928), pp. 107–109; J. Adhémar, *Graveurs du seizième siècle*, 2 vols. (Paris, 1938), *passim*; R. Mortimer, *Catalogue of Books and Manuscripts, Harvard College Library, Department of Printing and Graphic Arts: Part 1, French Sixteenth-Century Books*, 2 vols. (Cambridge, MA, 1964), *passim.*; E. Eisenstein, *The Printing Press as an Agent of Change: Communications and Cultural Transformations in Early-Modern Europe*, 2 vols. (Cambridge, 1979), vol. 2, pp. 544–545.

4 Evidence for Fine's employment as a supervisor of printing derives from the dedicatory letter in Charles de Bovelles's *L'Art et practique de geometrie* (Paris, 1542), which Fine apparently saw through the press. For an overview of his work with Parisian publishing houses see Ross, 'Studies on Oronce Finé', chapters 2 and 4.

5 Fine's printed instrument may be compared to a late fifteenth-century French manuscript equatorium in the Museum of the History of Science, Oxford, Inv. No. 49847. Notably, Fine's father published a treatise on this type of device (see below). On Fine's paper instruments, see Eagleton in this volume, and for paper instruments in general see S. Karr-Schmidt, 'Art – a User's Guide: Interactive and Sculptural Printmaking in the Renaissance', unpublished PhD. diss. (Yale University, 2006).

6 The historiography of the relationship between scholarship and craft in the Renaissance and Early Modern period is rich. See e.g. E. Zilsel, 'The Sociological Roots of Science', *American Journal of Sociology,* vol. 47 (1942), pp. 544–562; A. C. Keller, 'Zilsel, the Artisans, and the Idea of Progress in the Renaissance', *Journal of the History of Ideas*, vol. 11 (1950), pp. 235–240; J. V. Field and F. A. J. L. James (eds.), *Renaissance and Revolution: Humanists, Scholars, Craftsmen and Natural Philosophers in Early Modern Europe* (Cambridge, 1993); P. H. Smith, *The Body of the Artisan: Art and Experience in the Scientific Revolution* (Chicago and London, 2004).

compiled by Hillard, Poulle, and Ross in the 1970s list some seventy-four books by (or edited by) Fine on topics such as practical geometry, arithmetic, gnomonics, optics, music, astronomy, and cosmography.[7] This prolificacy ensured wide dissemination – Fine, who published mainly in Latin, was read throughout France and was well known in Italy, Spain, England, and the Holy Roman Empire – as well as praise from contemporaries.[8] For example, his first biographer, Antoine Mizauld, celebrated Fine as the first scholar to have 'brought these buried [mathematical] arts back to light' in France.[9] The cosmographer André Thevet, who included Fine in his *Les vrais pourtraits et vies des hommes illustres* (1584), was similarly laudatory, noting that through his mathematical teaching Fine

> [...] revived the School of Plato at the University of Paris, where mathematics was refined by the skill, vigilance, and lessons of this Dauphinois. With such lively expressions and demonstrations in his lessons, he greatly advanced these sciences, illuminated even further by his writings and by the invention and manufacture of sundry beautiful instruments and maps, for he possessed not only the ingenuity (*esprit*) to invent these devices, but also the hand to

7 See Hillard and Poulle, 'Oronce Fine et l'horloge planétaire', pp. 335–349; R. P. Ross, 'Oronce Fine's Printed Works. Additions to Hillard and Poulle's Bibliography', *Bibliothèque d'Humanisme et Renaissance*, vol. 36 (1974), pp. 83–85. As Poulle notes, 'The list of books published by Fine is difficult to establish, for it involves sorting out many reprintings, some of them only partial, and a number of translations.' Poulle, 'Fine, Oronce', p. 156. The total given above does not include Fine's maps, works to which he contributed only images or liminal text (the 'Imprimés divers' of Hillard and Poulle's list), and the works listed by Mizauld in his 1556 edition of Fine's *De rebus mathematicis* that remain unidentified.

8 For Fine's reception abroad see the brief comments in Ross, 'Studies on Oronce Finé', pp. 368–374 and Mosley in this volume. It should be noted that Fine occasionally published in French and some of his Latin works were translated into the vernacular – French, English, and Italian – after his death. He was well known to Oxford scholars in the second half of the sixteenth century. See M. Feingold, *The Mathematicians' Apprenticeship: Science, Universities and Society in England, 1560–1640* (Cambridge, 1984), pp. 116–118. For his reception in Italy and Portugal see Dupré and Leitão in this volume. A collection of verse eulogies was printed immediately following Fine's death, indicating his fame in France: *Funebre symbolum virorum aliquot illustrium, de optimo & doctißimo viro Orontio Finaeo, regio mathemat n professore & illustratore* (Paris, 1555).

9 'Mathematicae artes bene docentur: quas omnium / primus, sepultas misit in lucem'. Antoine Mizauld, 'Vita Orontii', in O. Fine, *De rebus mathematicis hactenus desideratis libri IV* (Paris, 1556), sig. v[r]–vi[r], at vi[r]. Mizauld knew Fine personally. See Ross, 'Studies on Oronce Finé', p. 363. That Fine's fame endured in France for some time after his death is evinced by his inclusion in Guy Lefèvre de la Broderie's *Galliade ou la Revolution des arts et sciences* (1578) and in the list of renowned mathematicians compiled by Claude Milliet de Chales towards the end of the seventeenth century. See (for de la Broderie) I. D. McFarlane, *A Literary History of Renaissance France, 1470–1589* (London and New York, 1974), p. 386 and (for Milliet de Chales) Leitão in this volume.

> make and design them as well. This led to his being greatly admired, and caused him to be honoured by Kings, Princes, Cardinals, Ambassadors and others […].[10]

But Fine's intellectual and artisanal energy also drew criticism. He was attacked in his own era for serious scholarly failings – principally sloppiness brought about by excessive production – and for wildly exaggerated claims, most notoriously the boast that he had managed to square the circle in more than a hundred different ways.[11] Bernardino Baldi's comments in the 'Life of Oronce Fine' that forms part of his *Vite de' matematici*, probably reflect the prevailing scholarly opinion of Fine and his works at the turn of the sixteenth century: 'He was a man of infinite industry, but with an intellect lacking that discernment required of a perfect mathematician; to this end he wrote very many books, but they are filled with many errors.'[12]

Later assessments of his achievements have been equally disparaging. Emanuelle Poulle, one of the few scholars to have devoted sustained attention to Fine in recent years, has dismissed his contribution to science as 'encyclopaedic, elementary, and unoriginal'.[13] But this is to miss the point entirely. As Adam Mosley explains in his contribution to this volume:

> The apologetic tone of such remarks reflects historians of science and mathematics's continuing thrall to narratives of progress and stories of revolution-

[10] '[…] par luy elles fussent ressucitées et que de nouveau l'echole de Platon fut reveillée dans l'université de Paris, où que les Mathemates fussent affinées par la dexterité, vigilance et leçons de ce Dauphinois. Que si de bouche et vive voix, ensemblé par demonstrations en ses leçons il avançoit grandement ces sciences, encore plus les illustrioit il par ses labeurs particuliers tant par ces escrits que par invention et frabricature de plusieurs beaux instrumens et cartes, come ayant la main non moins apte et duite à fabriquer et dresser, tels organes, et les peindre, que l'esprit à les inventer. Ce qui le rendit plus admirable, et le feit honorer par les Roys, Princes, Cardinaux, Ambassadeurs et autres […]'. A. Thevet, *Les vrais pourtraits et vies des hommes illustres Grecz, Latins, et Payens…* (Paris, 1584), fol. 565^{r}. Thevet's information derived from Fine's son, Jean. Given Thevet's comments about Fine's combination of intellectual subtlety and manual dexterity, he seems a perfect early example of the embodied ingenuity recently examined in L. Roberts, S. Schaffer and P. Dear (eds.), *The Mindful Hand: Inquiry and Invention from the Late Renaissance to Early Industrialisation* (Amsterdam, 2007).

[11] On which see Leitão in this volume. See also Pantin in this volume for the critiques of the so-called *Orontiomagistes*, who were apparently envious of Fine's success.

[12] 'Fu huomo d'infinita fatica, ma d'intelletto mancante di quella perspicacia che a perfetto matematico si richiede; laonde scrisse moltissimi volumi, ma ripieni di moltissimi errori'. B. Baldi, 'Life of Oronce Fine', in *Le vite de' matematici. Edizione annotata e commentata della parte medievale e rinascimentale*, ed. E. Nenci (Milan, 1998), pp. 442–456, at p. 444. Baldi provides a lengthy critique of Fine's errors in *De quadratura circuli*.

[13] Poulle, 'Fine, Oronce', p. 157.

> ary change. [...] Fine was much more representative of the mathematical culture of the period than the extraordinary figures who have tended to monopolise historians' attention. Arguably, indeed, he did much more to shape the nature of that culture than many of those who are now more famous than him.

The teacher of Petrus Ramus, Jacques Peletier and Pierre Forcadel, Fine educated an entire generation of French mathematicians, who went on to prepare the ground for the work of canonical figures such as Descartes and Viète.[14] He was instrumental in disseminating and raising the status of mathematics in France and beyond, not only through the volume of his publications but also by virtue of his bold arguments concerning geometry's ability to provide certain demonstrations.[15] Alongside his own, numerous works he edited important mathematical texts, including Siliceus's *Arithmetica theorica et practica* (1519), Peurbach's *Theoricae novae planetarum* (1525) and the first six books of Euclid's *Elements* (1536).[16] Through his instruments, Fine made a substantial contribution to the material culture of mathematics – Thevet, for instance, noted that contemporaries revered him especially for having 'recovered the ingenious devices made by the renowned Saragossian, Archimedes'.[17] Indeed, Fine should properly be regarded as one of the progenitors of a French renaissance of mathematics, a renaissance

14 On Fine's teaching see, in addition to Ross, J.-C. Margolin, 'L'Enseignement des mathématiques en France (1540–1570): Charles de Bovelles, Finé, Peletier, Ramus', in P. Sharratt (ed.), *French Renaissance Studies 1540–1570: Humanism and the Encyclopedia* (Edinburgh, 1976), pp. 109–55; Pantin and Cifoletti in this volume.

15 Indeed, Fine is an important early contributor to this theme, on which see Axworthy in this volume and A. de Pace, *Le matematiche e il mondo: ricerche su un dibattito in Italia nella seconda metà del Cinquecento* (Milan, 1993). Notably, Fine's comments about the certainty of mathematics and its propaedeuticity to other subjects, especially the work of Aristotle, were taken up later in the century by Federico Commandino, although the Urbinate mathematican criticized Fine for not knowing Greek. See P. L. Rose, *The Italian Renaissance of Mathematics: Studies on Humanists and Mathematicians from Petrarch to Galileo* (Geneva, 1975), p. 205.

16 Fine also edited scholastic and philosophical works: Thomas Bricot's paraphrase of Aristotle's natural philosophy and Joannes de Bassoli's commentaries on Lombard's *Sentences* in 1517, Gregor Reisch's *Margarita philosophica* in 1535. See Ross, 'Studies on Oronce Finé', pp. 78 and 89. For Fine's attitude towards Aristotle see Axworthy in this volume. For his anti-scholasticism see Ross, 'Studies on Oronce Finé', p. 361.

17 'La plus-part d'iceux se saisoit entendre qu'il avout recouvert les ingenieux artifices qui avoyent tant faict renommer le Saragoussois, Archimede'. Thevet, *Vrais pourtraits*, fol. 565^r. Amongst these devices was the burning mirror, which Archimedes reputedly used against the fleet of Marcellus at the Siege of Syracuse, and to which Fine devoted a treatise: *De speculo ustorio*, discussed in Dupré's contribution to this volume. As Pantin notes, Fine's instrumental ingenuity was the key to his success: 'What guaranteed his international reputation, and the support of his protectors in France, were his talents as an inventor of instruments and guides for their use in practice'. See below, p. 21.

that consisted as much in objects and face-to-face tuition as in the printed books that poured from Parisian presses in the sixteenth century.

Born to a respected Dauphinois family in 1494, Fine's background played a major part in shaping his career. His father, François, was a physician and a student of astronomy who published a treatise on the equatorium, *De coelestium motuum indagatione sine calculo*, in the year of Oronce's birth. Fine's uncles, Pierre and Antoine, were painters, and have been described as 'serious students of perspective and astronomy'.[18] Thus, from an early age Fine was exposed to mathematics, art and the world of the book – in fact, Pierre and Antoine's only surviving work is an illuminated book of hours, which perhaps helps to explain Fine's celebrated skills as a book illustrator. It is generally assumed that Fine's very early education was supervised by his father, but François Fine died when Oronce was still only a child, and his estate passed to the eldest son, Antoine. Oronce was thus obliged to move to Paris to continue his education and forge a career, arriving in the French capital sometime ca. 1510.

Fine's first few years in Paris were passed under the supervision of a family friend, Antoine Silvestre, professor of *belles-lettres* at the Collège de Montaigu, who enrolled him at the Collège de Navarre.[19] There he studied grammar and philosophy, and was awarded the Bachelor of Medicine degree in 1522.[20] However, he seems also to have pursued mathematical studies, indeed it is generally agreed that he was a junior member of Jacques Lefévre d'Etaples's intellectual circle, which included Josse Clichtove and Charles de Bovelles, both of whom published extensively on mathematics in the first few decades of the century.[21] Lefévre's circle was particularly eager to harness the new medium of print in order to advance scholarship, so it is unsurprising that Fine's earliest employment was in Parisian print shops. Amongst his earliest book illustrations were those undertaken for Petit and Chaudiere's 1515 edition of Peurbach's *Theoricae novae planetarum*,

18 Ross, 'Studies on Oronce Finé', p. 10. The biographical sketch presented here is based principally on Ross's account (pp. 8–31), notable for its judicious evaluation of contemporary and later biographies of Fine. See also Hillard and Poulle, 'Oronce Fine et l'horloge planétaire', pp. 320–324; C. K. Aked, 'Biography of Oronce Finé', in *Oronce Finé's First Book of Solar Horlogy*, ed. P. I. Drinkwater (Shipton-on-Stour, 1990). Like Ross, I have found no evidence to confirm the oft-made claim (supposedly based on one of Fine's letters, now apparently lost) that Fine was employed as a military engineer by François Ier, nor that he worked in this capacity in Italy in the early 1520s. For the letter, see M. Fournier, *Histoire générale des Alpes Maritimes ou Cottiènes* (Paris, 1890), pp. 512–513.

19 Ross thinks it likely that Fine exited with a Masters degree in 1516, as in his 1517 edition of Bricot's paraphrase of Aristotle he signed himself 'magister'. Ross, 'Studies on Oronce Finé', p. 17.

20 Poulle, 'Fine, Oronce', p. 154.

21 Fine and Bovelles certainly knew one another, for which see Dupré in this volume.

for which he produced approximately two hundred figures, as well as a full-page illustration adapted from Regiomontanus's *Epitoma ... in Almagestum Ptolemei* (1496, Fig. 1.3). In later versions of this image, Fine presented himself as a scholar surrounded by books and instruments, conversing with Urania, Muse of Astronomy (Fig. 7.2).[22]

At around the time of his employment as an illustrator, Fine began working as a private tutor in mathematics at his alma mater, the Collège de Navarre, and as a public lecturer at the Collège de Maître Gervaise, but his promising career was brought to the brink of ruin by a period of imprisonment, sometime in the early 1520s.[23] The precise reasons for Fine's incarceration – which lasted from at least October 1524 to early 1525 – are not clear, but Ross has argued convincingly that the mathematician fell foul of Parisian theologians for having practised judicial astrology. This was certainly the opinion of Heinrich Cornelius Agrippa, who in 1526 wrote that he was wary of accepting an offer to become the King of France's astrologer since 'Oronce the Parisian, a worthy mathematician and astrologer [...], despite the fact that he foretold the future as truly as he was able, was punished by a long imprisonment.'[24] Fine, who corresponded with and may have met Agrippa, was certainly interested in the occult arts. He published an edition of the cabalist Augustinius Ricius's, *De motu octavae sphaerae* (1521) and composed several astrological treatises of his own.[25]

By early 1525 Finc had been freed, perhaps as a result of a petition from the Faculty of Arts of the University of Paris, presented to the king's mother in October 1524.[26] His release was followed by a period of intense activity in which he attempted to re-establish his reputation through a flurry of publications. Alongside editions of works by leading mathematical authors he produced treatises on

22 The depiction of the author is notably similar to the portrait printed in Thevet's *Vrais pourtraits*, suggesting that both are authentic likenesses of Fine. It is sometimes suggested that the Thevet engraving is after a lost Clouet pourtrait, but this cannot be confirmed.

23 On Fine's teaching see Pantin in this volume.

24 H. C. Agrippa to Capellanus, from Lyon, 3 November 1526. Quoted in Ross, 'Studies on Oronce Finé', p. 387. For a full account of the evidence about Fine's incarceration, see *ibid.*, pp. 386–397.

25 These include *Les canons et documens très amples touchant l'usaige et pratique des communs Almanachs* (1543) and *De duodecim caeli domiciliis* (1553).

26 Cited in Ross, 'Studies on Oronce Finé', p. 386. Ross also notes a letter written to William Farel by Joannes Angelus on 1 January 1525, which reads: 'Oronce, with whom I have spoken three times in jail, that he might be freed has sent the king as petitioners two Swiss who had sought him to teach them mathematics. But I do not know how the business has turned out. Wonderful it is how in these days the theologians, who rave against every type of learning, would not be difficult to conquer if faith were strong and constant in those things in which it ought to be.' Ross, 'Studies on Oronce Finé', p. 387.

instruments and planetary theory, and published his first map – the celebrated *Nova totius Galliae descriptio*.[27] In 1530 he composed a letter in verse: the *Epistre exhortative*, dedicated to François Ier, which urged the king to add a chair of mathematics to the recently founded Collège Royal.[28] Fine's suggestion bore fruit, for in 1531 he was named the first Royal Lecturer in Mathematics, a position he held until his death.

The status associated with the post of Royal Lecturer granted Fine entry into the upper echelons of French society. Indeed, according to his contemporary biographers, Fine's Paris town house became a meeting place for *virtuosi* curious about mathematics and its products.[29] According to Mizauld,

> He so impressed the royal family that they did not decline to enter his house many times; similarly many dukes, cardinals, numerous noblemen, and ambassadors came to converse with him and to see what, by his own hand, he had painted, or sculpted, or described – I say not only maps or books, but also a thousand mathematical instruments, and devices of other sorts.[30]

Although no inventory of Fine's possessions has come to light, his library and study must have been a veritable *wunderkammer* that contained, as Mizauld noted, at least some instruments made by his own hand (Fig. 1.4). Few of these seem to have survived (indeed the only instrument securely attributed to him is a small ship-shaped dial now in the Poldi Pezzoli Museum, discussed by Eagleton in

27 Hillard and Poulle list three later editions of this map, printed in 1538, 1553, and 1557 respectively. Fine also produced two cordiform world maps, a map of Dauphiné, and a map of the Holy Land published in Bernard de Breydenbach's *Grant voyage de Jherusalem* (1517). See Hillard and Poulle, 'Oronce Fine et l'horloge planétaire', pp. 342–343.

28 See Pantin and Axworthy in this volume.

29 In this regard, Fine's house must have been very similar to that of his friend John Dee, in London. See e.g. W. H. Sherman, *John Dee: The Politics of Reading and Writing in the English Renaissance* (Amherst, 1995), chapter 2. The relationship between Dee and Fine deserves further study. The two men met in Paris in the early 1550s, and in his famous 'Mathematicall Praeface' to the first English edition of Euclid's *Elements* (1570), Dee noted having seen a thaumaturgical wonder with Fine in the Abbey of Saint-Denis. See my 'Understanding Automata in the Late Renaissance', *Journal de la Renaissance*, vol. 2 (2004), pp. 205–221, at p. 208. Some of Fine's comments in the *Epistre exhortative* – for example, '[the mathematical arts are vilified] and their supporters are the victims of false accusations, as if it was heresy to know them' (see below, p. 27) – are very similar to Dee's complaints in the 'Mathematical Praeface'.

30 'Is Regibus tantùm fuit mirabilis, / Ut eius aedes ingredier saepissime / Non sunt recusati: pariter multi Duces, / Et cardinales, et numerosi nobiles, / Necnon Legati gentium atque Principium, / Ut cum viro tanto loqui percommodè / Possent: adhaec videre quae manu propria / Vel pinxerat, vel sculpserat, vel descripserat, / Non dico chartas, aut libros, sed mille organa / Mathematica, vel alterius artificij.' Mizauld, 'Vita Orontii', fol. vi[r].

her contribution to the present volume), but we know that Fine offered mathematical devices for sale because he says as much in his treatise on sundials, *De solaribus horologiis*.[31] We know also that he presented instruments as gifts – a standard route to patronage for ambitious mathematicians – for, as Anthony Turner notes, Fine unequivocally states that he presented an example of his water-clock to François Ier.[32] It is highly likely that instruments also accompanied the finely illustrated manuscripts of his mathematical works that Fine presented to various dedicatees, the king included (see below, Fig. 4.3).[33]

Despite securing royal patronage, Fine's later years were beset by financial difficulties. Having married in around 1535, Fine's wife produced a 'countless brood' of children, which he struggled to support.[34] Although the Royal Lectureship carried a generous stipend, Fine's numerous pleas for funds indicate that payments were made only sporadically. Indeed, the fact that he persisted in working for print shops (notably those of Guillaume Cavellat and Simon de Colines), as a surveyor, and as an instrument salesman after 1531 may be attributed in no small part to the negligence of his illustrious patron. As Oronce's son, Jean, explained in the preface to a posthumous edition of his father's treatise on sundials:

> After three decades and more spent and devoted to restoring and explaining mathematics, not only by lecturing but also by writing, all this time waiting and begging for payment for his efforts and being mocked and put off with courtly pittances, all this time watching his family shrink and old age come upon him while the number of his published works continued to mount, having borne such indignity as unworthy, he died cheerfully and steadfastly in the Lord in his sixtieth year from a fatal disease. My beloved mother, who had sailed in the same deplorable ship of hopes and troubles, followed a short while after, leaving behind six little sheep to wander among starving wolves without a shepherd.[35]

While the present volume cannot remedy the plight of Fine or his children, it can at least bring this much neglected polymath to the attention of a new audience. The essays gathered here aim to cast fresh light on Fine and his myriad activities,

31 See Eagleton in this volume. In ca. 1553 Fine constructed a new panel, showing his ingenious conception of unequal hours, for Charles de Lorraine's astronomical clock, now in the Bibliothèque Sainte-Geneviève. See Hillard and Poulle, *Science et astrologie au XVIe siècle: Oronce Fine et son horloge planétaire* (Paris, 1971).

32 See Turner in this volume.

33 On gift-giving in this period see N. Z. Davis, *The Gift in Sixteenth-Century France* (Oxford, 2000). For a list of the surviving manuscript works by Fine, See Hillard and Poulle, 'Oronce Fine et l'horloge planétaire', pp. 346–347.

34 Mizauld, 'Vita Orontii', quoted in Ross, 'Studies on Oronce Finé', p. 25. For Fine's financial difficulties see Pantin in this volume.

35 Jean Fine, preface to O. Fine, *De solaribus horologiis* (Paris, 1560), quoted in Ross, 'Studies on Oronce Finé', p. 31.

placing him within the broad socio-intellectual context of Renaissance Europe and demonstrating his important contribution to the worlds of mathematics, instruments, and print. Isabelle Pantin considers Fine's role as Royal Lecturer, charting the evolution of mathematics teaching at the Collège Royal, the circumstances (and vicissitudes) of his appointment, and the type of programme he instituted. She shows that Fine's vigorous promotion of mathematics as a subject of high intellectual standing was instrumental in firmly embedding this discipline in sixteenth-century French intellectual culture. Fine's arguments for the significance of mathematics are explored further by Angela Axworthy, who examines in detail his comments on the epistemological status of mathematics in works such as the *Epistre exhortative* and the *Protomathesis*, Fine's masterful mathematical compendium. Axworthy demonstrates that Fine presented his discipline as propaedeutic to most others, and certainly to the study of Aristotle. Drawing on the work of authorities such as Plato and Cicero, Fine argued that mathematics was a necessary first step along the path to true understanding, an intermediary between Christian theology and natural philosophy, or between metaphysical and physical knowledge.

However, as Pascal Brioist illustrates in his contribution, Fine was equally concerned with practice as with theory. Many of his texts focused on practical geometry and instrumentation, such as the *Quarré geometrique* and parts of the *Protomathesis*. Brioist places Fine squarely within a tradition of 'useful' mathematics stretching from Antiquity to his own age, which encouraged the use of geometry in activities such as surveying, a domain in which Fine himself worked.[36] Brioist investigates the ways in which Fine negotiated theory and practice, encouraging the use of instruments while maintaining a level of abstraction – culled from deep immersion in the Euclidean tradition – in his printed works. The potential tension between practice and theory is a key theme of Sven Dupré's study of Fine's much neglected treatise on the burning mirror, *De speculo ustorio*, one of the first optical treatises published in France. In a thorough analysis of Fine's sources and treatment of the subject, Dupré argues that although concerned with an actual object, the work partook thoroughly of bookish culture. Indeed, all of the instructions related to craft that Fine provided in his text were mathematized. As Dupré explains, 'Fine's practical mathematical knowledge is to be differentiated from material knowledge, [his] optics was an affair of paper, print and drawings.'

Instruments – paper or otherwise – are the subject of Catherine Eagleton's study of Fine's treatise on sundials, *De solaribus horologiis*. Like Dupré, Eagleton illustrates Fine's deep commitment to bookish culture, in particular his familiarity with medieval and early Renaissance manuscripts on instruments, which he drew upon heavily when writing about devices such as the navicula or 'ship-shaped' dial.

36 For example, he worked as a surveyor for the University of Paris in a land dispute of the late 1540s. See Ross, 'Studies on Oronce Finé', p. 29.

Yet where previous writers had written about sundials in a somewhat haphazard fashion, Fine gathered, selected, and ordered his material to create a new genre: the sundial book, which became a staple of later mathematical authorship. Indeed, Eagleton shows that Fine exerted considerable influence on important later mathematicians such as Christoph Clavius.

Cosmography and cartography are major aspects of Fine's output. The contributions of Jean-Marc Besse, Adam Mosley and Jean-Jacques Brioist investigate various aspects of his work in these fields. Beginning with Fine's visual, quasi-emblematic representation of cosmology, Besse argues that Fine perceived this discipline as part history, part mathematics. Turning to Fine's texts, Besse proceeds to tease out the relationship between geography and cosmology present in the mathematician's works, placing Fine's treatment of these subjects within the wider context of Renaissance learned culture. In a close reading of the content, context, and influence of Fine's principal cosmological work, the *Sphaera mundi*, Adam Mosley argues forcefully that Fine's cosmology is not 'the blend of descriptive geography and proto-ethnography that the discipline has sometimes been taken to be'. Rather, he shows that Fine's cosmography – although peppered with natural philosophical, astrological and religious reflections – was largely mathematical. In fact, studying Fine's cosmography helps us to arrive at a more nuanced understanding of what kind of mathematician Fine was. Mosley suggests that Fine's cosmology was part of what he calls 'theoretical practical mathematics', for although ostensibly useful, *Sphaera mundi* was concerned more with providing instruction in theoretical principles than with solving specific problems. However, as Jean-Jacques Brioist shows, Fine was adept at overcoming the practical challenges inherent in cartography, for he produced a series of important, elegant and (for their age) highly accurate regional and world maps. By tracing carefully Fine's instructions, in his printed books, for map projection, and by studying the resulting artefacts themselves, Brioist reconstitutes Fine's complex cartographical methods. Intriguingly, Brioist concludes that in this artisanal activity Fine cast aside mathematical consistency, opting instead for a mixture of differing techniques – some of which derived from Ptolemy – depending on what kind of map he was making.

Fine's reception and legacy are the subjects of contributions by Henrique Leitão, Giovanni Cifoletti and Anthony Turner. In his study of the Portuguese mathematician Pedro Nunes's critique of Oronce Fine – *De erratis Orontii Finaei* (1546) – Leitão evaluates why and how Fine was castigated by contemporary scholars for a series of intellectual failings. Leitão shows that Nunes's book was a carefully crafted piece of self-promotion, designed to entrench his position as one of Portugal's foremost mathematical authorities, and he traces Nunes's demolition of Fine's work on squaring the circle, the *Quadratura circuli*. While Fine was lambasted abroad, his reception in France was largely favorable. In her essay, Giovanna Cifoletti assesses Fine's legacy in the French algebraic tradition. Starting with Natalie Zemon Davis's assertion that Fine's most important contribution lay in educating an important generation of French mathematicians, she looks for traces of Fine's influence in the works of his students and followers, Jacques Peletier du Mans, Petrus Ramus and Guillaume Gosselin. Cifoletti argues that alongside his impact as a teacher, Fine's importance lies in his promotion and dissemination of mathematics in print, through accessible, affordable text books – a programme expanded by his followers into a 'vernacular scientific book project'. Anthony Turner considers Fine's legacy in technology and horology by tracing the fortunes of one of his most famous inventions, the water clock or *clepsydra*. A largely overlooked device in the history of horology, Fine's water clock was based on ancient and medieval prototypes, indeed the French mathematician probably first encountered it in the writings of Hero of Alexandria. Turner shows precisely how Fine used earlier descriptions of the *clepsydra* to make his water clock, before examining the persistence of interest in such devices up to the eighteenth century.

Finally, in the Epilogue, Stephen Clucas draws together the various strands of Fine's life and work examined in the preceding essays. He points out that while many of the contributors to the present volume share previous scholars' opinion of Fine's work in terms of its originality, his impact as a pedagogue was considerable. Moreover, far from dismissing Fine as a lacklustre popularizer, we should instead appreciate him as a representative of the broad sweep of sixteenth-century mathematical culture. With his text books to his maps, his lectures and his instruments, Fine was 'one of the most typical and characteristic figures of this phase of the development of Renaissance mathematical culture in Europe'.

Oronce Fine's Role as Royal Lecturer

ISABELLE PANTIN

I. *The Collège Royal and its Uncertain Beginnings*

The origins of the Collège Royal seem, at first glance, sufficiently well known to modern scholarship that they do not need to be recapitulated – provided one does not wish to enter into detail. François 1er was ever ready to make his authority felt on the University of Paris, a powerful conservative institution.[1] After the defeat at Pavia (1525) and with the rise of religious tensions, his reign entered a difficult phase and it mattered to him all the more to affirm himself as the *Père des Lettres* and the hero of *L'Ignorance Chassée*, as he is portrayed in the celebrated fresco in the gallery at Fontainebleau.[2] Influenced by the humanists close to the court, in particular by Guillaume Budé, in 1530 he established the institution of the *Lecteurs Royaux*, paid for out of his own treasury and charged with teaching subjects unknown to the scholastics (or badly handled by them), namely Greek and Hebrew initially, followed by mathematics, which did not have quite the same status as the languages. Later he had other disciplines added, such as Latin eloquence, philosophy and medicine.

The Royal Lecturers had a complex relationship with the University.[3] They were outside this institution, since they did not confer degrees, delivered only public lectures, and were placed under the special protection of the *Grand Aumônier du Roi*.[4] But at the same time they were part of it, as is shown by the

1 See J. K. Farge, *Orthodoxy and Reform in Early Reformation France: The Faculty of Theology of the University of Paris* (Leiden, 1985).

2 François 1er dressed as a Roman warrior and crowned with laurels, carries a sword in his right hand, and a book in his left. He enters the doorway of a magnificent temple from which shines a brilliant light, while in the foreground a crowd of people, with their eyes blindfolded, feel their way in the darkness with desperate gestures. The Abbé Goujet was the first to show the connection between this painting and the institution of the Royal Lecturers. See C.-P. Goujet, *Mémoire historique et littéraire sur le Collège Royal de France*, 3 vols. (Paris, 1758), vol. 1, p. 80; A.-M. Lecoq, 'La fondation du Collège Royal et *L'Ignorance chassée* de Fontainebleau', in M. Fumaroli (ed.), *Les origines du Collège de France (1500–1560)* (Paris, 1998), pp. 185–206.

3 See J. K. Farge, 'Les lecteurs royaux et l'Université de Paris', in A. Tuilier (ed.), *Histoire du Collège de France. I: La création 1530–1560* (Paris, 2006), pp. 209–232.

4 Jacques Colin (from 1534 to 1537); Pierre du Chastel from 1534 to 1552.

title they held: *Lecteurs Ordinaires du Roi en l'Université de Paris.*[5] Because they did not enjoy the use of any particular building, they taught in different Paris colleges, and, from 1546, they were submitted to the jurisdiction of the Parliament.[6] In 1568, an *ordonnance du roi* ordered that:

> All those who teach and will teach or deliver lectures in the said University, either in private or in public schools [...] notably those who teach and deliver lectures publicly by His Majesty's appointment, will be Roman Catholics and will comply with the laws, statutes and decrees of the said University.[7]

Thus, the general situation can be seen quite simply. However, the particular conditions of the profession of Royal Lecturer are not well known. The first reason for this is that the Collège was founded without statutes or rules. This lack is clearly recognized in a parliamentary decree of March 1566, concerning Jacques Charpentier, accused by Ramus of not fulfilling his duties as a Royal Lecturer in Mathematics: no text defined what was expected from the incumbents of the different chairs.[8] In February 1574, the directorship of the Collège was conferred by *lettres patentes* upon Jacques Amyot, the king's *Aumônier*, with the task of putting an end to the abuses which had shockingly been taking place:

> We have been informed that some people, having obtained from Us, by using indiscrete intrigue or otherwise, Letters that designated them as Royal Lecturers in Our University of Paris, and having introduced themselves in [the said institution] without having been submitted to test and examination [...], after they have thus usurped, by such schemes, lectureships assigned to the Mathematical Sciences, have converted them into lectureships in other subjects, as they pleased, without Our permission, that some teach when they like, with long intermissions, teaching what they like and when they like, because they carry on professions incompatible with the lectureship.[9]

5 *Lettres patentes*, 9 November 1532, quoted in A. Lefranc, *Histoire du Collège de France* (Paris, 1893; reprint, Geneva, 1970), pp. 394–395.

6 Farge, 'Les lecteurs royaux', p. 225. The University had been granted this privilege for one hundred years. See F.-A. Isambert, *Recueil général des anciennes lois francaises depuis l'an 1420 jusqu'à la révolution de 1789*, 29 vols. (Paris, 1821–33), vol. 9, no. 173, pp. 138–144.

7 'Tous ceux qui enseignent et enseigneront ou feront lectures, soit en escholes privees ou publiques en ladite Université [...] mesmes ceux qui ont fonction et gages de sa majesté pour faire lecture et exercice publique, seront de la religion catholique et romaine, observeront les loix, statuts, et ordonnances de ladite université.' C. E. Du Boulay, *Historia Universitatis Parisiensis*, 6 vols. (Paris, 1665–1673), vol. 6, p. 661.

8 See C. Waddington, *Ramus, sa vie, ses écrits, ses opinions* (Paris, 1855), pp. 176–178.

9 'Nous avons esté advertys que quelques uns ayant impetré de Nous par importunité ou autrement Lettres de Lecteurs du Roy en Nostre Université de Paris, s'y estant ingerez sans avoir esté esprouvez et examinez et apres [...] ayant obtenu par telles surprises places de lecteurs affectées aux Sciences Mathematiques les ont à leur plaisir sans per-

The new director had to restore discipline, according to the original intentions of François 1er:

> We want to regulate the said lectures, thus following in the traces of the will and good intentions of Our forebear of glorious memory, the Great King François who was the first to create these lectureships, you have been chosen and elected by Us so that hereafter you will govern, administer, and control the said lecturers, order them in what way and at what times they have to deliver their lectures, and examine those who will be appointed to the vacant chairs to certify their ability.[10]

All these uncertainties re-enforced the role of custom and tradition. When confronted with a problem, the deans of the Collège took up the habit (from the end of the sixteenth century) of referring to past precedent, by recapitulating the careers of their colleagues, in order to give weight to their requests.[11] The name of Oronce Fine recurs in these writings, in laudatory mentions, though generally brief and sometimes inexact. He was clearly the founder of a line of mathematical lecturers and thus an essential figure in the history of the institution. Taking account of this fact, I will consider the career of Fine in the somewhat larger context of the history of the Collège Royal in the sixteenth century. Can studying his role throw some light on the beginnings of the institution? And to what extent did this first *Lecteur Mathematicien du Roy en L'Université de Paris* create a tradition?

---

mission de Nous convertyes en lectures d'autres qualités que plusieurs lisent quand bon leur semble faisant de grandes et longues intermissions, lisant ce que bon leur semble et quand bon leur semble, exerceans aucunes vacations lesquelles ne peuvent pas bonnement compatir avec la lecture'. *Lettres patentes* of 7 February 1574, published in S. A. Sédillot, 'Les professeurs de mathématiques et de physique générale au Collège de France. Deuxième période. Les derniers Valois 1547–1589', *Bullettino di Bibliografia e di Storia delle Scienze Matematice e Fisice*, no. 2 (1869), pp. 387–448, at p. 409. From the beginning, the *Aumônier du Roi*, who was generally also *Lecteur particulier du Roi*, exercised authority over the Royal Lecturers (this was the case for Jacques Colin and Pierre du Chastel under François 1er).

10 '[...] Nous voulant [...] régler les dites lectures et par mesme moyen suivre les erremens de la volonté et bonne intention de Notre ayeul de glorieuse mémoire, le Grand Roy François qui le premier institua lesdits lecteurs, vous avons choisy et esleu pour doresnavant régler, scindiquer et controoler lesdits Lecteurs, leur ordonner comment et quand ils devront lire, examiner et approuver ceux qui debvront succeder aux places vacantes'. Sédillot, 'Les professeurs de mathématiques', p. 409.

11 H. de Monantheuil, *Pro suo in regiam cathedram reditu oratio* (Paris, 1585); J.-E. Girot, 'La notion de lecteur royal: le cas de René Guillon (1500–1570)', in M. Fumaroli, *Les origines du Collège de France*, pp. 43–108, at pp. 87–88. See also H. de Monantheuil, *Oratio qua ostenditur quale esse deberet Collegium professorum regiorum, ut sit perfectum atque absolutum, habita 18 cal. dec. in auditorio regio* (Paris, 1595).

II. *The Entry of Mathematics into the Collège*

The circumstances of Fine's appointment are not well known. He was probably chosen because he had already shown proof of scholarly competence in his previous publications. But, chiefly, he must have been favoured at court. The original plates that he had drawn for the second edition of Nicole Le Huen's *Grant voyage de Jherusalem* (1517)[12] illustrated precisely a theme current to royal propaganda at the time: they showed the hoped-for arrival of the Royal Army in the Holy Land, a motive repeated in the ballad which Fine signed with an acrostic.[13] Shortly after, Fine struck up a close friendship with Albert Pigghe, as Mizauld was to recall in his *Vita Orontii*.[14] Significantly, Pigghe was a close friend of François Desmoulins, François 1er's old tutor who had become one of the principal inventors of fictions and allegories conceived for the purpose of embellishing the king's magnificence. Pigghe had dedicated to Desmoulins his *De aquinoctiorum inventione* (1520) and collaborated on the superb manuscript started by the said Desmoulins to establish an analogy between François 1er and Julius Caesar: the *Commentaires de la guerre gallique*.[15] At any rate, Oronce Fine became, in 1528 or 1529, an *escolier* at the *Collège de Maître Gervais*, which means that he benefited from a scholarship paid for by the king on the condition that he gave courses in mathematics and astronomy.[16] He owed this to his patrons and it was a decisive step towards his future status as a Royal Lecturer.

But the most important question remains: why did the king decide to add a chair of mathematics, a quite traditional discipline, to those of Greek and Hebrew, which were humanist novelties? At the beginning of 1532, Oronce Fine, now a Royal Lecturer for about a year, published an epistle in verse that he had delivered

12 Nicole Le Huen's book was a French adaptation of the *Sancta peregrinatio* of Bernard von Breydenbach (1st ed., Mayence, 1486); it had been first printed at Lyons, by Jacques Heremberck and Michel Topié in 1488, under the title: *Pérégrination de Oultre-Mer en Terre Saincte au tres glorieux et Saint Sepulchre nostre Seigneur ihesucrist en iherusalem*. This second edition, printed by Nicolas Hygman for Françoys Regnault, carried a supplement: the account of the death of Louis XII and the beginning of the reign of François 1er, with an appeal for a new Crusade led by the victor of Marignan.

13 A.-M. Lecoq, *François 1er imaginaire. Symbolique et politique à l'aube de la Renaissance française* (Paris, 1987), pp. 259–262.

14 See A. Mizauld, *Funebre symbolum virorum aliquot illustrium de optimo et doctißimo viro Orontio Finaeo ... Ejusdem Orontii Vita, carmine per Anton. Mizaldum paucis expressa: cum catalogo omnium librorum, quos ille idem Orontius partim emisit, partim moliebatur* (Paris, 1555). Pigghe is quoted in Ricius, ed. Fine, 1521, preface.

15 The manuscript was completed in 1519–1520. Book 1 is now in the British Library, Book 2 in the Bibliothèque Nationale de France, and Book 3 in the Musée Condé at Chantilly. See Lecoq, *François 1er imaginaire*, pp. 229–244.

16 See J. Dupèbe, *Astrologie, religion et médecine à Paris. Antoine Mizauld (c. 1512–1578)*, unpublished PhD. diss. (Université de Paris X – Nanterre, 1998), p. 538.

before an audience, perhaps in the form of an inaugural lesson. This *Epistre exhortative*, supposedly pronounced by Philosophy and addressed to the king, contained an encomium of mathematics and ended with this invitation: François 1er should engage a lecturer to teach publicly these disciplines for the greater good of the country:

> I ask of you, so that they endure for ever, / the appointment of one who will by night and day endeavour / To teach them [the mathematical arts] to all publicly; / And by that means, we shall see without delay / France flourishing and surpassing in sciences / The surrounding nations and peoples.[17]

This allegorical vision of things tells us very little about how the events took place. The historical sources show that Fine's appointment was made a few months after that of his colleagues. On 27 March 1531, Pierre Danès, Jacques Toussain, François Vatable, and Agacio Guidacerio, professors of Greek and Hebrew, each received 200 *écus*, while Oronce Fine received only 150.[18] Moreover, in this list Fine was the only one of his kind, while the other chairs were double, which made his discipline the poorest in the Collège. Only in 1540, when Pascal Duhamel was appointed, was Fine assisted in his onerous task, for he was expected to teach the four branches of the quadrivium, as well as optics, cosmography, and the use of instruments.

Later on, this initial imbalance was seen as an anomaly. The *lettres patentes* of 1574, sought, as we have seen, to recover the original spirit of the Collège: 'following in the traces of the will and good intentions of [...] the Great King François'. Still inspired by the recently deceased Ramus's repeated injunctions, Charles IX then emphasized that the presence of two lecturers in the three principal disciplines (Hebrew, Greek, and mathematics) was a fundamental necessity:

> And as soon as these lectureships will be made vacant, by the death of their incumbents, We desire that they should be reduced to the number and destination that they had at their creation, which was intended to introduce in Our said University the matters which appeared to be missing, that is the knowledge of Greek and Hebrew and of Mathematical Sciences. By such means, We desire that there should be two Lecturers in Greek, the one to teach the beginners grammar and the most accessible authors, and the other to read the poets and orators to more advanced students. The same in Hebrew, the one to teach grammar and the most accessible books in the Old Testament, and the other to read the Prophets, the Psalms of David, the Canticles, and other books in verse. And two lecturers in Mathematics, the one to

17 'Plaise toy donc affin que leur cours dure, / Ordonner un qui, jour et nuit procure, / Les demonstrer à tous publiquement; / Et ce faisant lon verra briefvement / France florir et passer en doctrines, / Les nations, et gens circumvoisines.' Fine, 1532*a*, B4r.

18 Archives Nationales de France, Paris, AE II 1817, 27 March 1531. Quoted in J. Linet, D. Hillard and E. Poulle, *Science et astrologie au 16e siècle. Oronce Fine et son horloge planétaire* (Paris, 1971), p. 25 (no. 41).

> teach practical Arithmetic up to Algebra, the theory [of planets] and Music, and the other to lecture on geometry, optics and catoptrics, and astrology.[19]

The need for a double chair of mathematics had been a recurrent theme in Ramus's writings on teaching reform. Without this essential condition, a nation, according to him, could not hope to raise its standards in these disciplines.[20] It is probably for this reason that, at least from Henri de Monantheuil's 1595 *Oratio*, the lists of the Royal Lecturers gave Fine a colleague: Juan Martinez Población, a doctor of medicine to Queen Eleanor.[21] Población certainly at one time or another enjoyed the right to the title of *Lector regius*.[22] However, this title was not exclusively reserved in the sixteenth century for members of the Collège Royal. It could extend to all those who gave courses at the king's request, or whose salary was, at least in part, paid from the royal treasury.[23] Moreover, Población is never referred to as a 'Royal Lecturer in Mathematics' on the title pages of his publications, (something which always occurs with Fine), and the royal accounts never mention him under this heading. It is thus certain that Población did not belong to the Collège, although some preferred to believe this to be the case, in order to better argue for the legitimacy of a double chair in mathematics and to confirm the appointment's original aim, namely promotion of the mathematical sciences.

III. *What Mathematics for the Collège?*

This will to promote mathematics certainly existed, for the idea of increasing the technical competences and grasp of arithmetic and geometry of artisans, merchants, and sailors was in the air at the time (Fine refers to it, as one would expect, in his *Epistre exhortative*). In the large countries of Europe, there began a move

19 '[…] et à mesure que telles places de Lecteurs viendront à vacquer par mort, Voullons qu'elles soient reduittes au nombre et à l'intention qu'elles furent premierement instituees, qui fut pour mettre en notre dite Université les choses qui sembloient y defaillir, à scavoir la connoissance des langues Grecque et Hebraïque et des Sciences Mathematiques. Au moyen de quoy Voulons qu'il y ayt deux Lecteurs en langue Grecque l'un pour lire la Grammaire et les Autheurs plus faciles pour ceux qui commencent, et l'autre pour lire les poëtes et orateurs pour ceux qui seront plus advancés. Semblablement en la langue Hebraïque, l'un pour lire la Grammaire et les livres plus facilles de l'ancien testament, et l'autre pour lire les Prophètes, les Psaulmes de David, les cantiques et autres Livres écrits en Carmes. Et deux pour lire les Sciences Mathématiques, l'ung l'Arithmetique pratique jusqu'à l'Algebre, la théorique, et la Musique, l'autre pour enseigner la Géométrie, optique et catoptrique et l'astrologie.' Quoted in Sédillot, 'Les professeurs de mathématiques', p. 409.

20 See I. Pantin, 'Ramus et l'enseignement des mathématiques', in K. Meerhoff and M. Magnien (eds.), *Ramus et l'Université* (Paris, 2004), pp. 71–86.

21 H. de Monantheuil, *Oratio qua ostenditur…* (Paris, 1595).

22 See M. Bataillon, 'Les lecteurs royaux et le Nouveau Monde', *Bibliothèque d'Humanisme et Renaissance*, vol. 13, (1951), pp. 231–240.

23 See Girot, 'La notion de lecteur royal', pp. 49–54.

to use printing for this purpose, by publishing little manuals in the vernacular destined for a different public to that of the scholars.[24] The Collège Royal, however, had not been founded to provide courses of popular education, such as those that existed in the nineteenth century. Its mission consisted in embedding a humanist program and its first duty was rather in promoting rare and difficult disciplines (such as were Greek and Hebrew). And yet humanism, above all in France in this period, had a double face: elitist on one hand, it was a populariser on the other. Whether it was a case of the bible or the medical corpus of Galen, it was held necessary to establish correct editions through the erudite labour of philology, but also to publish translations and to spread useful and salutary knowledge as widely as possible.[25] Lefèvre d'Etaples, Budé, and their friends, whose influence was decisive in the founding of the Collège, all shared this double preoccupation.

In the literary disciplines, this ambivalence was only slightly felt. But the chair of mathematics was profoundly marked by it. From the end of the fifteenth century, Jacques Lefèvre d'Etaples and his disciples, Josse Clichtove and Charles de Bovelles, had sought to increase the role of mathematics in teaching. They played a major part in the plan to found a royal chair in this discipline, of which they held a complex view. They knew of the hopes which the Italian and German humanists, from Regiomontanus on, had placed in the re-discovery of Greek mathematics. However, as far as they were concerned, they were engaged in another type of research, influenced by Pythagoreanism, by neoplatonism and by Nicolas of Cusa.[26] Since God had created the world according to 'weight, number and mesure' (Book of Wisdom, 11:12), mathematical notions should symbolically

24 See N. Z. Davis, 'Sixteenth-Century French Arithmetics on the Business Life', *Journal of the History of Ideas*, vol. 21, no. 1 (1960), pp. 188–222. See also Cifoletti in this volume.

25 See I. Pantin, 'Alessandro Piccolomini en France, le problème de la langue scientifique et l'évolution du genre du traité de la sphère', in A. Perifano (ed.), *La Réception des écrits italiens en France à la Renaissance: ouvrages philosophiques, scientifiques et techniques (Colloque, Paris, Institut Culturel Italien, 1997)* (Paris, 2000), pp. 9–28; 'La traduction latine des *Œuvres* d'Ambroise Paré', in E. Berriot-Salvatore (ed.), *Ambroise Paré (1510–1590). Pratique et écriture de la science à la Renaissance (Colloque, Pau, 1999)* (Paris, 2003), pp. 315–336.

26 The treatises of Boethius, filled with Pythagoreanism, remained practically the only manuals of mathematics all through the Middle Ages. See V. F. Hopper, *Medieval Number Symbolism* (New York, 1969); G. Beaujouan 'Le symbolisme du nombre à l'époque romane', *Cahiers de civilisation médiévale,* vol. 4 (1961), pp. 159–169. In his *Oratio de hominis dignitate*, Giovanni Pico della Mirandola mentions the 'Ancient institution of philosophizing through numbers which had been developed by the Ancient Theologians, especially Pythagoras, [...] by Plato and by the first Platonists' ['per numeros philosophandi institutio antiqua, illa quidem et a priscis theologis a Pythagora praesertim (...) a Platone prioribusque Platonicis observata'], G. Pico della Mirandola, *Oratio de hominis dignitate*, ed. E. Garin (Florence, 1942), p. 146.

express certain mysteries.[27] Lefèvre d'Etaples, Clichtove and Girard Roussel published or commented on Boethius and composed works of arithmetic.[28] Moreover Bovelles, who published a *Liber de duodecim numeris* (1511), spoke of mathematics as a 'mystical science' allowing the principles of things to be scrutinized.[29] God had 'wanted numbers to reveal the traces of the intelligible world and the quantities that of the sensible world'.[30] Mathematics was thus capable of lifting the mind to divine realities of which numbers, with their properties, were the symbols.[31] At the same time, these 'Pythagorizing' minds were popularizers, careful to publish treatises that were easy to grasp as well as useful for teaching purposes, and when they could, they put them into French. Their evangelical convictions prevented them from hiding knowledge of supreme value under a bushel, especially when the latter could bring together practical usefulness and religious values.

With Fine, they found an excellent focus for their concerns. Fine had already shown his worth as a popularizer and as a teacher. As early as 1515, he had produced an edition of Peurbach's *Theorica planetarum*, at a time when this manual was still less widely known than it was later to become, and the following year an edition of Sacrobosco's *De sphaera*. In 1528, he had published a paraphrase in French of Peurbach's *Theoriques*.[32] On the other hand Fine, in his publications, revealed a conception of the value of mathematics which was close to that of Lefèvre d'Etaples and of Bovelles. His (somewhat questionable) attempts in the

27 See I. Pantin. 'La représentation des mathématiques chez Jacques Peletier du Mans: cosmos hiéroglyphique ou ordre rhétorique?', *Rhetorica*, vol. 20, no. 4 (2002), pp. 375–389.

28 In 1503, Lefèvre d'Etaples published an introduction to Boethius's *Arithmetica* and Josse Clichtove a commentary on this text. The edition of Boethius, printed by Simon de Colines in 1521, was accompanied by a commentary by Girard Roussel who developed the *mystica numerorum applicatio.* Clichtove published in 1513 a *De mystica numerorum significatione*, which in particular sought out the meaning of the numbers found in the bible.

29 For example, in the preface to his *Art practique de geometrie* (Paris, 1542). On the *Liber de duodecim numeris* see J. Céard, 'Bovelles et la numérologie', in J.-C. Margolin and M. Laporte (eds.), *Charles de Bovelles en son cinquième centenaire (1479–1979)* (Paris, 1982), pp. 211–228.

30 C. de Bovelles, *Physicorum elementorum ... libri X* (Paris, 1512), quoted in Céard, 'Bovelles et la numérologie', p. 213.

31 On this trope, which appears in Fine's writings, see Axworthy in this volume. See also, for example, Lefèvre d'Etaples's preface to his edition of Euclid's *Elements* (Paris, 1517): 'Where can analogies more appropriate, more abstract, more pure to lift one to the divine be found [...] than mathematical characters?' ['Sed quae (...) promptiores, abstractiores, puriores ad divina surgendi praebere possint analogias (...) quam litterae mathematicae?] Quoted in E. F. Rice, *The Prefatory Epistles of J. Lefèvre d'Etaples* (New York, 1972), p. 380.

32 Peurbach, ed. Fine, 1515; Sacrobosco, ed. Fine, 1516; Fine, 1528.

field of speculative mathematics, in particular his treatise on the squaring of the circle (*De quadratura circuli*, 1544), went in the same direction.

Fine was probably less talented in all that dealt with erudite philology, a domain nevertheless essential if he wished to appear on the same level as his colleagues teaching Greek and Hebrew. He did his best in publishing an edition of Euclid,[33] in which the definitions of the propositions appeared in Greek, carefully re-copied from Simon Grynaeus's Basel edition of 1533, whereas the preceding Paris edition of 1517, established by Lefèvre d'Etaples, had merely reproduced existing Latin translations – the medieval version of Campanus and the humanist one of Bartolomeo Zamberti.[34] [Fig. 2.1] However, Fine's edition could not deceive the eyes of experts. The corrections to Zamberti's text, 'ad fidem geometricam', as the title claims, were limited to slight alterations. Fine often replaced Zamberti's terminology with that of Campanus, which was clearer and more familiar. In doing this, he placed himself in the tradition of Lefèvre d'Etaples, who had, in his work on Aristotle, recognised the inconvenience of undertaking too much reformulation. Furthermore, although the title appeared to attribute to him the authorship of the demonstrations, Fine had contented himself with slightly modifying those of Zamberti's translation.[35] His personal contributions were thus strictly for teaching purposes. He had added introductions to several of Euclid's books, and brief commentaries to the most important propositions.

Fine's lack of erudition probably mattered little to his students and patrons. What guaranteed his international reputation, and the support of his protectors in France, were his talents as an inventor of instruments and guides for their use in practice, as well as his composition of practical works teaching knowledge indispensable for cosmographers and astrologers. With only one or two exceptions, all his extant manuscripts deal with these fields (to which should be added those on alchemy).[36] No doubt this was not by accident.

However, there remained a problem for his successors. No candidate as polyvalent, as active and so endowed with similar talents seems to have followed him.

33 Fine, 1536.

34 Campanus's version was printed for the first time in Venice in 1482; that of Zamberti, done from the Greek, in Venice in 1505. Comparing traditional and modern versions was typical of Lefèvre d'Etaples's method: he did the same with Aristotle.

35 That is to say the demonstrations which should have been Euclid's, but which were at the time attributed to Theon.

36 See entries 81–90 in Hillard and Poule's bibliography, especially *Carré geometrique* (1538, no. 85); *Astrolabium organum novum* (1539/1546/1555, no. 84) ; *Æquatorium astronomicum* (no. 88); *L'art et manière de trouver certainement la longitude de tous lieux ... Item la composition et usaige d'ung singulier Metheoroscope geographique* (1543, nos. 83 and 89) ; *De speculo ustorio* (1548, no. 90); *De astrolabio sive planispherio* (1553, no. 84). These various disciplines are treated at length in the other contributions to this volume.

Obviously the level of competence in philology required of a Royal Lecturer did improve, while that in technical competence declined. The publishing, translation and commentary on Greek mathematicians (Euclid, Archimedes and Ptolemy) remained important. Yet this aspect of the lecturer's role produced results which never rose above mediocrity, despite being maintained all through the century. Maurice Bressieu, after being appointed, evoked the great task expected of him (and which he never sought to accomplish):

> There remain hidden in the Royal Library and elsewhere, buried in dirt and dust, many works written in Greek by the most excellent mathematicians [...]; they still long for a translator who would give them back light and voice, and they beg for help.[37]

Still, for Guillaume du Val in the seventeenth century there was no doubt about the issue: a Royal Lecturer must have a 'precise and perfect knowledge of the Greek, Hebraic and Arabic languages, and of Latin eloquence, because they are not only very useful but even necessary to understand the founders of the sciences, who were mostly Greeks'.[38]

The result was that, after Fine's death, the 'standard' profile for applying for a chair was that of being a doctor of medicine, with the necessary humanist training and a sufficient knowledge of mathematics and astrology. All the lecturers of this type, from Jean Magnien to Henri de Monantheuil, did more or less what was expected of them, without being brilliant in any way. The other way of becoming a Royal Lecturer, the one illustrated by Fine, remained a possible option, but that was rarely followed. Only Ramus seriously tried to defend it when he had two of his disciples elected: Jean Pena (who was a good humanist, though above all a specialist in optics) and Pierre Forcadel.[39]

IV. *The Duties of a Royal Lecturer: Did Fine found a Tradition?*

The duties of a Royal Lecturer can be empirically deduced from the practices observed throughout the century, or else on the other hand from the complaints brought before Parliament, if there were shortcomings. Courses were required everyday for an audience as large as possible. In the dedication to the king of his

37 'Multa adhuc summorum Mathematicorum scripta Graeca, et in Regia bibliotheca, et alibi, pulvere et squalore obsita [...] delitescunt: quae interpretis adhuc lucem, vocemque desiderant, et auxilium implorant.' M. Bressieu, *De mathematica professione a P. Ramo instituta...oratio* (Paris, 1576), p. 34.

38 '[...] exacte et parfaicte connoissance des langues Grecques, Hebraiques, Arabiques, et de l'Eloquence Latine, comme tres-utiles, voire necessaires pour entendre les Autheurs des Sciences, Grecs pour la pluspart'. G. du Val, *Le Collège Royal de France ou institution, establissement et catalogue des lecteurs et professeurs ordinaires du Roy* (Paris, 1644), p. 17.

39 On which see I. Pantin, 'Teaching Mathematics and Astronomy in France: the *Collège Royal* (1550–1650)', *Science and Education*, vol. 15, nos. 2–4 (2006), pp. 189–207.

edition of Euclid, Fine affirmed this, acknowledging what he owed 'to [royal] munificence and to the public, I did [my] best to fulfill the obligations of [my] office and, *besides [my] daily lectures*, leave some legacy to posterity'.[40] For a wider dissemination of knowledge, publications were encouraged, notably those in the form of manuals, as well as the more specialised and prestigious works, whether they concerned Greek philology or technical inventions, of which mention has been made above. What is important to notice here is that Fine, who worked bravely on all these fronts, set up for his successors a sort of canonical model, which they were never able to equal. Fine for his part never ceased to emphasize the gap between his outstanding efforts and the mediocre benefits which came his way. It is sufficient to quote the dedication to Henri II in his *Sphère du monde* of 1551:

> I have known the singular love the late King your father [...] bore the said mathematical arts, of which he eventually ordered that I should be public expositor in the University of Paris. There I made every effort to restore them and teach them, by delivering ordinary lectures as well as by writing books, for more than thirty years. And as for the benefits that came from it, and which come every day, I leave it to the testimony of all wise and well-intentioned persons, either foreigners or living in your kingdom. In return for [these labours] I am still waiting to be rewarded.[41]

This was not all. It seems that a Royal Lecturer should also assume a sort of 'ideological function'. He had to put forth themes and even sorts of programmes. He was obliged to play a role in various important publications and above all in the long dedications which were addressed to famous figures. On this level one may ask whether Fine's contribution had a durable effect. I shall limit myself to briefly examining two of the themes or undertakings which marked the teaching of the quadrivium in sixteenth century France. These were the defence of the discipline in the context of the university and the growth of German influence.

40 '[...] ut igitur pro mea virili parte, tum erga munificentiam tuam, tum erga ipsam rempublicam debito fungar officio, et præter *quotidianas lectiones*, aliquod hominis vestigium [...] posteris relinquam [...].' [emphasis mine]. Fine, 1536, sig., ij^{v}.

41 '[...] je congneu le feu Roy vostre pere (auquel Dieu doint repos eternel) [...] porter singuliere affection ausdittes mathematiques: desquelles il me ordonna finablement publique interpreteur en l'université de Paris, ou j'ay fait mon devoir, tant par leçons ordinaires, que par oeuvres escrittes, les remettre sus, et icelles demonstrer l'espace de trent' ans et plus: et quant est du fruit qui en est advenu, et advient tous les jours: je m'en rapporte à la relation de toutes gens de bon vouloir et jugement, tant estrangers que de vostre royaume: dont je suis encores attendant la recompense'. Fine, 1551*c*, dedication. See also the beginning of the notice to the reader in Fine, 1542, or Antoine Mizauld's *Orontii Vita* (quoted above, n. 14), pp. 5–6.

V. *Defending Mathematics*

The defence of mathematics is a vast issue that concerns the whole of Europe in the Renaissance. I shall thus limit myself here to a few particular aspects. In the sixteenth century the most energetic defenders of mathematics at the University of Paris were Lefèvre d'Etaples and Ramus. Fine, who taught Ramus, stands in between. One wonders whether he provided a link between the other two, and whether he made a personal contribution.

On the whole, the defence of mathematics developed in France along the same lines as elsewhere; that is to say on the Neoplatonic grounds laid down by Proclus, by Boethius and by others.[42] To praise the primordial nature of mathematics, the infallibility of the order it produced and its perfect certitude were all commonplace. The Parisians, however, distinguished themselves by emphasizing, in particular, the practical usefulness of these sciences and their value to religion, and by placing their defence in the context of a ferocious struggle against certain of their colleagues in philosophy.

The bitter way in which Lefèvre d'Etaples, right from his beginnings at the *Collège du Cardinal Lemoine*, attacked the Scotists, still very influential in the University, is well known. He called them 'sophists', reproaching them for their love of empty abstractions and for their way of cultivating logic with excessive subtlety, applying it to all the disciplines, including theology.[43] For Lefèvre d'Etaples, logic was but the art of emptiness and unreality, an instrument for thinking that was purely rationalist and thoroughly impious. He defended the mathematical arts for their quite opposite aptitudes. On the one hand, they were perfectly compatible with an approach to reality that was intuitive, spiritual and authentically Christian, as Nicolas of Cusa had recently shown. On the other hand, they were quite naturally to be used for concrete and useful applications. These disciplines could thus legitimately assume certain functions which had been perniciously usurped by logic. Just by remaining in their place, they could constitute the foundation of knowledge. In his dedication to Robert Fortuné (Principal of the *Collège du Plessy*) in his edition of translations from the *Metaphysics* (1515), Lefèvre stated, for example, that philosophy has three levels, first mathematics, then physics, finally theology, there obviously being a progression from one to the other.[44] Mathematics was placed right at the bottom, serving as an introductory preparation.[45] However,

42 See Axworthy in this volume.

43 See E. F. Rice, 'Humanist Aristotelianism in France. Jacques Lefèvre d'Etaples and his Circle', in A. H. T. Levi (ed.), *Humanism in France at the End of the Middle Ages and in the Early Renaissance* (Manchester, 1970), pp. 132–149.

44 J. Lefèvre d'Etaples (ed. and trans.), *Opus metaphysicum Aristotelis* (Paris, 1515). This edition was the first to give Bessarion's translation (*ca*. 1450) which closely followed that of William of Moerbeke. To it was joined, in the form of a synopsis, that of Argyropoulos (Books 1 to 12), done about 1460 and already printed at Venice in 1496 and in 1507.

Mathematics was also, in its own right, a mediatory science. In presenting his commentary on Jordanus Nemorarius's *Arithmetica*, Lefèvre d'Etaples deplored (in his dedication to Jean de Ganay) the fact that the students in philosophy at the University of Paris were deprived of the science of numbers, which offered a 'road so necessary to rise to divine things and as well as to descend to human ones'.[46]

In the second half of the century, Ramus developed similar themes: first in 1544 and 1550, when the sentence, to which he had been condemned, obliged him to fall back on teaching mathematics, first at the *Collège de l'Ave Maria*, and then at the *Collège de Presle* (it was on this occasion that he profited from the courses he had taken with Fine); then again later in the 1560s, when he took up the defense of the discipline at the Collège Royal.[47] In his texts can be found bitter complaints, with polemical intentions, against the decadence of the University brought about by degenerate scholastics, complaints as forthright as were those of Lefèvre d'Etaples. For obvious reasons, religious themes were less present; on the other hand, useful and practical considerations came to the fore. As with Lefèvre d'Etaples, Ramus used mathematics to denounce the inanity of scholastic logic, but he did it in a different way. For Ramus these disciplines were indeed meant to be perfect models of method, but for this they still needed to be somewhat reformed, to fit the way in which his own dialectic was ordered. During the last part of his life he endeavoured to do all he could to achieve this.[48]

Between Lefèvre d'Etaples and Ramus, Fine stood as a good intermediary. One can easily see him as a transmitter, to the succeeding generation, of the ideas of early French humanism. His prefaces show his conviction of the Christian value of his role. In teaching mathematics, he cultivated the gifts given him by God, to the greater advantage of his neighbour, of his king and of his nation, while, on a speculative level, he developed a way of attaining religious truth. The idea presented by Lefèvre d'Etaples of the both fundamental and mediatory character of mathematics, is one of Fine's most frequent leitmotifs. In the *Epistre exhortative* of 1532, Philosophy sees mathematics 'as the true roots of all knowledge and of

45 'Mathematics is at the bottom, because it refers to accidents; then we go up to the natural and physical level which deals with the substance, but only the mobile and median one; after physics if we ascend, there only remains the divine kind of philosophy [...].' ['Infimum mathematicum, ut quod de accidente sit; post quod ascendendo naturale ac physicum, ut quod de substantia est, sed mobili ac media. Post physicum vero sursum vergendo solum restat divinum philosophiae genus, (...).'] Quoted in Rice, *Prefatory Epistles*, p. 356.

46 '[...] non ab re dolebas hanc numerosam hujus almi parisiensis studii philosophantium turbam et bonarum litterarum cupidam tam necessaria semita tum ad divina assurgendi tum descendendi ad humana, esse destituta[m].' Quoted in Rice, *Prefatory Epistles*, p. 18.

47 See P. Sharratt, 'La Ramée's Early Mathematical Teaching', *Bibliothèque d'Humanisme et Renaissance*, vol. 28 (1966), pp. 605–614.

48 See Pantin, 'Ramus et l'enseignement des mathématiques'.

the noble disciplines'.[49] A note in the margin emphasises that they lie between natural philosophy and theology: 'Mathematica media inter naturalem et divinam philosophiam'. The rest of the poem illustrates this:

> For as the angelic spirit / Stands beneath the omnipotent Deity / And above human creatures, / Exactly so, I say, the mathematical order / stands between theology and natural sciences. / It is linked to both of them, in which it participates, / And it is not above human understanding, / Because of which it deserves even greater praise.[50]

These verses are the prototype, rather clumsy it must be said, of other comparable passages. In his preface to Euclid, Fine states that a mathematical training is indispensable to philosophy, in terms sometimes close to those of Boethius, and sometimes to those of Proclus, although he could hardly have known directly the latter's commentary on Euclid.[51] Geometry, he argues, applies a perfect method of reasoning, 'founded on the most potent precepts of dialectic (*potissimis dialectices innixa praeceptis*)', proceeding from axioms and simple elements to compounds, which it 'eventually reduces to their proper principles (*ad propria tandem singula resolvendo principia*)'. Moreover, 'although it deals with intelligible and abstract objects, as does also theology, we see that at the same time it concerns sensible and material objects, like physical reasoning'.[52] In 1551, the same idea recurs: mathematics 'agrees with natural philosophy as far as it concerns material bodies, and it participates in theology because it deals with the said bodies in the abstract, as if they were spiritual beings, separated from their movement and matter'.[53] As for Lefèvre d'Etaples and for Ramus, these praises were developed in a context that was strongly polemical. In his *Epistre exhortative*, Fine vituperated against the 'sophists' and 'stupid lunatics' who had forced philosophy into exile by destroying mathematics.[54]

49 '[...] comme vraie racines / De tout scavoir et nobles disciplines'. Fine, 1532*a*, sig. A3^{v}.

50 'Car tout ainsi que lesprit angelique / Est au dessoubz du pouvoir deifique / Et surmontant lhumaine creature, / Ainsy je diz lordre mathematique / Estre moyen du droit theologique / Et du scavoir qui concerne nature, / Participant des deux par ligature / Sans exceder lhumain entendement / Dont on le doibt priser plus amplement'. *Idem.*.

51 See Boethius, *Institutio arithmetica*, Prologue. Proclus's commentary on the first book of the *Elements* was published for the first time in Grynaeus's edition of Euclid in Greek (1538). A Latin translation by Francesco Barozzi was printed in Padua in 1560.

52 'Quanquam insuper circa intellectilia et abstracta, quemadmodum et divina versetur philosophia: sensilia tamen et ipsi materiae subjecta, veluti physica ratiocinatio, simul attingere comperitur [...].' Fine, 1536, sig. ij^{v}. See also Proclus, *Les commentaires sur le premier livre des Eléments*, trans. P. Ver Eecke (Bruges, 1948), pp. 1–2.

53 '[...] convenans avec la naturelle, d'autant qu'elles concernent les corps materiels: et participans de la divine, pource qu'elles traictent desdits corps abstractivement, comme des choses spirituelles separees de leur mouvement et matiere'. Fine, 1552*c*, sig. aa2^{r}.

54 Fine, 1532*a*, sig. A2^{v}.

These complaints and accusations were rendered even more bitter by constant allusions to persecution. It is known that Oronce Fine had spent a long term in prison, from the end of 1523 to the beginning of 1525, for obscure reasons which were probably linked to his practising astrology.[55] The verses in the *Epistre* particularly echo this. The mathematical arts, it is said, are vilified 'and their supporters are the victims of false accusations, as if it were heresy to know them'.[56] Until the death of the Royal Lecturer and even in posthumous praise of him, there were allusions to envious ones who had attacked him, those *Orontiomastiges* who sought to ruin his reputation and prevent him from cultivating the gifts he had received from God, using 'whispered lies' and 'impostures'.[57]

To conclude the list of the themes which he shared with Lefèvre d'Etaples and Ramus, Fine always kept in view the 'practical usefulness' of mathematics. His declarations reveal this as well as his publications, a fair number of which deal with practical instruments for surveyors and astrologers. Furthermore, his *Cosmographia*, the first version of which appeared in the *Protomathesis* of 1532, marked a decisive change in the tradition of treatises on the Sphere through a considerable development of technical, astrological and geographical aspects. It was perhaps through the influence of Fine that Ramus became obsessed with introducing teaching in practical disciplines derived from mathematics (notably geography and optics), while the University of Paris hardly succeeded in assuring an elementary training of the students in arithmetic and geometry.[58] Already in his *Epistre* Fine had expressed the hope that geography and perspective might make a brilliant entry into the University which had banned them.[59]

VI. *Influence from Germany*

However, it is just as revealing to compare the praise of mathematics by Fine with that which had been developed by Regiomontanus, three quarters of a century earlier, which leads us to examine the question of the penetration of German influence in France. This influence became very important from 1550 on, mainly due to the Royal Lecturers, as I have tried to show elsewhere.[60] Matters are not so clear in the preceding generation, but it does seem that Fine played a significant role in this context. From his formative years, he seems to have paid attention to German

55 See Marr in this volume.

56 'Et leurs suppoz faussement on accuse / Comme s'il fut den scavoir heresie'. Fine, 1532*a*, A3^r.

57 Fine, 1542, sig. A2^r. André Thevet, in his *Vrais portraits*, mentions these envious ones who had made Fine suffer.

58 For example he pushed Jean Pena and Frederich Risner to work on optics. See also his *Prooemium mathematicum*. The *lettres patentes* of 1574, quoted above, echo his concerns when they evoke the subjects taught by the two Lecturers.

59 Fine, 1532*a*, sig. B3^v.

60 See Pantin, 'Teaching Mathematics'.

publications. As early as 1519, he drew up a cordiform world map and dedicated it to the king. This map, which is now lost, was updated and printed in 1534/1536, with a note mentioning the earlier version.[61] Most probably, Fine had been inspired by the geometrical description of three variants of the cordiform projection published by Johannes Werner in his *Libellus de quatuor terrarum orbis in plano figurationibus*. This work had been printed in Nuremberg in 1514, together with Werner's translation of the first book of Ptolemy's *Cosmographia* and with Regiomontanus' *Epistola ad ... Bessarionem ... de compositione et usu Meteoroscopii*.[62]

The Municipal Library at Meaux holds Fine's copy of Regiomontanus's *Epitome* (Venice, 1496), the preface of which strongly deplores the decadence of mathematics and above all of astronomy, the most sublime and holiest of all the sciences. It is not surprising that Fine owned this fundamental work. He probably also acquired Regiomontanus's other works, which progressively appeared up to the 1540s, since they dealt with all the subjects that interested him: trigonometry, surveying and cartography, instruments, astronomical tables, the establishing of celestial figures for horoscopes and even the squaring of the circle.[63] At least two of Fine's treatises, the *Metheoroscope* (which remained in manuscript)[64] and the *De duodecim caeli domiciliis* (Paris, 1553) should be placed in relation to two treatises by Regiomontanus published in 1514 (the *Epistola* mentioned above) and in 1536 which bear similar titles.[65] There is still much to be said on this point.

61 On which see J.-J. Brioist in this volume.

62 See G. Kish, 'The Cosmographic Heart: Cordiform Maps of the Sixteenth Century', *Imago Mundi*, vol. 19 (1965), pp. 13–21; R. Shirley, *The Mapping of the World. Early Printed World Maps 1472–1700* (London, 1993), no. 66, pl. 60, and no. 69; M. Pelletier, 'Die herzförmigen Weltkarten von Oronce Finé', *Cartographica Helvetica*, vol. 12 (1995), pp. 27–37; F. Lestringant and M. Pelletier, 'Maps and Descriptions of the World in Sixteenth-Century France', in D. Woodward (ed.), *The History of Cartography Volume 3: Cartography in the European Renaissance*, 2 vols. (Chicago and London, 2007), vol. 2, pp. 1463–1479.

63 See especially these editions by Johann Schöner: *De triangulis omnimodis ... pleraque D. Nicolai Cusani de quadratura circuli deque recti ac curvi commensuratione, itemque Joh. de Monte Regio eadem de re* (Nuremberg, 1533); *Problemata XXIX Saphaeae* (Nuremberg, 1534); *Rudimenta astronomica Alfragani ... Item oratio introductoria in omnes scientias mathematicas Joannis de Regiomonte ... Ejusdem utilissima introductio in elementa Euclidis* (Nuremberg, 1537); *Tabulae sinuum* (Nuremberg, 1541).

64 'La composition et usaige d'un singulier méthéoroscope géographique ... nouvellement inventé ... par Oronce Fine ... 1543'. Bibliothèque Nationale de France, Paris, MS Français 1337, fols. 15–22 and MS Français 14760.

65 J. Regiomontanus, *Tabulae astronomicae ... Ratio ... duodecim domorum caeli* (Nuremberg, 1536). In his own *De duodecim*, Fine criticizes Regiomontanus's method of domification.

All the posthumous editions of Regiomontanus, in addition to the significance of their specific contents, also played a role in spreading the author's ideas about the need to restore mathematics. These ideas were repeated and enriched by Melanchthon and his disciple Johann Schöner, who ensured the publication of Regiomontanus's work in the 1530s and 1540s. From 1536 on, one finds Fine writing on themes which were especially present in the *Oratio de disciplinis mathematicis*, which Regiomontanus had pronounced in 1464. Whereas up until then, when he wanted to quote the prestigious ancients who had honoured his discipline, Fine always quoted Plato and Cicero, he now emphasized the case of Aristotle, whose philosophy he said, presupposed a knowledge of geometry: '[...] thus, it appears obscure and difficult to those who ignore it'.[66] Regiomontanus had said that no one could be considered capable of understanding Aristotle without knowing the quadrivium.[67] Furthermore, Fine not only praised the order and certainty of mathematics, but also its eternal character, inviolable and incorruptible,[68] qualities which Regiomontanus had celebrated in lyrical terms.[69]

It is not even necessary to suppose that Fine had read Regiomontanus's *Oratio*, printed in 1537 together with an edition of Alfraganus and Albategnius. The ideas that the latter work expressed had begun to spread in Italy from the end of the fifteenth century (Regiomontanus had pronounced his discourse at Padua in 1464), and Lefèvre d'Etaples had perhaps picked up something of it during his travels in 1491–1492. Concerning Fine more particularly, he was probably informed of the importance of the German mathematical renaissance from 1520, through his friendship with Albert Pigghe. An encomium of this renaissance fig-

66 '[...] geometricum praesupponere videtur auditorem: hinc sit, ut iis qui Geometriam ignorant, subobscurus difficilisque videatur Aristoteles'. Fine, 1536, sig. iij^{v}. On the importance of mathematics for Plato and Aristotle, see the dedication of Fine, 1551*c*, sig. aa2^{v}.

67 See Regiomontanus, *Oratio*, sig. b3^{v}, facsimile of the 1537 edition in *Joannis Regiomontani Opera collectanea*, ed. F. Schmeidler (Osnabrück, 1972) p. 50. In the subsequent passage, Regiomontanus linked, as did Lefèvre d'Etaples later, the decadence of philosophy to the abandoning of mathematics in favour of the sophisms of the Scotists.

68 See the dedication to Chancellor Guillaume Poyet in Fine, 1542.

69 'Theoremata Euclidis eandem hodie quam ante mille annos habent certitudinem. Inventa Archimedis post mille secula venturis hominibus non minorem inducent admirationem, quam legentibus nobis jucunditatem. O perpetuae mortalium comites, non prius cessaturae quam mundus ipse desinat. O divina philosophorum numina summis prosequenda honoribus [...].' ['Euclid's theorems possess now the same certitude that they had a thousand years ago. A thousand centuries hence, Archimedes's inventions will inspire future men with an admiration in no way inferior to the pleasure they give us when we read them. O perpetual companions of mortals, who will not make default until the end of the world itself! O sacred goddesses of philosophers to whom the highest honours must be paid (...).'] *Ibid.*, sig. b4^{r}.

ures notably at the beginning of Pigghe's treatise on the points of the equinox, to which Fine made an allusion in the dedication to his edition of Ricci.[70]

Even if there is nothing to show that Fine played a role in the republication (in Paris) of the editions (prepared at Wittenberg) of the *Computus* and of Sacrobosco's *Sphaera* (1543), it is certain that he followed with great attention that which was being published by the 'Melanchthon circle'. When, in 1546, Gervasius Marstaller (one of Melanchthon's disciples, who was persuaded of the urgency in promoting a Christian astrology), took refuge in Paris, he was welcomed at the *Collège de Beauvais*, where Fine gave courses. Marstaller dedicated to Fine a collection in praise of astrology, containing, among other things, texts by Melanchthon and Joachim Camerarius.[71] It was also perhaps Marstaller who drew Fine's attention to a course of cosmography given at Wittenberg about 1533, the *Elementale cosmographicum* of Martin Borrhaus, also called Cellarius. This text had already been printed at Strasburg in 1536 (by Crato Mylius), through the good offices of Achille Pirmin Gasser. Fine made corrections to it and added notes and figures so that it could be published by Cavellat at the beginning of 1551.[72]

One thus sees in these few examples, how the personality and activities of Oronce Fine deeply marked the role of Royal Lecturer in mathematics. At least for the sixteenth century (for later profound changes took place), this role constituted a sort of model and example, never forgotten even it was not always followed. The enterprises which he undertook were continued after him (if not by all his successors, at least by Ramus and his disciples). These were the defence of the discipline of mathematics in the university, the progress and spreading of technical knowledge, and the introduction into France of the best foreign innovations, and even, where it was possible, an increase in the prestige of French science beyond the frontiers of the kingdom.

Were all these actions carried forward by Oronce Fine on his own initiative? The answer must be strongly nuanced. It goes without saying that the Royal Lecturer followed the leads established by his great patron, the king. More fundamentally still, Fine reflected the major tendencies of the *milieu* from which he came, and which had pushed for his election. This *milieu* was that of the earliest French humanism, deeply marked by its particular ethical and religious orientations.

70 A. Ricci, *De motu octavae sphaerae* (Paris, 1521).

71 See Dupèbe, *Astrologie, religion et médecine*, pp. 325, 352–355. See also I. Pantin, 'La lettre de Melanchthon à Symon Grynaeus: avatars d'une défense de l'astrologie', in J. Céard (ed.), *Divination et controverse religieuse au XVIe siècle (Colloque, Paris, 1986)* (Paris, 1987), pp. 85–101.

72 Borrhaus, ed. Fine, 1551; I. Pantin, *Imprimeurs et libraires parisiens du XVIe siècle d'après les manuscrits de Philippe Renouard: Cavellat – Marnef et Cavellat* (Paris, 1986), no. 27.

The Epistemological Foundations of the Propaedeutic Status of Mathematics according to the Epistolary and Prefatory Writings of Oronce Fine

ANGELA AXWORTHY

François Ier's decision to establish a royal chair of mathematics shortly after the creation of the institution of the Royal Lecturers in March 1530 aimed, first of all, to compensate for the insufficiency of the mathematical teaching that was then provided in the Faculty of Arts of the University of Paris.[1] Thus, the task of Oronce Fine, as the first lecturer appointed to this chair was, first, to propose a course of instruction which would be able to rekindle the University students' interest in mathematics; second, to set forth the dignity and fruitfulness of the mathematical disciplines, as well as their unity and autonomy from the other subjects taught in the Faculty of Arts. In his *Epistre exhortative touchant la perfection & commodite des ars liberaulx mathematiques* – a text published in 1532 but written for François Ier before 1531 – but also in most of the prefaces of the mathematical treatises he published after his appointment to the newly created chair, Fine provided his reassertion of the value of mathematics on the idea that the study of this discipline is necessary to reach the most perfect degree of knowledge in any field of investigation.[2] Hence, in the *Epistre*, he starts his praise of mathematics by stating that mathematics corresponds to 'the keys to all perfect knowledge' and concludes it by stating that 'no man gained pure intelligence of any knowledge without their teaching'.[3] To support these two assertions, Fine sets forth, within this text, an argument which would serve as a model for most of the later texts in which he aimed to prove the necessity of studying mathematics. In particular, it is the

1 A. Tuilier, 'L'entrée en fonction des premiers lecteurs royaux', in A. Tuilier (ed.), *Histoire du Collège de France. I: La création 1530–1560* (Paris, 2006), p. 154. See also Pantin in this volume.

2 Fine, 1532*a*. This text corresponds to a verse eulogy of mathematics, written in French around 1530 for François Ier, although it is not certain whether Fine wrote this epistle before or after the king's decision to appoint him to the first royal chair of mathematics.

3 'Ce sont les clefz de tout perfaict scavoir'. Fine, 1532*a*, sig. Aiii^{v}; 'Onc homme n'eust l'intelligence pure / D'aucun scavoir sans leur enseignement'. *Ibid.*, sig. Biv^{r}.

basis for the general preface of his *Protomathesis* (1532)[4] – a compendium offering a fundamental and elementary course in different mathematical disciplines – as well as for the preface of the *Sphère du monde* (1551), a French edition of the cosmographical part of the *Protomathesis*.[5] This argument aimed to show, by presenting the fact that the mathematical object may be known perfectly and by revealing the eminent certainty of mathematical demonstrations, why mathematics was considered by the ancients (and mostly by Cicero and Plato, to whom Fine often refers his texts) to be the necessary path towards knowledge, and why, therefore, it should be valued as such by all those who wish to reach truth about any subject.

The notion that the study of mathematics should constitute the foundation of all perfect knowledge was frequently put forward by sixteenth-century mathematicians in order to assert the value of their discipline. These mathematicians generally related this idea to the Platonic theory of education, according to which the philosophical contemplation of eternal and divine truths is only accessible to those who have properly studied mathematics beforehand, as was supposedly indicated on the pediment of the Academy.[6] Indeed, according to the *Meno* and to the *Republic*, mathematics, thanks to the fact that it allows one to contemplate things which are immaterial, but nevertheless accessible through the senses, has the power to assist the human soul in going beyond the imperfect stage of empirical and probable knowledge to reach the ultimate and most perfect degree of knowledge, which deals with eternal and divine realities.[7]

During the Renaissance (indeed, since late Antiquity), the propaedeutic status of mathematics had often been connected to the very etymology of the word 'mathematics'. In fact, according to its etymological origin, the term *mathematica* (most commonly *mathematicae* in the plural, to indicate the plurality of the mathematical arts) derives from the Greek terms *mathesis* and *mathema*, which had the general meaning of 'instruction' or 'teaching' in Ancient Greek. As such, the arts which were called *mathemata* (as were arithmetic, geometry, astronomy and music in Plato's *Republic*), were given the status of *propaideia*, that is a condition of knowledge.[8] The theme of the propaedeutic nature of mathematics was taken

4 Fine, 1532*b*. This work comprises a compendium offering an elementary course on different mathematical disciplines gathering together four books of arithmetic, two of geometry (theoretical and practical), five of cosmography and four of gnomonics. Although it was published very shortly after Fine started teaching mathematics in the name of François Ier, this work was probably the most important of his whole career as Royal Lecturer. For further discussion of the *Protomathesis* see P. Brioist in this volume.

5 Fine, 1551*c*.

6 Fine clearly refers to this in Fine, 1532*a*, sig. Aiii[v] and in the preface of Fine, 1532*b*, sig. AA2[r].

7 Plato, *Meno*, 81b–d; *Republic*, 522c–532e.

up from the Platonists by Boethius and widely contributed (at least until the twelfth century) to the placing of these disciplines among the liberal arts, the main function of which was to lead the human soul towards the contemplation of noble and divine things; in other words, to prepare men for philosophy and wisdom.[9] In the Middle Ages, this theme remained present through a literary tradition rooted in the mathematical writings of Boethius, but did not allow the mathematical arts to be properly considered as propaedeutic, or even to be considered as an autonomous subject within the curriculum of the medieval and Early Modern universities. However, during the sixteenth century, thanks to the rediscovery of the works of Plato and the commentary of Proclus on the first book of Euclid's *Elements*, the reassertion of the propaedeutic status of mathematics went hand in hand with the *restauratio mathematicarum*, which was undertaken especially in Italy in the mid-sixteenth century.[10] The growing influence of Platonism, and the desire to return to a model of teaching that was inscribed in the continuation of the Platonic tradition, was also one of the determining factors in François Ier's decision to found a royal chair of mathematics following his initial creation of chairs in Greek and Hebrew.[11]

Thus, it is not surprising to find that Fine established his own promotion and defence of mathematics on the notion that the study of mathematics should constitute the first and necessary step towards the acquisition of knowledge. It is, however, particularly interesting to see the extent to which this idea holds a central place in Fine's discourse, since it not only allowed him to explain the finality, the dignity and the fruitfulness of these arts, but also allowed him to legitimise his own status of Royal Lecturer in mathematics (not to mention the many other aspects of his mathematical practice, notably cartography and mathematical instrument making). We may therefore understand what is at stake behind the choice of the title *Protomathesis* for his own mathematical encyclopaedia, as this word is none other than the Latin transcription of a Greek word literally meaning 'first instruction'. Thus, in order to show why the ancients considered arithmetic, geometry, music and astronomy[12] to be the only arts which are 'rightfully called mathemat-

8 Plato, *Republic*, *idem.*

9 See Boethius, *De institutione arithmetica libri duo; De institutione musica libri quinque; accedit geometria quae fertur Boetii*, ed. G. Friedlein (Leipzig, 1867), pp. 7–10.

10 On the importance of the *Elements* and Proclus's commentary for sixteenth-century mathematics, see Cifoletti in this volume. For the 'renaissance of mathematics' see P. L. Rose, *The Italian Renaissance of Mathematics: Studies on Humanists and Mathematicians from Petrarch to Galileo* (Geneva, 1975).

11 Tuilier in his *Histoire du Collège de France*, p. 155. See also Pantin in this volume.

12 According to the tradition instituted by Boethius, in his *De institutione arithmetica*, Fine chose, as did most of his contemporaries, to follow the Pythagorean division of mathematics. We see this in the *Epistre exhortative*, but also in the prefaces of the *Pro-*

ics, that is true disciplines' (the term *disciplina* implicitly corresponding to the Greek term *mathema*), Fine developed (in the three above-mentioned texts) a discourse which not only presents the main aspects of his philosophy of mathematics, but also his definition of knowledge.[13]

As we shall see, in the texts mentioned above, Fine's definition of the propaedeutic status of mathematics is mainly founded on the idea that the study of this discipline offers the intellect a form of investigation which is able, by itself, to lead to irrefutable judgments (that is judgments of the highest degree of truth) and which is able, from there, to prepare the mind to seize truth outside the field of mathematics. In the *Epistre exhortative*, in the general preface of the *Protomathesis*, and in the preface of the *Sphère du monde*, this assertion is advanced through two main arguments.[14] The first is based on the traditional notion that mathematics holds an intermediate place between metaphysics and physics in the hierarchy of speculative disciplines. The second is founded on the higher certainty of mathematical demonstrations as opposed to the demonstrations of other disciplines. This second argument, as we shall see, allowed Fine to define the procedures through which mathematics is said to offer the human intellect its first experience of certainty and, from there, enable it to produce truthful judgments in non-mathematical domains.

If, in order to explain why mathematics should be studied before any other discipline, Fine starts by explaining in these texts why it must be considered as intermediate between metaphysics and physics, it is because this allows him to put forward the notion that the things which are studied by the mathematician are, as for their nature, partly sensible and partly intelligible and are therefore fully accessible to the human intellect, as opposed to things which are completely sensible or intelligible. From this, he advances the notion that mathematics presents the firmest and most certain demonstrations of all and thus represents the first step towards the production of all truthful judgments. As my aim is to show on what grounds Fine founded the propaedeutic status of mathematics, I shall follow the order of his argument and therefore start my analysis by examining in greater detail how the assertion of the intermediate position of mathematics between metaphysics and physics enabled him to demonstrate the epistemic superiority of his discipline and, from there, its ability to lead to a fully accomplished knowledge in all fields of investigation.

---

tomathesis and of the *Sphère du monde*. Interestingly enough, in his *Protomathesis* Fine displays the theory of four different mathematical disciplines, thus evoking the quadripartite division of the Boethian quadrivium, though in this text music is replaced by gnomonics.

13 'Quales sunt ueteres illae, fideles, ac divinae artes: quae solae Mathematicae, hoc est, uerae disciplinae, haud immeritò uocitantur.' Fine, 1532*b*, sig. AA2^{r}.

14 Fine, 1532*a*, sig. Aiii^{r-v}; Fine, 1532*b*, sig. AA2^{r-v}; Fine, 1551*c*, sig. [Ai.^{r}].

I. *The Intermediate Status of Mathematics as First Proof of its Ability to Lead to Knowledge*

In the *Epistre exhortative*, the notion that mathematics should be situated at an intermediate level between metaphysics and physics is introduced through the presentation of the different divisions of 'Lady Philosophy' (in fact the narrator of the epistle):[15]

> Coming to the point that must be understood first, / I can be taken in three ways / Or separated in three principal parts: / For I allow the secrets of God to be understood, / And then afterwards bring the mind down / To the living, vegetating, mineral, / Between which are the liberal arts, / Named above, as the true roots, / Of all knowledge and noble disciplines.[16]

In this passage, we can clearly perceive the underlying influence of the Aristotelian division of theoretical sciences that was traditionally present within medieval and Early Modern classifications of sciences and arts.[17] Founded on the distinction of the different types of beings according to their degree of abstraction from matter, this division includes theology (which is here presented as the contemplation of the 'secrets of God'), natural philosophy (which here corresponds to the study of the 'living, vegetating, mineral') and mathematics (which Aristotle defined as the study of abstract quantity, but which is interestingly defined in this passage by its propaedeutic status rather than by the nature of its subject matter, being here referred to as 'the true roots of all knowledge and noble disciplines').

The hierarchical order which Fine establishes here between the three parts of philosophy (placing mathematics at an intermediate level between the superior level of theological or metaphysical knowledge and the inferior level of natural philosophy) evokes the hierarchy of knowledge defined by the Platonists, and mainly by Proclus, in the prologue of his commentary on the first book of Euclid's *Elements*.[18] In this text, Proclus explains that mathematics, inasmuch as it concerns things which are immaterial and divisible, must be regarded as inferior – on the

[15] Fine's decision to place his defence of mathematics in the mouth of the personification of Philosophy seems to stem from the will to inscribe his discourse in the tradition of Boethius's *Consolation of Philosophy*, thereby manifesting the highly philosophical value of his project to restore mathematics.

[16] 'Venant au poinct il faut premier entendre / Que l'on me peust en troys manieres prendre / Ou separer en trois pars principales / Car les secretz de dieu je fais comprendre / Et puis apres l'entendement descendre / Sur les vivans vegetans minerales / Entre lesquelz sont les ars liberales / Dessus nommez, comme vrayes racines / De tout scavoir et nobles disciplines.' Fine, 1532*a*, sig. Aiii[r].

[17] Aristotle, *Metaphysics*, 1026a13–20. For the history of this division during the medieval period, see O. Weijers, *Le maniement du savoir: Pratiques intellectuelles à l'époque des premières universités (XIIIe–XIVe siècles)* (Turnhout, 1996), pp. 187–196.

[18] Proclus, *In primum Euclidis elementorum librum commentarii*, ed. G. Friedlein (Hildesheim, 1967), pp. 3–5.

scale of dignity and ontological reality – to the contemplation of intelligible or metaphysical entities (which are immaterial and indivisible) but nevertheless as superior to the study of sensible or physical beings (which are material and divisible). Thus, following the binary representation of reality of his master Plato, Proclus considered mathematical things to be the necessary ontological intermediary between the radically distinct and unconnected worlds of intelligible and sensible beings.[19] Indeed, while mathematical things represent, according to this conception, the multiple and divisible images of the immaterial and indivisible principles of all things, they constitute at the same time the immaterial models of material and divisible beings. As such, mathematical things were conceived as those things to which sensible things (described as material and divisible) may imitate and partake of intelligible principles (believed to be immaterial and indivisible) according to the double mode of materiality and multiplicity. Now, as mathematical beings (which were considered to be immaterial and divisible) were therefore regarded as ontologically intermediate between the intelligible and the sensible, Proclus, following Plato, conceived of their study as the necessary means through which the philosopher (whose soul was thought of as limited and weakened by its conjunction with the matter of the body) may learn to look beyond the imperfect and untruthful level of corruptible and complex realities to progressively grasp the perfect and imperishable truth of eternal and undivided beings.

Though it is not certain whether Fine was familiar with this part of Proclus's writings, the fact that, in this text, he describes mathematics as the condition of all perfect knowledge at the same time as proposing that it should be placed at an intermediate level between theology and natural philosophy, could very well appear to be – at first sight – related to an underlying Platonic influence on his conception of knowledge.[20] However, if the association of these two themes is indeed generally ascribed to the Platonic tradition, the notion that mathematics holds an intermediate place between metaphysics and physics is more likely to have been transmitted to Fine through the prologue of Ptolemy's *Almagest*. Indeed, in this text (which Fine knew of, since his cosmological teaching is practically entirely based on the Ptolemaic system), this assertion appears in a way which is similar to that in which it appeared in Fine's own prefaces and prologues fourteen centuries later. According to Ptolemy, if mathematics must be considered as intermediate

19 Plato, *Republic*, 510c–511d.

20 Though the manuscript of this text had been rediscovered at least by 1501 (as parts of it may be found in Giorgio Valla's *De expetendis et fugiendis rebus opus* [Venice, 1501]), this text was only published as a whole in 1533 by Simon Grynaeus, along with the *editio princeps* of Euclid's *Elements*. If it is therefore unsure whether Fine new this text directly, it is however certain that he had read Valla's *De expetendis et fugiendis rebus opus*, since he mentions the treatise of geometry which was part of it in his *Quadratura circuli* (Fine, 1544c).

between metaphysics and physics, it is because the knowledge of its objects may be reached both through the senses and without the senses (in other words, through the intellect alone), therefore embracing material as well as immaterial things.[21] Interestingly, in his prologue, Ptolemy draws from this statement a conclusion which (as we shall see) was to play an essential part in Fine's demonstration of the propaedeutic status of mathematics. According to it, mathematics should be regarded, on one hand, as more accessible to human beings than metaphysical knowledge (of which the object is so simple and so pure that it may not be directly understood by mortal beings) and, on the other, as more necessary and certain than physical knowledge (which aims at contemplating things which are material, and therefore constantly changing and contingent).[22]

In the following passage, Fine gives further details about the ways in which he considers mathematics to be intermediate between metaphysical and physical knowledge, and about how this intermediate position determines its power to lead to true knowledge:

> For just as the angelic spirit / Is inferior to divine power / And is superior to human beings: / Thus I say mathematical order / Is intermediate between theological law / And the knowledge which concerns nature / Partaking in both, through a chain / Without exceeding the human understanding, / For which it must be more greatly prized. // It is therefore clear that [the most noble / mathematics] is perfect, authentic, / And the mirror of all certainty: / For all the noble or mechanical arts, / Even those which are the most magnificent / Have drawn from it their course and habit.[23]

In this passage, Fine uses a comparison with the intermediate status of the angel between God and man to illustrate the fact that he considers mathematics to be intermediate between theology and natural philosophy insofar as it is related in a certain way to each of these two forms of speculative knowledge. Immediately after, he affirms that this double relation occurs according to a mode which does not exceed the limits of human understanding, adding, in the following verse, that it is for this reason that mathematics (here referred to as 'mathematical order') should be considered valuable. Having said this, he then asserts that mathematics is 'perfect, authentic, and the mirror of all certainty', in other words, that this dis-

21 See Ptolemy, *La composition mathématique*, trans. M. Halma (Paris, 1988 [facsimile of the edition of 1813]), pp. 2–3.

22 Ptolemy, *La composition mathématique*, p. 3.

23 'Car tout ainsi que l'esprit angelique / Est au-dessoubz du povoir deifique / Et surmontant l'humaine creature / Ainsy je diz l'ordre mathematique / Estre moyen du droit theologique / Et du scavoir qui concerne nature / Participant des deux, par ligature / Sans exceder l'humain entendement / Dont on le doibt priser plus amplement. // Il est donc cler que les mathematiques / Tresnobles sont parfaictes, authentiques / Et le miroer de tout certitude / Car tous les ars nobles ou mechaniques / Mesmement ceux qui sont plus magnifiques / D'elles ont prins leurs cours et habitude.' Fine, 1532*a*, sig. Aiii^{v}.

cipline not only is the most perfect as regards certainty and truth, but also has the power to reveal the degree of certainty and truthfulness of all types of investigations. In order to justify this, Fine pleads the fact that all arts, be they of a technical or of a speculative nature, draw their disposition and their way of proceeding from mathematics.

It should be noted, before going any further, that, according to this passage, the propaedeutic power of mathematics has a properly universal scope, leading to 'mechanical' or utilitarian arts as much as to noble or speculative knowledge. As such, this text seems to reveal Fine's intention to contribute to the evolution of the status of technical knowledge, which progressively took place during the sixteenth century, thanks to the efforts of humanists and mathematicians to reassert the certainty and the reliability of mathematics and, through this, the dignity of its fields of application.[24]

Returning to the general purpose of this passage, what Fine appears to say here is that, thanks to the manner in which mathematics is, on one hand, connected to the study of divine things and, on the other, related to the study of natural things, this discipline offers a form of contemplation which conforms perfectly to the requisites set by the limitations of the human intellectual faculties. Indeed, as we shall see, mathematical things are partly sensible, partly intelligible, and, for this reason, may be reached by the intellect through the senses. From there, the mathematician's enquiry may be considered as fully able to satisfy the requirements of *scientific* knowledge.[25] Thereafter, mathematics may be conceived as able to display the procedures through which the human mind may produce a discourse which is truthful and certain about all subjects of investigation.

To confirm this hypothesis, let us first examine (thanks to a passage from the preface of the *Protomathesis*) how Fine presents, through a justification of the ontological intermediacy of mathematics between metaphysical and physical knowledge, the proper nature of mathematical things:

> Mathematics is intermediate between the natural or physical investigation and the supernatural or metaphysical investigation (which deserve to be

24 See G. Cifoletti, 'L'utile de l'entendement et l'utile de l'action: discussions sur l'utilité des mathématiques au XVIe siècle', *Revue de synthèse*, vol. 4 (2001), pp. 505–519; A. Keller, 'Mathematical Technologies and the Growth of the Idea of Technical Progress in the Sixteenth Century', in A. G. Debus (ed.), *Science, Medicine & Society in the Renaissance: Essays to Honor Walter Pagel*, 2 vols. (London, 1972), vol. 1, p. 11–27; P. Rossi, *Les philosophes et les machines 1400–1700* (Milan, 1962), p. 60–65.

25 Here, the word 'scientific' must be understood as a means to distinguish the type of knowledge which is firmly founded on irrefutable propositions (in accordance with Aristotle's definition of *episteme* [*scientia*] in the *Posterior Analytics*) and which is regarded as the highest point on the scale of necessity and truth, from a knowledge which is only founded on probable propositions and, as such, considered as unnecessary and less truthful.

> called conjectures [*coniecturae*] rather than knowledges [*scientiae*]), taking part, with the natural, in matter, and joining the supernatural in the fact that it considers the same things as if they were separated from matter.[26]

According to what is said in this passage, if the investigation of the mathematician is here related to that of the natural philosopher, it is insofar as they both aim at knowing things which are material. Having said this, if mathematics is said to partake of the investigation of the metaphysician, it is inasmuch as the mathematician, in order to examine the nature of his objects, considers them as if they were separated from matter, just as are the divine and supra-sensible things which the metaphysician investigates. Thus, if Fine compares mathematics, on one hand, with physics and, on the other hand, with metaphysics, it is because of the nature of its objects of investigation, which are at the same time connected with matter (as are natural things) and able to be considered without any matter (as are metaphysical things).

This definition may seem contradictory, as natural things, with which mathematical things here seem to be identified, can never be separated from matter, not even in the imagination. However, if Fine can say, in this context, that the mathematician aims at knowing material things and, at the same time, that the mathematician considers these things as if they were deprived of matter, it is because – according to a conception which was commonly held during the Renaissance (which followed Thomas Aquinas's interpretation of Boethius's *De Trinitate*) – he implicitly distinguishes two types of relationship which mathematical objects have with matter.[27] The first is established *secundum esse*, that is according to the 'being' of the known thing (which here designates the nature or the mode of existence which it presents in reality, before any apprehension by the intellect), and the second *secundum intellectum*, that is according to the way it is

26 'Sunt enim Mathematicae mediae inter naturalem seu Physicam auscultationem, & supernaturalem siue Metaphysicam (quae coniecturae potius, quàm scientiae dici meruerunt) cum naturali participantes in materia, & cum supernaturali conuenientes in eo, quoniam res easdem, perinde acsi forent à materia seiunctae considerant.' Fine, 1532*b*, sig. AA2r. In the preface of the *Sphère du monde*, Fine expresses this idea according to a similar formulation: '[...] lesdittes mathematiques sont moyennes entre la divine & la naturelle philosophie: convenans avec la naturelle, d'autant qu'elles concernent les corps materiels: & participans de la divine, pource qu'elles traictent desdicts corps abstractivement, comme des choses spirituelles separees de leur mouvement & matiere.' ['Mathematics is intermediate between divine and natural philosophy, joining the natural inasmuch as it concerns material bodies, and partaking of the divine insofar as it deals with these bodies in an abstract manner, as spiritual things separated from their movement and matter'] Fine, 1551*c*, sig. [Ar].

27 T. Aquinas, 'Opusculum LXX, Super Boetio De Trinitate' in A. Pizamini (ed.), *Thomae opuscula* (Venice, 1490), sig. FFiv. See also Aristotle, *De Anima*, III.7, 431b12–16; *Physics*, II, 193b22–35 and 194a4–7; *Metaphysics*, VI.1, 1026a13–17 and XI.7, 1064a31–35.

seized by the intellect. Therefore, according to their existential condition, numbers and magnitudes (which are assimilated with the different types of quantity and which properly correspond to the objects of mathematical investigation) cannot exist outside sensible matter and are therefore necessarily encountered within physical bodies. Now, according to the intellect, this connection between mathematical things and matter is not imposed by any logical necessity and therefore must be conceptually represented as if they were actually separated from matter.[28] From there, if the object of the natural philosopher may be distinguished from that of the mathematician it is insofar as it is inseparable from matter according to facts, as well as according to the intellect. Reciprocally, if the object of the metaphysician may be distinguished from that of the mathematician, it is inasmuch as it is separated from matter not only according to the intellect, but also according to facts. Thus, if Fine may assert, without any contradiction, that mathematics partakes of both physics and metaphysics, it is because its subject matter has the same relationship with matter as that of the former *secundum esse* and as that of the latter *secundum intellectum*.

According to the above-described distinction, we can see clearly that Fine follows the abstractionist conception of Aristotle, who, in *De anima*, states that mathematical objects – though they are inseparable from the matter of natural beings – are considered as separate when they are conceived by the intellect.[29] According to this conception, mathematical things do not, in fact, have a separate being from that of sensible things and are not, from an ontological point of view, considered as proper substances, but rather as properties of physical substances.[30] As such, mathematical things may only be regarded as independent from natural beings as a result of the process led by imagination, through which the quantitative elements of bodies are separated from their matter, movement and individual characters. As we shall see, the capacity of numbers and magnitudes to be separated from matter by the imagination is, in Fine's conception, the condition of the

[28] If this particular status of mathematical things is generally illustrated by the condition of the geometrical object in the discourse of Aristotle and his followers, as is also often the case in Fine's texts, it must also be noted that Fine does consider numbers to bear the same ontological duality as magnitudes, as indicated by the preface of the 1544 edition of the *Arithmetica practica*: 'Inter ea quae in rerum universa natura (…) non possunt absque materia subsistere, & nihilominus extra materiam facilè sunt intelligibilia: maximè est ipsa unitas, & qui ex eadem unitate procreatur numerus.' ['Among those things which, in the whole of nature, cannot subsist without matter and which are nevertheless easily seized outside matter, there is most of all unity and number which is produced from the same unity'.] Fine, 1544*a*.

[29] Aristotle, *De anima*, III.7, 431b15–16.

[30] See references in n. 27, above, and *Metaphysics*, M, 2, 1076b12–13, 1076b40–1077a19, 1077a31–b18.

higher knowability of the mathematical objects, from the point of view of the human intellect.

II. *The Higher Knowability of the Object of Mathematics*

As indicated above, the implicit definition of the ontological status of mathematical objects is clearly what allows Fine to put forward the idea that the nature of mathematical things is perfectly compatible with the condition of the human intellect. Indeed, in this context the human intellect is traditionally conceived as dependent upon the senses to reach its objects but, as opposed to the senses, represents a purely rational mode of apprehension, aimed at knowing that which is unchanging and deprived from individual properties (in other words, that which is immaterial, universal and necessary). As such, mathematical things are regarded as perfectly accessible to the human intellect. Indeed, because (as we have seen) Fine considered mathematical things to be present within natural things but separable by the intellect from matter, he therefore perceived them as accessible to the senses, but also as proper to be seized as things which are immaterial, universal and necessary. It is in this sense that we should understand Fine's words when he says, in the *Epistre*, that mathematics is linked to theology and natural philosophy in a manner which does not 'exceed the human understanding'.[31] As we shall see, it is first of all on the basis of this compatibility of the nature of mathematical things and of the human intellect that Fine considered (notably in the *Protomathesis*) that the mathematician, as opposed to the metaphysician and the natural philosopher, is able to reach by himself properly irrefutable conclusions in his own domain of investigation and is consequently fully able to satisfy the requisites of *scientia*.

In a passage from the prologue of the *Geometria libri duo*, Fine presents the mode through which the geometer, and, by extension, the mathematician in general, apprehends and examines his objects. On this occasion, Fine explains more precisely the reasons why the abstract nature of mathematical things determines the epistemic perfection of mathematical investigation:

> Geometry [...] as it examines intelligible things while being in contact with sensible things, has knowledge of the cause (*propter quid*) and of the fact (*quia*). For the soul, as it feebly embraces the properties of these things by the sense of vision, aims at separating their knowledge from the senses by conceiving a figure that is different from that which is seen, and by displaying demonstrations about this other figure.[32]

[31] Fine, 1532*a*, sig. Aiii^{v}.

[32] 'Cognoscit enim propter quid, & quia est, circum intellectilia versans, sensilia tamen attingendo. sententia namque animi, cum suas aspectu debiliter amplectatur rationes, à sensibus cognitionem ipsarum tentat abducere: aliam ab ea inspicitur concipiendo figuram, & circum aliam demonstrationes ostentans.' Fine, 1532*b*, sig. G2^{r}.

Using, in this passage, the expressions *propter quid* and *quia* to describe the types of information which can be obtained from the study of geometrical figures, Fine clearly assimilates the demonstration of the geometrician to the firmest and most necessary form of demonstration, as defined by the scholars of the medieval period according to Averroës's interpretation of Aristotle's theory of scientific demonstration. While Aristotle distinguished, in the *Posterior Analytics*, two types of demonstrations (demonstration of the fact – subsequently designated as *demonstratio quia* – and demonstration of the cause or the reason why – *demonstratio propter quid*), Averroës defined a third type, traditionally referred to as *demonstratio absoluta* or *simpliciter*, which is at the same time *quia* and *propter quid* (giving the fact and at the same time its proper cause), and which is considered as the most 'powerful' in terms of production of knowledge.[33] Averroës (in his sixteenth commentary on the second book of Aristotle's *Metaphysics*) attributed this third type of demonstration to mathematics in order to show that this form of knowledge is first in the order of certainty.[34] By asserting that the geometer knows the *quia* and the *propter quid*, Fine therefore followed Averroës's opinion according to which mathematical demonstrations are the most certain of all and are the most likely to reach a properly scientific knowledge.[35]

According to this interpretation of Aristotle's theory of demonstration, a demonstration is truly knowledge-producing if it shows not only that something is so (as does the *demonstratio quia*), but also why it can never be otherwise (as does the *demonstratio propter quid*). Now, what is seized within a demonstration

33 Aristotle, *Posterior Analytics*, I.13, 78a22–79a16; Aristotle-Averroës, *Aristotelis Opera cum Averrois commentariis*, 11 vols., 3 suppl. (Venice, 1562–1574), reprinted (Frankfurt am Main, 1962), vol. 4, p. 4r. Shin'ichiro Higashi, 'Penser les mathematiques au XVIe siècle. La Philosophie des mathematiques chez Marcantonio Zimara', in *Historia Scientiatum*, 11, no. 2, 2001, pp. 143–167.

34 Aristotle-Averroës, *Aristotelis Opera cum Averrois commentariis* '[C]ontra id, quod fit in ipsis Mathematicis, in quibus scilicet fiunt vt plurimum demonstrationes simpliciter, et absolutae.', vol. 4, p. 4r. Thus, Averroës started a debate concerning the certainty of mathematics that lasted at least until the end of the sixteenth century, and which culminated with Alessandro Piccolomini's *Commentarium de certitudine mathematicarum scientiarum* which was published with his *In mechanicas quaestiones Aristotelis paraphrasis* (Rome, 1547).

35 Again, if Fine here only describes the type of knowledge reached in geometry, and does not mention the other parts of mathematics, it is first of all because he is introducing his treatise on geometry and, secondly, because the general form of mathematical demonstration is traditionally assimilated with the demonstrations presented in Euclid's *Elements*, which principally deals with geometry. Thus, this does not mean that Fine considers that such demonstrations do not occur in other parts of mathematics. As we shall see, the fact that geometry is able to present demonstrations which are at the same time *quia* and *propter quid* is founded on the abstract nature of its objects which is the same for all mathematical things (see also above, n. 28).

which is both *quia* and *propter quid* is the essential properties of a given object of investigation, which explain not only what a thing is, but also why it is universally and necessarily so (thus constituting the proper subject of *scientia* according to Aristotle).

Here, the ability of the geometer to produce demonstrations of the 'reasoned fact' (that is, demonstrations which display at one and the same time the studied fact and its proper cause) is clearly related to him investigating things which, though they cannot in fact exist without matter, may be apprehended in the intellect as if they were independent from matter. In other words, thanks to their abstract nature, magnitudes are apprehensible by the intellect as things which are invariable and inalterable, thus revealing their essential properties and proper causes or, in other words, that which in them may be regarded as universal and necessary. As such, thanks to the fact that they can be seized by the intellect as invariable and inalterable, geometrical things (and with them all mathematical abstractions) may constitute the subject of the firmest and most certain demonstrations and, from there, lead to truthful judgments.[36]

As presented in the above-quoted passage, if the geometer is able to investigate things in a manner which is proper to *scientia*, it is however thanks to the dual nature of magnitudes, which (as was suggested in the *Epistre*) is fully compatible with the limitations of the human intellect. Indeed, because the intellect is itself conjoined with the matter of the body, it is forced at first to reach all of its objects of investigation thanks to the senses, separating them (if possible) from matter in a purely intellectual manner and exploring their essential properties by contemplating the product of this process of abstraction. Thus, because the mind is 'weakened' by its connection with the body, only mathematical things, which are accessible through the senses but potentially separable from matter by and within the intellect, are able to be seized in their universal and necessary form. In this sense, we can see clearly that if mathematics, and not only geometry, is considered by Fine as the 'mirror of all certainty' and as able to reach at the same time the fact and the cause, it is indeed thanks to the compatibility of the nature of its objects with the requirements set by the nature of the human intellect.

III. *The Higher Certainty of Mathematical Demonstrations*

If the human intellect depends upon the senses to apprehend its objects of investigation, metaphysical or divine things, which are separated from matter not only *secundum intellectum*, but also *secundum esse*, must therefore be considered to be out of its reach. As such, the essential properties and proper causes of divine things may not at all be embraced and properly known by the human mind (weakened as it is by its connection with the body). On the other hand, despite the fact

36 This argument is present in Averroës's commentary to Aristotle's *Metaphysics*. See Aristotle-Averroës, *Aristotelis Opera cum Averrois commentariis*, vol. 8, fol. 35^{v}.

that natural things are accessible to the intellect through the senses, the fact that they cannot, even in the imagination, be separated from matter (and from the variations induced by their material condition) implies that they can never present the invariability and inalterability required of all subjects of scientific investigations. Thus, since metaphysics and physics aim at knowing things which are either completely separated from matter (and thus inaccessible to the senses) or absolutely inseparable from matter, not even by the intellect (and hence inconceivable as universal and necessary), none of these forms of investigation 'deserve to be called *scientiae*', but should rather be assimilated to *coniecturae*.[37] In other words, neither physics nor metaphysics, as opposed to mathematics, present truly necessary demonstrations and are able to lead to judgments other than probable. Therefore, if the mathematical object, thanks to its partial materiality and its partial immateriality, is properly compatible with the limitations of the human intellect, it is clear that mathematics (and not only geometry) may be conceived as the only form of investigation which is capable of leading by itself to true *scientia*.

In the preface of the *Protomathesis*, in the passage that follows the presentation of the relationship of mathematics with metaphysics and physics, Fine asserts the epistemic superiority of mathematics by appealing this time to the very form of the mathematical demonstration:

> [Mathematics] also has the first degree of certainty among all of the disciplines of the very liberal philosophy, given that it draws its origin from principles which are first and immediately known by themselves.[38]

In this passage, Fine intends to show that mathematics is more certain than physics and metaphysics insofar as it presents demonstrations which are founded on non-demonstrable and self-established premises.[39] To be precise, Fine implicitly establishes the higher certainty of mathematics on the conformity of its demonstrations to the definition of scientific demonstration, as set out by Aristotle in the *Posterior Analytics*. Indeed, according to the Aristotelian theory of knowledge, the highest form of demonstration (through which are reached the most certain and truthful judgments) is defined as a deductive reasoning thanks to which the knowledge sought is thoroughly and firmly drawn from principles which are immediate, non-demonstrable, self-evident, and thanks to which are

37 Fine, 1532*b*, sig. AA2^r.

38 'Primum quoque certitudinis gradum, inter omnes liberalioris Philosophiae disciplinae obtinent: quoniam ex primis, ac immediatè per sese notis principijs, suam ducunt originem.' *Idem.*

39 In this context, the disciplines referred to as *liberalioris Philosophiae disciplinae* evidently correspond to the three divisions of philosophy which we first encountered in the *Epistre exhortative*. So, after having been compared to metaphysics and physics as for the nature of its objects, mathematics is here compared to the same disciplines regarding the degree of certainty of its demonstrations.

displayed the proper causes of that which is known.[40] According to this definition, the scientific character of the knowledge thereby produced is founded on the fact that it may be apodictically established on particular premises, the necessity of which immediately imposes itself on the intellect. These premises, which are necessarily more evident than all the propositions obtained by demonstration and therefore known before all demonstration, may, as such, ensure the irrefutability of the conclusion which is deduced from them and, hence, constitute the proper foundation of a knowledge-producing demonstration. Therefore, by asserting, in the above-quoted passage, that mathematics 'draws its origin from principles which are first and immediately known by themselves', Fine surely intends to found the eminent certainty of mathematical knowledge on the conformity of its demonstrations to the Aristotelian definition of scientific reasoning. In so doing, Fine makes explicit the message which underlies the parenthesis included in the preceding passage of the *Protomathesis*, namely that if metaphysics and physics should be regarded as *coniecturae*, only mathematics may properly be assimilated to *scientia*.

IV. *The Foundation of the Ability of Mathematics to Lead us to Knowledge*

One of the main reasons Fine considered mathematics to be more certain than the other parts of philosophy is that, thanks to the abstract nature of its objects and to the necessity of its demonstrations, the mathematician, as opposed to the metaphysician and the natural philosopher, is able to reach necessary and irrefutable conclusions by the sole power of his own intellectual faculties. As such, the mathematician does not need to have studied any other discipline beforehand in order to reach irrefutable judgments in his own field of investigation. Now, Fine did not consider that it is impossible to produce a properly scientific knowledge in the framework of a physical or metaphysical investigation, but rather that, in order to do this, one must have beforehand acquired the knowledge thanks to which he would afterwards be able to discern truth from falsehood in any discipline. Therefore, if metaphysics and physics, as opposed to mathematics, are considered in the preface of the *Protomathesis* as *coniecturae* rather than *scientiae*, it is because (for the reasons we have seen) these forms of investigation are not *by themselves* able to lead to a knowledge which is universal and necessary, and hence properly irrefutable. This point is fundamental for Fine's demonstration of the propaedeutic nature of mathematics, as it explains not only why this discipline should be considered to be more certain than the other types of speculative knowledge, but also why it must be conceived as the necessary condition of the acquisition of all types of knowledge. Thus, after having shown, in the preface of the *Protomathesis*, that mathematics represents the only type of investigation which may (thanks to the

[40] Aristotle, *Posterior Analytics*, I.2, 71b17–22.

abstract nature of its subject matter and thanks to the firmness of its demonstrations) reach *by itself* a knowledge which is certain and conforms to the traditional definition of *scientia*, Fine affirms that this discipline may therefore *rightfully* be regarded as the necessary propaedeutic to all disciplines:

> Any learned man must consider that only mathematics holds an intermediate place between intelligible and sensible things and is pure, certain, inviolable and eternally invariable as for its essence. Its excellent worth, order, firmness of reasoning and invariability of contemplation show the path towards the knowledge of all things and towards erudition.[41]

As a consequence, that which allowed Fine to show evidence of the higher certainty of mathematics is also that which allowed him to prove that this type of study is the only one which has the power to lead anyone to truthful knowledge in all fields of investigation. Indeed, if, as is said here, only mathematics may represent the proper propaedeutic to all disciplines, it is because only its object is at the same time accessible through the senses and conveys the purity, the inalterability and the invariability which conditions the production of all universal and necessary judgments.

The mathematical object's potential to be known in a way which is proper to *scientia* is based, as we have seen, on its ability to found invariable speculations and to lead to demonstrations which are firmly established on self-sufficient and self-evident premises. Having said this, if, according to Fine, mathematics is able to assist us in the learning and the discovery of all types of truths (and not only of mathematical truths) it is firstly thanks to the fact that its demonstrations represent the model of all demonstrations. Indeed, in the texts we have presented, mathematical demonstration is assimilated to the most perfect type of demonstration. As Fine says in the *Epistre*, mathematics is perfect, authentic and the 'mirror of all certainty: / For all the noble or mechanical arts, / Even those which are the most magnificent / Have drawn from it their course and habit'.[42] In this passage, Fine advances the idea that mathematics fulfils its propaedeutic function towards the other disciplines by unveiling the method, in other words the disposition and the procedures, which must be followed in all investigations in order to reach the desired knowledge.

Now, if the demonstration of the mathematician clearly displays, according to Fine, the criteria of the highest form of demonstration, what properly allows mathematics to show us the path towards knowledge is mainly the fact that, thanks

41 'Adeò quidem ut solae Mathematicae, medium inter intellectilia sensiliaque locum adeptae, purae, certae, inuiolabiles, ac stabilis semper essentiae, ab quouis censendae sint erudito : quarum excellens decor, ordo, rationum firmitudo, ac inspectionum stabilitas, ad uniuersorum scientiam uiam praebet, & eruditionem.' Fine, 1532*b*, sig. AA2^{r}.

42 See above, n. 23.

to the thorough study of mathematical things, this discipline gives individuals the opportunity to exercise and develop their natural abilities to know. In this sense, the very recognition of the certainty of mathematical demonstrations is itself dependent upon the exercise which mathematics offers to the human intellect and which afterwards enables one to discern truth from falsehood in other disciplines. Thus, more than the formal structure of mathematical reasoning, it is the very *disposition* of the human cognitive faculties which is fashioned by the study of mathematical things, which seem to be properly knowledge-producing. Therefore, if Fine regards mathematics as the necessary condition of the acquisition of universal and necessary knowledge in all disciplines, it is insofar as it enables one's intellect to reach the very disposition through which it may discern truth.

In the preface of his edition of the first six books of Euclid's *Elements*, Fine therefore says about geometry (after having presented the perfection of its demonstrations and the dual nature of its objects) that there is no discipline which 'favours, develops and enriches more the powers of the mind, or which leads the same mind, *carried by its own nature*, to the purest studies and to the invention of the noblest discoveries'.[43] In this passage, Fine implies that if geometry (and with it all the other mathematical disciplines) may lead us to the knowledge of all things, it is indeed inasmuch as it has the power to reveal the intrinsic nature and manner of proceeding of human reason, thus allowing one to develop and to operate one's intellectual faculties in a way that is most proper to reach truly scientific knowledge.[44] Thus, in the preface of the *Sphère du monde*, to conclude an argument which is structurally identical to that put forward in the *Epistre* and in the preface of the *Protomathesis*, Fine says:

> And this being so, we could not name an author, who is excellent and highly considered in whichever profession, who has not been a mathematician: and, on the contrary, no man has despised mathematics who has not been found dull-witted, and not only incapable of reaching mathematics, but also all philosophy. As fire is the proof of gold, mathematics is the confirmation of the understanding, drawing all good minds from all things to itself, and compelling them to go through the rigor of truth: which does not happen in other arts, dependant as they are upon the will and the invention of men.[45]

43 'Nulla etiam quae vires ingenij magis foveat, augeat, locupletétque: *vel quae ingenium ipsum* ad puriora studia, omniùmque ingenuarum adinventionum excogitationem, adeò facile reddat, ac suapte suapte natura propensum.' Fine, 1551*a*, sig. *ij^{r} (emphasis mine).

44 In the preface of the *Sphère du monde*, a similar idea is set forth concerning all mathematical disciplines, that is arithmetic, music and astronomy, as well as geometry.

45 'Et qu'il soit ainsi, on ne sçauroit nommer autheur excellent & approuvé, en quelque profession que ce soit, qui n'ait esté mathematicien: & au contraire, jamais homme ne desprisa lesdittes mathematiques, qui ne fut trouvé lourd d'esprit, & non seulement incapable de la mathematique, ains de toute philosophie. Car tout ainsi que le feu est

In this passage, Fine starts by putting forward the notion that there is a true relation of causality between the fact of possessing knowledge and the fact of having successfully studied mathematics and, reciprocally, between ignorance and the refusal to instruct oneself in this discipline. Thus, according to Fine, it is necessary that those who have been recognised as fully accomplished investigators and teachers in any discipline have studied mathematics beforehand. In return, it is necessary for those who have rejected mathematics to be found ignorant and absolutely incapable of properly reaching the principles of any part of philosophy, including its mathematical part.[46]

Since Fine, in this passage, regards mathematics as the means through which the intellectual faculties of human beings may be developed, he therefore considers that mathematics is proper to show if a man is able or not to apprehend truth and, from there, to reach the highest degree of knowledge in any discipline. In order to describe this, Fine evokes an alchemical metaphor (which also appears in Tartaglia's preface to his Italian edition of Euclid's *Elements*) which compares the capacity of mathematics to reveal man's natural powers to know with the ability of fire to reveal the degree of nobility of a particular metal.[47] As fire, by burning a metal, gradually reveals its quality and strength, mathematics, by directing the intellect to things which are pure and inalterable, although linked in reality to matter, gradually reveals the degree of perfection of the individual's cognitive facul-

---

la preuve de l'or, aussi sont les mathematiques la probation de l'entendement, tirans tous bons esprits du tout à elles, & les contraignans passer par la rigueur de verité: ce qui ne peult advenir des autres ars, dependans de la volunté & invention des hommes.' Fine, 1551*c*, sig. [A]^r. The first sentence corresponds almost exactly to the translation of a part of the general preface of the *Protomathesis*: 'Quem enim excellentem, probatùmue cuiuspiam facultatis authorem dabimus, non Mathematicum ? quem uersa uice apertum Mathematicarum offendemus inimicum, non stupidum, uel ingenio crassum, & qui non Mathematicam tantummodò, sed omnem prorsus ignoret Philosophiam?' Fine, 1532*b*, sig. AA2^v.

46 Let us note that in the original French version of this text (as well as in its Latin model found in the preface of the *Protomathesis*), Fine seems to distinguish – by the alternate use of the plural and of the singular in writing the word 'mathematics' ('les mathématiques' / 'la mathématique'; 'mathematicae' / 'mathematica') – mathematics as a set of different particular disciplines and mathematics as an undivided form of knowledge (which is at the same time a part of philosophy). The former represents the path towards the latter as well as towards all philosophy. Indeed, if mathematics, at first, corresponds to the study of a specific type of being (as defined through the division into three of speculative knowledge), it is afterwards recognised, thanks to the ontologically intermediate position of this type of being, as having the power to lead to the knowledge of all types of beings, thus being at one and the same time a part of philosophy and the way towards philosophy.

47 'Sicut aurum probatur igni, & ingenium Mathematicis.' N. Tartaglia, *Euclide megarense philosopho, solo introduttore delle scientia mathematice* (Venice, 1565), sig. A3^r.

ties. Accordingly, since mathematical demonstrations lead the human mind to produce irrefutable judgments, those men who, faced with such demonstrations, do not recognize the necessity of their conclusions are therefore shown to be unfit to reach any truth.

As Fine adds at the end of this passage, if only mathematics has the power to show the value of the minds of human beings (by revealing those who are truly fit for *scientia*), it is because it is the only type of investigation which has an object at the same time accessible to the senses and intellectually separable from matter, and which presents absolutely firm and necessary demonstrations. Indeed, because other types of investigation, as opposed to mathematics, aim at knowing things which are inaccessible to the senses or completely inconsistent, and are therefore unable to present apodictic demonstrations, they are forced (without the assistance of mathematics) to rely upon the decisions and the intuitions of men whose minds – weakened as they are by their link with the matter of the body – may not transcend the contingency and uncertainty of empirical knowledge.

According to the argument which is repeatedly set out by Oronce Fine in his main prefatory texts about mathematics, if the study of mathematics constitutes, as the ancients thought, the necessary condition of knowledge, it is insofar as it is able to reveal and develop our natural abilities to know. Mathematical studies thus allow all those who are naturally endowed with a sane mind to reach the theoretical principles of all disciplines, be they of a speculative or of a technical nature. As we have seen, if, according to Fine, mathematics has this ability, it is because the rigor and the necessity of its demonstrations enable us to produce conclusions which are firmly established on self-evident principles. Consequently, Fine asserted that mathematical demonstrations allow us to have a first experience of certainty, that is an experience through which we understand not only the nature of truth, but also the process through which it may be reached in spite of the limitations of our intellect.

If, in order to prove this, Fine took the time to explain why mathematics should be considered as intermediate between metaphysics and physics, it is because this allowed him to set forth the specific nature of the mathematical object and, from there, to justify its full knowability as regards the human intellect. This argument implicitly enabled him to explain why mathematics is able to present apodictic demonstrations and produce universal and necessary judgments. Eventually, this gave him the opportunity to advance the notion that mathematics truly has the capacity to lead us to properly scientific knowledge and thus to show us the path towards the contemplation of all truths.

Therefore, it appears that Fine's conception of mathematics is the result of the intertwining of a certain Platonism, which considers mathematics as the privileged intermediary between the sensible and the intelligible worlds, with elements of Aristotelian theory of being and knowledge. Indeed if, on one hand, Fine chose to interpret the Platonic assertion of the intermediate status of mathematical things according to Aristotle's abstractionist conception, it is to better set forth the correspondence (originally presented in Plato's *Republic*) between mathematics and dianoetic or demonstrative knowledge, a correspondence which is founded on the harmony between the nature of mathematical things and the condition of the human soul.[48] Now, instead of considering, as Plato and Proclus did, discursive or dianoetic knowledge, and thus mathematics, as a mere step towards a higher type of knowledge (that is *noesis*, which corresponds to the supra-rational or non-discursive contemplation of ideal and divine beings), Fine, following Aristotle and his medieval commentators, regarded demonstrative knowledge as the higher degree of knowledge directly accessible to the human intellect. As such, from the assertion of the compatibility of mathematical things with the requirements of *scientia*, Fine drew the conclusion (in accordance with the Averroist interpretation of Aristotle's theory of demonstration) that mathematics has the firmest and most certain demonstrations. As such, in the texts considered in this essay, mathematics emerges as a methodological model for all disciplines to follow and, more particularly, as bringing about the conditions required for grasping all truths by the intellect. In this, Fine stayed faithful to Plato's representation of mathematics as the means to lead the human soul from the imperfect level of probable and perishable knowledge to the ultimate level of necessary and eternal truths.

Therefore, if Fine's definition of knowledge seems to have been mainly influenced by conceptions derived from Aristotle's epistemology, it could however be said that his vision of mathematics indirectly contributed to the introduction of certain elements of Platonic thought within the representation of this discipline which was proposed at the University of Paris. Indeed, not only did Fine's conception of mathematics openly establish the mathematical arts as the foundation of the university curriculum, it also led to a certain representation of mathematics as the means to seize the rationality, that is the order and harmony, which govern sensible realities. This last element may explain the popularity of Fine's works within certain universities of northern Europe during the second half of the sixteenth century (notably Pisa, in which Galileo had taught in his youth, as noted by Charles Schmitt) and may allow us to measure the importance his works had for certain ulterior developments in the field of applied mathematics and natural philosophy.[49] To conclude, we could say that, from this reinterpretation of the Platonic assertion of the propaedeutic character of mathematics through arguments which belong to the peripatetic tradition, Fine proposed a definition of mathe-

48 Plato, *Republic*, VI, 510c–511e.

49 See Charles B. Schmitt, 'The Faculty of Arts at Pisa at the Time of Galileo', *Physis*, vol. 14, no. 3 (1972), pp. 243–272.

matics which was in full harmony with the preoccupations of his contemporaries. This definition, in the sense that it coincided (in a certain way) with the humanist desire to establish a model of universal knowledge, was to be searched for not only in 'noble' and speculative disciplines, but also in arts which are 'mechanical' and which deal with matter for practical purposes.

Oronce Fine's Practical Geometry

PASCAL BRIOIST

> He was called Professor in mathematical sciences by that great restorer of Letters François Ier. In this profession, he was so accomplished that [...] he gave a new life to the splendour of Mathematics in the University of Paris that was until then very degenerated. [...] He greatly advanced mathematical learning [...] as much by his writings as by his invention of sundry beautiful instruments and maps. His hand was no less fit to build and draw such things, and to paint them, than was his ingenuity to invent them [...].[1]

This description of Oronce Fine by the cosmographer André Thevet in the *Histoire des plus illustres et scavants hommes de leurs siècles* (1584), illustrates the importance of mathematics in Fine's life, but also insists on the professor's genuine manual or artisanal qualities – his 'subtlety'. The manual arts in which Fine was (according to Thevet) 'expert' were closely bound up with a particular branch of mathematics that he treated extensively in print: practical geometry. Practical geometry played an important role in the history of Renaissance culture, for it connected intellectual knowledge to numerous fields of practice, which all engaged in their own way with artisanal practices of the day: painting, architecture, fortification, topography, shipbuilding, carpentry, stonecutting, etc. According to the twelfth-century mathematician Hugh of St Victor, practical geometry may be opposed to theoretical geometry in that it is 'done by means of certain instruments and makes judgements by proportionally joining together one thing to another.'[2]

Throughout his life, Fine engaged in activities that had direct need of geometrical practice – such as his work as a surveyor and as a practising astrologer.[3]

1 '[...] il fut appelé par ce grand restaurateur des lettres le Roy François premier, Professeur ès sciences mathématiques. En cette profession il versa si bien, qu'au gré & contentement des gens de bien & malgré l'envie [...] il ressuscita en l'Université de Paris la splendeur des Mathématiques, qui pour lors étoient par trop abastardies. [...] Il avançoit grandement ces sciences [Mathématiques] [...] tant par ses escrits que par invention de plusieurs beaux instruments & cartes, comme ayant la main non moins propre à fabriquer & dresser de tels organes, & à les peindre, que l'esprit à les inventer. [...].' A. Thevet, *Histoire des plus illustres et scavants hommes de leurs siècles* (Paris, 1671 [first edition, Paris, 1584]), vol. 7, pp. 564–566.

2 Quoted by L. R. Shelby in 'The Geometrical Knowledge of Medieval Master Masons', *Speculum*, vol. 47, pp. 379–421.

More importantly, practical geometry was essential for Fine's work on cartography (which, by the end of the 1520s, had made him famous), cosmography, mirrors, and dialling, and played a crucial part in the lessons he delivered in his capacity at Royal Lecturer.[4] Between 1532 and 1542, Fine wrote a number of texts which can be identified as 'practical geometries':

1. The second part of the book on 'Geometry' in the *Protomathesis* (composed in 1530, first printed 1532).

2. The *Composition et usage du quarré geometrique* (1538), which is a translation into French of one chapter of the *Protomathesis*.[5]

3. The *Liber de geometria practica, sive de practicis longitudinum, planorum et solidorum* (1544), which is an upgraded version of Book 2 of the second part of the *Protomathesis* (although it lacks the pages from the *Protomathesis* concerning the use of the geometrical square).[6]

A fourth text could perhaps be added to this corpus, namely Charles de Bovelles's *Géométrie practique* (1551), as this important work seems to have resulted from a collaboration between the famous grammarian and Fine.[7] To a certain extent, Fine's practical geometries do not seem to be addressed to a scholarly audience.[8] They dismiss academic demonstrations and employ instead case studies with numerical examples, as in the *Protomathesis*:

3 It has been claimed, moreover, that in the 1520s he took part in the French army's expedition into Italy and tradition says he was consulted as a technical adviser during the siege of Pavia of October 1524, contributing that same year to the strengthening of the fortifications of Milan. In February of 1524 he was apparently taken by the enemy while instructing pioneers to build a bridge to cross the river Thezin. The evidence for this is supposedly a letter written by Fine that no longer exists, quoted in M. Fournier, *Histoire générale des Alpes Maritimes ou Cottiènes* (Paris, 1890), pp. 512–513. See also R. P. Ross (who dismisses the letter), 'Studies on Oronce Finé (1494–1555)', unpublished PhD. diss. (Columbia University, 1971), pp. 393–395.

4 As demonstrated in the other contributions to this volume.

5 The first version of this text is a manuscript by Fine dated 1538 (Bibliothèque Nationale de France, Paris, MS Français 1334). It was first printed independently of the *Protomathesis* in 1556 (Fine, 1556*a*).

6 This work was reprinted in 1556 as *De re et praxi geometrica libri tres*, and again in 1558, 1564 and 1586. A French translation by Pierre Forcadel was published under the title *La practique de géométrie d'Oronce* in 1570, 1585 and 1586.

7 C. de Bovelles, *Géométrie practique, composée par le noble Philosophe maistre Charles de Bovelles, & nouvellement par luy reveur, augmentée, & grandement enrichie* (Paris, 1551). A liminal piece in this work – a Latin poem written by Bovelles in Noyon in 1542, addressed to the Abbot of Ourscamp – states that Fine frequently came to the small Picardy town, and 'assisted' Bovelles in his task. It seems that it was Fine who took the manuscript to a Parisian printer, and he may also have helped illustrate the book with accurate engraved figures. On this issue, see A. F. Johnson, 'Oronce Finé as an Illustrator of Books', *Gutenberg-Jahrbuch*, vol. 3 (1928), pp. 107–109.

> Let us propose, for instance, a triangle with acute angles and unequal sides called LMN, let its left side LM be 7½ cubits, its right side LN 7 cubits and its basis MN properly 7 cubits. Let us multiply then the first thing 6 and ½ corresponding to the left side with itself and you shall have 42 [...].[9]

It is worth noting that these numerical examples are more or less canonical within the genre and that, sometimes, that canon was established as early on as Antiquity.[10] In particular, it should be observed that material objects are generally integral to these case studies, such as mathematical instruments, steeples, towers, wells or barrels. But despite the 'everyday' content of the works, the fact that two of Fine's texts on practical geometry were written in Latin, and that in the *Protomathesis* he addressed the 'studious reader', suggests a degree of ambiguity in terms of audience. In fact, Fine clearly wrote for students as well as for the newly enlarged public that could be reached through the *Institution des Lecteurs Royaux*. This aspect of the Collège Royal was intended to be useful not only for humanist intellectuals but also for aristocrats, mariners, topographers, and traders qualified by experience.[11]

In publishing practical geometries in the sixteenth century, Oronce Fine followed a tradition inherited from Antiquity and embedded in Western culture for many centuries prior to the Renaissance. It is nevertheless clear that practical geometries grew in strength and number during the fifteenth and sixteenth centuries. When Fine started his career, two types of treatise on practical geometry existed. One, which belonged to the tradition beginning with Hugh of St Victor, is characterised by an emphasis on the use of instruments in solving practical geometrical problems, and mainly addressed the professional world of masons and master builders. For example, in the fifteenth-century lectures given in Montpellier by Robertus Englicus, on the way in which to handle the quadrant and the knowledge belonging to the building yards of the cathedral (such as Villard de Honnecourt's solution to duplicating the square), were transcribed by Italian scholars such as Paolo Gherardi or Paolo dal' Abbaco. who disseminated them in Tuscany.[12] The second type originated with Italian abacus books and dealt with

8 It is notable that Fine translated his *Liber de geometria* into French but never published it, as if the mathematician felt some kind of tension between his mundane practice as a lecturer addressing the Parisian public and his scholarly status.

9 Fine, 1532*b*, fol. 70.

10 Traditionally, for instance, a circle has a diameter of 7 or 14, the equilateral triangle a side of 10, or a multiple of 10.

11 The lectures were free, open to the Parisian public and so popular that when Ramus (Fine's successor) delivered his first lecture, two members of the audience collapsed after being suffocated by the crowd. See Pantin, in this volume, and M. Fumaroli, *Les origines du Collège de France 1500–1560* (Klincksieck, 1998). For the story about Ramus see A. Chastel, *Culture et demeures en France au XVIe siècle* (Paris, 1985), p. 37.

12 See e.g. the *Libro di ragioni*, Biblioteca Nazionale, Florence, MS. Magl. XI 87.

problems of lines, surfaces and volumes.[13] Texts such as Leon Battista Alberti's *Ludi rerum mathematicarum* (1452), Nicolas Chuquet's *Géométrie* (circa 1480), and Luca Pacioli's *Tractatus geometriae* in the *Summa de arithmetica* (1494) (all of which built upon the abacus tradition) determined a pattern of basic knowledge characterized by the necessity of measuring distances, surfaces or volumes, or by the constructivist approach of medieval master builders.[14] This type of treatise addressed, through its *ragioni*, a world of merchants and educated lay men who required quick solutions to everyday problems.[15] For example, the series of practical questions presented by Alberti indicates the sort of tasks addressed in practical geometries of the period: measuring the height of a tower, the breadth of a river, the depth of a well, or the areas of differently shaped fields or bodies; constructing (with ruler and compass) a square angle and regular figures (normally the pentagon and heptagon); squaring the circle; finding the centre of an arc; and transforming an equilateral triangle with a given side (c) into a square with sides of length 2c/3. Most of these questions use only the properties of similar triangles and a rule of thumb, although in Alberti's *Ludi* some problems show the author's acquaintance with the Greco-Arabic learned tradition. This is the case, for instance, when Alberti treats measuring the height of a tower knowing neither its distance nor the dimension of any of its parts, or the issue of squaring the circle using chords and segments of circles.[16] At the end of the fifteenth century, a synthesis of the two types of practical geometry was effected. Dürer's *Underweysung der Messung* (1525) is a good example of this synthesis, his particular contribution being his answers to the specific concerns of artists (how to draw rope-moulded columns, how to draw a bishop's crozier, how to draw a spiral staircase, etc.).[17]

Fine's practical geometry should be placed squarely in this tradition, but is it at all possible to determine with which authors he was familiar and which sources

13 For the abacus tradition and school mathematics see e.g. R. Franci and L. Toti Rigatelli, *Introduzione all'algebra mercantile del medievo e del Rinascimento* (Urbino, 1982); G. Cifoletti, 'Mathematics and Rhetoric: Jacques Peletier, Guillaume Gosselin and the Making of the French Algebraic Tradition', unpublished PhD. diss. (Princeton, 1993), pp. 71–102.

14 See e.g. W. Van Egmond, 'The Commercial Revolution and the beginnings of Mathematics in Renaissance Florence, 1300–1500', unpublished PhD. diss. (Indiana University, 1976); *Practical Mathematics in the Italian Renaissance: A Catalogue of Italian Abacus Manuscripts and Printed Books to 1600* (Florence, 1980).

15 On these two audiences, see the introduction to Chuquet's *Géométrie* by Hervé l'Huillier in N. Chuquet, *La Géométrie. Première géométrie algébrique en langue française (1484)* (Paris, 1979), pp. 10–26.

16 See Pierre Souffrin, 'La geometria practica dans les *Ludi rerum mathematicarum*', *Albertiana*, vol. 1 (1998), pp. 87–103.

17 See J. Peiffer, 'Projections Embodied in Technical Drawings: Dürer and his Followers', in W. Lefèvre (ed.), *Picturing Machines, 1400–1700* (Cambridge, MA and London, 2004), pp. 245–275.

he used in his works? Because the contents of Fine's library are not known, and as his personal papers are scarce, it is very difficult to state with any great certainty which sources he used, but some hypotheses can nevertheless be formulated. First, during his formative years on the bench of the Collège de Navarre, it is highly probable that Fine was made aware of the tradition represented by the French author Chuquet; second, there exists some evidence for his connection to German practical mathematicians: he contributed in 1523 to the publication of Reisch's *Margerita philosophica,* to Georg Peurbach's *Theoricae novae planetarum* in 1525, and later, in 1544, he let the Strasburgian printer Knobloch publish his *Geometria practica* suggesting that he was familiar with the work of authors from the Holy Roman Empire.[18]

However, despite certain similarities between Fine's works and those of his contemporaries and immediate predecessors, his style certainly differs from that of authors such as Alberti and Dürer. One reason for this is that he constantly manages to keep abstraction present within his texts (except perhaps in the *Quarré geometrique*, discussed below). The *Protomathesis*, for instance, starts with a first part which is purely theoretical and which names what Fine calls the principles of *Geometria*: definitions (nature and properties of things), questions, theorems and axioms.[19] He seems to have used this progression, using a specialized vocabulary, to familiarize students with the kind of Latin and Greek terms that they would need if ever they wanted to return *ad fontes*, i.e. to classical geometers such as Euclid and Archimedes.[20] Very soon, though, Fine forgets to make use of this terminology and goes straight to the practical exercises necessary for a topographer's or an engineer's career. In a text like the *Quarré geometrique*, for instance, he attempts to reintroduce another type of pedagogical progression, dividing his chapters into definitions, examples, corollaries, and differences. These are less elements of a theoretical framework than classification devices – an art of memory for men of action. Nevertheless, even in the texts obviously written for the benefit of topographers, Fine expresses the need to defend the virtue of abstract mathematics. This is clearly the case in the tenth chapter of *Quarré geometrique*, entitled 'Of the mathematical rationality of the previous rules and operations', where the engravings of valleys, mountains and towers are replaced by purely geometrical schemes made of points and lines. Here, the author wants to convince his reader of the simplicity of the mathematical rationality lying beneath the calculations he has performed. Teaching mathematics through games was only acceptable for him if due respect was paid to the noble discipline of the quadrivium, which lent him

18 On this issue see also Pantin and J.-J. Brioist, in this volume.

19 Fine, 1532*b*, fol. 50v.

20 In fact, Fine frequently quotes Euclid's *Elements*. For instance the *Liber de geometria practica* cites Proposition VI.4 of Euclid's *Elements* (Fine, 1544*b*, p. 44), while the *Quarré geometrique* refers to, 'the penultimate proposition of the said first book of Euclid' (Fine, 1556*a*, p. 32).

social status.

The first important mathematical treatise published by Fine was the *Protomathesis*. Given the publication date of this work (1532), it is highly likely that its content roughly corresponds to that of the lectures he delivered at the Collège Royal. The work (from which, as we have seen, the *Liber de geometria practica* derives), starts with an introduction dealing with the definition and the principles of geometry, that is with the theoretical basis of the subject. Book 1 contains definitions of figures (points, lines, surfaces, angles), general properties and axioms. It also deals with sinuses, chords and arcs (including a table). Book 2, which is properly practical, addresses the art of measuring lines, surfaces and bodies according to Euclid, before going on to elucidate the method of building a geometrical square or quadrant. The remaining chapters explain how to measure distances using the quadrant, staffs, Jacob's staff and the geometrical square; how to measure surfaces (triangles, parallelograms, figures with multiple angles such as the pentagon or hexagon, and circles); how to demonstrate (according to the method of Archimedes) the relation between the circumference and the diameter of the circle; how to square the circle; and, finally, how to measure solid bodies (columns, pyramids, spheres, dodecahedrons etc., the rhomb or 'almond' [that is, barrels]). Notably, the structure of this part of the *Protomathesis* follows the threefold structure of the medieval geometer's treatise: *altimetria*, *planimetria* and *steriometria*.[21]

Instruments are much in evidence in the *Protomathesis*, indeed a notable feature of many of Fine's published works is their engagement with instrumentation. Fine in fact authored an early example of the 'manufacture and use' genre: the *Composition et usage du quarré geometrique* (1556), which opens with instructions for the manufacture of the 'geometrical square' but which contains also numerous examples of his practical geometrical method. [Figs. 4.1 and 4.2] The second part explains how to measure all manner of things on earth with the said instrument: length from the ground (specific secondary examples, such as measuring the back of a mountain or of a wall, are here called 'corollaries'); length from a summit, height of a building, or a wall, on condition that it can be approached; height of buildings that cannot be approached; length of a wall that cannot be approached; heights of two unequal buildings from the summit of one of them; height of a building erected on an inaccessible mountain; depth of wells or valleys; height of the sun or stars. Chapter 10 explains the principles at stake in this art of measuring: essentially, the properties of similar triangles and the rule of thumb.[22]

21 This threefold structure can be found, for example, in Hugh of St Victor's *Practica geometriae* (Bibliothèque Mazarine, Paris, MS 717 or Bibliothèque Nationale de France, Paris, MSS 14506 and 15362) or in Chuquet's *Géometrie* (see above, n. 17), both of which consider ways of measuring heights, depths and lengths.

22 In fact, the theoretical basis of Fine's *Quarré geometrique* is Proposition VI.3 of

For example, in Chapter 3 (part 2), Fine addresses the problem: 'By which means we can measure the height of any wall, or other thing rightly erected on earth, for as much as it is possible to approach it'. His explanations reads as follows:

> If from point A, we would like to measure the height EH of the following figure, & that the square ABCD was located as described, the visual line AH passes rightly through the point C, as is shown in the said figure. I say that the said height H will cover no less and no more than the length or distance AE, well measured, let us say 30 feet. Then we can with certainty conclude that the said height EH is also 30 feet long.[23]

Now, if the section IB measures 40 parts of the side BC, to find the height EK, one multiplies 30 by 40 and divides the resulting 1200 by 60 (the shadow square of the instrument being graduated from 0 to 60). The result will be 20. This result is obtained because of a proportional rule: since BC (that is 60) – IC (that is half 40) = 20, and since AE=EH=30, so AE=EK (that is 20) plus half EK (10), which is 30. This example encapsulates the somewhat confusing way in which Fine explains his rule. He simply uses the similitude between the triangle AIB and AKE and compares it to the similitude between ABC and AEH: the proportion of IB to BC provides the required height: 40 is to 60 what the height is to 30. In modern algebraic (and anachronistic) terms, this could be expressed as: $h = BI \times AE/BC = 40 \times 30/60$ In his 'third difference', or third case, related to this example, Fine explains what to do when the visual line rises above the line AC: one should multiply the distance by 60 and divide the number obtained by the number of parts measured on the geometrical square. To prevent his reader becoming lost while executing the rule of thumb, Fine even gives special tips such as 'you should always put the numbers in such an order that the unknown and desired number would be the fourth'.[24] Comments such as this remind us of Fine's commitment to practical utility and efficiency (which is in evidence in every single text on geometry that he authored), but what exactly did this practicality mean?

Describing a mathematical book as a work of 'practical geometry' was intended to appeal to an audience of practitioners, but the content justifying the

Euclid's *Elements*, concerning the proportionality of the sides of equilateral triangles, known today as 'Thales's theorem'.

23 'Comme si du point a, l'on voulait mesurer la hauteur eh de la figure qui s'ensuit; & que le carré abcd étant colloqué ainsi qu'il a été dit ci-dessus, la visée ah, passe droitement par le point c, ainsi comme il est exprimé en ladite figure. Ie dis que ladite hauteur h, contiendra lors autant précisément & non plus ny moins que ladite longueur ou distance ae. Supposé doncques qu'icelle distance ae, bien mesurée, soit de 30 pieds. L'on peut conclure seurement que la susdite hauteur eh, est aussi de 30 pieds.' Fine, 1556*a*, p. 8^{v}.

24 '[…] il fault touious ordonner les nombres en telle manière que le nombre incongneu & désiré vienne estre le quatrième'. *Idem*.

use of the adjective 'practical' has yet to be clearly established. What is the difference, for example, between books that deserve such a description and Euclid's *Elements*? In this case, it is not sufficient to claim that 'theoretical' refers to contemplation while 'practical' refers to action, in accordance with the use of classical authors, since Euclid's *Elements* is often considered a work of both theory and practice.[25] However, it is worth noting that a number of the characteristics of Fine's treatises correspond to a general pattern that can be found in the mathematical literature of his age. For instance, practical geometries such as Fine's *Quarré géometrique* most generally adopt a prescriptive format and address (more or less directly) the reader:

> It is convenient to multiply the said length of 30 feet by the 40 parts, out of which multiplication you shall obtain 1200; then you have to divide the total by 60 [...].[26]

In Bovelles' *Géométrie practique*, we find similar phrasing:

> Position the foot of the compass on the said point H, and turn the round figure according to the quantity and spread of the lines or length HA, HB & HC, and you shall have the circle within which you shall easily draw the required pentagon.[27]

Thus, the text establishes through a (by this date) standard 'dialogue' format, the fictitious context of a classroom in which a master discusses practical geometry with his pupil, using common tools like a ruler and a pair of compasses.[28] Indeed, it should be observed that a usual criterion for defining practical geometries in this era is their systematic instructions about the manipulation of mathematical instruments. The *Protomathesis*, for instance, provides instructions on the use of ordinary staffs, mirrors, the Jacob's staff, the quadrant, and the geometrical square. These are very simple and traditional instruments that were either in use in Antiquity or in the Middle Ages (the quadrant, for instant is mentioned in Ptolemy's *Almagest*), or derived from the astrolabe, itself known for centuries.[29]

25 P. Guyot and F. Métin, 'Les ouvrages de géométrie pratique au XVIe siècle', in E. Hébert (ed.), *Instruments scientifiques à travers l'histoire* (Paris, 2004), pp. 251–293.

26 'Il convient donques multiplier ladite longuer de 30 pieds, par les 40 parties dessusdites, dont ils proviendront 1200 : qu'il faut diviser par 60 [...].' Fine, 1556*a*, p.3.

27 'Parquoi mets le pied du compas dessus ledict point H, & tourne le rond selon la quantité & ouverture des lignes ou longueurs HA, HB & HC, & tu auras le cercle dedans lequel parferas facilement le pentagone qu'on demande [...].' Bovelles, *Géométrie practique*, p. 20^{v}.

28 On the uses of the dialogue format to confer a certain social authority upon an argument, see V. Cox, *The Renaissance Dialogue. Literary Dialogue in its Social and Political Contexts, Castiglione to Galileo* (Cambridge, 1992).

29 Fine's predecessor Chuquet, for example, insisted on the use of the astrolabe and the quadrant in his *Géometrie.* See Chuquet, *La Géométrie*, pp. 10–26. The Jacob's staff

In fact, an important feature of practical geometries is their constructivism, that is the instructions they provide as to how one may draw a given figure with a ruler and a pair of compasses. For instance, in the *Protomathesis*, Fine explains how to create the figure of a regular pentagon:

> A line AB being given, according to its quantity, I draw two arcs ACD and BCE, as long as I want. These arcs will divide themselves on the point C, which will be the head of the isopleure ABC. I draw consequently on the said line AB two right angles ABF and BAG, by two perpendicular lines AG and BF [...].[30]

Charles de Bovelles also recommended, in a liminal poem to his *Géométrie practique*, that his book should be read with the basic mathematical instruments in hand:

> Hear this, and do not forget / The square, the rule and the compasses: / For, from these three depend art and practice / And the profit of geometrical knowledge.[31]

The limits of this approach are that some figures cannot be drawn with a ruler and compasses. These are precisely the issues at stake in the construction of the heptagon and of squaring the circle, another field of interest to Fine, which he dealt with in his *De rebus mathematicis hactenus desideratis* (1556).

Fine considered the geometrical square to be the most useful and precise instrument of all, as demonstrated in his treatise *Quarré geometrique* (Fig. 4.3). In actuality, the only reason Fine's geometrical square may be considered superior to the quadrant is because it contains a quadrant within it. It is thus one of the first 'composite instruments', a category of device that was to become very common in the later part of the sixteenth century. Notably, the author does not describe in his text an abstract construction but a very tangible object, defining the

---

(called in French *arbalestrille*), described in the *Protomathesis*, consists simply in a straight rule on which a piece of wood can slide to provide the angular distance between two points. The measures taken with the Jacob's staff at two different stations, a known number of feet apart, can thus determine a distance between two remote points by virtue of the theory of similar triangles. This theory, as demonstrated in the Arabic tradition, enables anybody in possession of a mirror to work out the distance of an approachable object, since the angle of incidence is equal to the angle of reflection. The quadrant (also described in the *Protomathesis*) is a graduated arc encompassing a quarter of a circle, with sights on one side. Its hanging line, a line which is weighted with a 'plummet', shows the inclination of the quadrant on a degree scale.

30 'Soit la ligne assignée AB, selon la quantité d'elle, je tourne deux arcs ACD, & BCE, si longs ie que vouldray. Lesquels se diviseront sur le point C, qui sera chef de l'isopleure ABC. Ie fai conséquemment sur ladicte ligne AB deux angles droicts ABF, & BAG, par deux perpendiculaires AG & BF [...].' Fine, 1532*b*, p. 20.

31 'Entens le donc, & si n'oublie pas / L'esquierre droict, la reigle et le compas: / Car de ces trois despend l'art & practique / Et le profict du scavoir géométrique.' Charles de Bovelles, *Géométrie practique,* dedicatory poem (not paginated).

materials that should be used: a strong sort of wood for the alidade, brass for the sights, metal for the screwis, and so on. This material technology required the author to familiarize his reading public with a specific vocabulary. Thus, in his *Quarré geometrique*, Fine defines such terms as the sights (*pinules*), or the fiducial line (that is the visual line passing through the said sights of the mobile straight rule), or even the 'parts' of the shadow square.[32] Although Fine allows the instrument-maker the liberty of choosing the decoration, he advises him to make the square large, for the sake of precision, and to make sure that the instrument can be disassembled for carrying – thus emphasising the fact that his eye was constantly trained on practicality and efficiency.

It should be observed, however, that using such instruments was not always straightforward (Fig. 4.4). Indeed, Frédéric Métin has pointed out with irony that the gestures of the measurers that employed these devices were sometimes very awkward.[33] In an image from the *Protomathesis*, for instance, the measurer is lying on his back on the battlement of a tower, jeopardizing his equilibrium so that he may measure the height of another tower. Similarly, another individual tries desperately to identify himself with the very ground he is lying on, crawling, so that his eye is exactly at the ground level to measure another tower, while another image shows two mathematicians holding their geometrical squares in a very uncomfortable position, prompting doubts as to the quality of the measurements taken in these conditions.[34] Such engravings emphasize the fact that one cannot easily place one's eye in the situation required for perfect observation – i.e. the instrumental demands of practical geometry render it effectively impractical. At the same time, the works also seem to state that geometry possesses the power to rectify not only nature, or man-made items, but also the bodies of the measurers. To take an accurate measurement is difficult and the body has to submit itself to the exigencies of mathematical rules in the same way as the mathematician has to submit himself to the will of God, secretly inscribed in the laws of the universe.

With this in mind, the illustrations that appear throughout Fine's work are especially significant. If, on occasion, Fine asked his engraver to draw only appropriately dimensioned lines and to name points with capital letters (this is mainly the case in the most theoretical books such as the *Protomathesis*), he also indulged in asking him to include details that would make solids appear more real: shadows on a polyhedron or on a barrel, for example (Fig. 4.5). This is even more obvious in the *Quarré geometrique* where it seems that the purity of the mathematical object has to respond to the 'annoyance' of reality. The iconic signs serve as some kind of rhetorical device. Birds, trees, pla nts are not useful in geometri-

32 Fine, 1556*a*, p. 3.

33 Frédéric Métin, 'Le début et la fin de l'histoire: d'Oronce Finé à Samuel Marolois', in E. Hébert (ed.), *Instruments scientifiques à travers l'histoire* (Paris, 2004), pp. 233–250.

34 Fine, 1532*b,* fols. 69r and 73v.

cal demonstration, but they act as tokens of the 'reducing' power of mathematics. The decorative details of the buildings themselves respond to certain semiotic issues: strength (symbolised by the stone apparel and the battlements), exoticism (stipples with bulbs, palm trees, etc.), or simply plain ordinariness (such as a church tower).[35] The underlying discourse seems to be that, wherever you are, whatever the force of your adversary, geometry will solve your problems.

Indeed, a characteristic common to most practical geometries is the materiality of the things they always measure: towers, walls, bell towers, wells, rivers or valleys. They tend to correspond to objects which were familiar to military men and topographers. The problems addressed are also concerned with the universal 'annoyance' of certain things: a ditch, a river, an uneven mountain or any other impediment that keeps the engineer from accomplishing his job.[36] Chapter 4 of *Quarré geometrique* is very explicit on the military issues at stake, as demonstrated when Fine starts to discuss the manner of measuring unapproachable heights:

> It occurs, sometimes, that it is impossible to approach the bottom of towers, curtains, or similar heights that you desire to measure, either through fear of the enemy, or because of the impediment of a river or ditch [...].[37]

The chapter ends by advertising the fact that this trick will be 'greatly useful in time of war, to judge the range of a cannon'.[38] The determination of volumes in Fine's works deals with somewhat different objects, but again with a certain preference for objects related to warfare, such as bullets (the dimensions of the spheres described by Fine, a diameter of ten or fourteen inches for instance, leave no doubt on the fact that he is alluding to small shot) or barrels.[39]

Oronce Fine's works on practical geometry are especially relevant for the cultural history of Europe in the Renaissance because they tied together different traditions from the past and offered to the renewed audience of the sixteenth century the basis, both theoretical and practical, of a new knowledge ready for application. The institutionalisation of the teaching of mathematics outside the university,

35 Fine, 1556*a*, pp. 8, 15, 11.

36 '[...] tout autre empêchement.' *Ibid.*, p. 10.

37 'Il advient aucunesfois, que l'on ne peut approcher du pied des tours, murailles ou semblables hauteurs, que l'on désire mesurer, soit par crainte des ennemis, ou par l'empeschement de quelque rivière ou fossé: et lors il convient procéder comme s'ensuit'. *Idem.*

38 '[...] qui peut grandement servir en temps de guerre, pour juger de la portée d'un canon.' Fine, 1556*a*, p. 11.

39 Gauging barrels was a standard problem for measurers and some solutions are frequently present in abacus books. See e.g. M. Baxandall, *L'oeil du Quatrocento*, *l'usage de la peinture dans l'Italie de la Renaissance* (Paris, 1985).

namely in the Collège Royal, created the conditions for the transmission of a very particular kind of intellectual resource. Lecturing in French, but never renouncing abstraction completely, Fine managed to render accessible academic learning to city people, to soldiers and to courtiers qualified by experience, but who did not possess the necessary skills to attend, say, the courses of the Collège de Navarre. This was novel in France but perhaps not in Italy, where engineers already taught their aristocratic masters the tricks they needed to understand the art of fortification or the rules of poliorcetics.[40] Was this, then, any great achievement? The answer is elusive, since Fine's audience in the *quartier latin* cannot be identified with any great precision. However, it is clear that Fine's writings outlived him and that, in publishing a number of fundamental practical geometrical textbooks, he helped to fashion the sixteenth-century French revival of mathematics.

Indeed, at the end of the sixteenth century, some military textbooks, such as Claude Flamand's, *Guide des fortifications* (1597), or Jean Errard's *La fortification réduicte en art et démonstrée* (1600) used Fine's printed lessons as an introduction to their own teachings.[41] Similarly, it is significant that a manuscript written in England, probably at the end of the sixteenth century, provide a translation of the *Protomathesis* and of the *De rebus mathematicis*, mixing these with other sources, and arguing that because Fine's works were not available in English, their content could be used freely.[42] The anonymous author (possibly connected to the Gresham College if we consider his practical mathematical interests) was not, however, the first Englishman to have read Fine's geometries: Thomas Digges clearly had access to Fine's work on the geometrical square, since some pages of his *Pantometria* (1571) plagiarize the French mathematician. Examples such as these testify to the continued currency of Fine's practical geometries into the late sixteenth century, and even if Fine was not the great mathematician he advertised himself to be, in his own time his pedagogy was highly valued and he probably deserves the description by Thevet with which we began: a gifted practitioner who 'greatly advanced mathematical learning [...] as much by his writings as by his invention of sundry beautiful instruments.'

40 See e.g. C. Cresti, A. Fara and D. Lamberini (eds.), *Atti del convegno di studi Architettura Militare nell'Europa del XVI secolo* (Siena, 1988), *passim*.

41 It is noteworthy that Errard also published *La géométrie et pratique d'icelle* (Paris, 1594) and that he was, like Fine, interested in the squaring of the circle. See, for example, his *Réfutation de quelques propositions du livre de M. de l'Escale de la quadrature du cercle par luy intitulé: Cyclometrica elementa duo* (Paris, 1594).

42 British Library, London. MS Sloane 2102: 'A short treatise on Geometrie'. The manuscript is divided into three books: 'The First booke of Oruntius Fine'; 'The secund booke of Oruntius Finé', and 'The third booke of Oruntius Fine touching certaine mathematical conclusions very profitable and necessary never as yet published.'

Printing Practical Mathematics: Oronce Fine's De speculo ustorio *between Paper and Craft*

SVEN DUPRÉ

In 1551 Oronce Fine published *De speculo ustorio*. It was one of the first works on optics to be published in France.[1] At this point the only optical treatise widely available in print in France was – if we exclude Jean Pélerin Viator's *De artificiali perspectiva* – the *Introductio in scientiam perspectivam* of the French humanist, Charles de Bovelles. First printed in 1503 in Paris, its availability was enhanced by Fine's reprint of the work as an appendix to his edition of Gregor Reisch's *Margarita Philosophica* of 1535. Bovelles's treatise was, however, quite unlike Fine's own work on burning mirrors. Bovelles's treatise was a basic introduction to the science of optics, structured around definitions of sight, light, shadow, colour, rays and mirrors, without demonstrations and with simple, even crude, small diagrams. As far as burning mirrors are concerned, Bovelles limited his comment to one single sentence, namely, that 'when turned towards the sun concave mirrors produce fire from the reflections of rays'.[2]

Fine's *De speculo ustorio* might have been new for the French local market, but it was far less original within the European mathematical culture. Only three years earlier Antonius Gogava, a mathematician who belonged to the circle of Gemma Frisius in Louvain, had published his own work on burning mirrors.[3] A comparison between Fine's and Gogava's treatises on burning mirrors is highly informative of the type of knowledge which Fine wished to see in print, as well as

1 R. P. Ross, 'Oronce Fine's *De Speculo Ustorio*: A Heretofore Ignored Early French Renaissance Printed Treatise on Mathematical Optics', *Historia Mathematica*, no. 3 (1976), pp. 63–70.

2 '[...] specula concava ad solem conversa, ignem ex radiorum reflexione gignunt'. C. De Bovelles, 'Introductio in scientiam perspectivam'. Reisch, ed. Fine, 1535, p. 1492.

3 Gogava studied mathematics with Gemma Frisius in Louvain before leaving for Italy at the end of the 1540s to study medicine at the University of Padua. Around 1550, he established a medical practice in Venice, but he also published a treatise on music, *Aristoxeni ... Harmonicorum Elementorum libri III* (1562). Thanks to his connection with the Gonzaga court in Mantua, he moved to Madrid shortly before his death. See V. Jacques, 'Gogava, Antoine-Hermann', in *Biographie Nationale* (Brussels, 1866), pp. 86–88.

of Fine's motivations to publish *De speculo ustorio*. Two differences between the two texts stand out. First, Fine's work on burning mirrors consisted of two parts, a section on the mathematics of the parabola, and a section on the 'mechanical' making of a parabolic mirror. There was no equivalent of the latter section in Gogava's earlier treatise. Should we, therefore, conclude that Fine's work was closely associated to the craft of mirror-making? In this essay I will try to be more precise about where we should place Fine's work between theory and practice, or between paper and craft. Did it reflect knowledge of workshop practices or bookish knowledge? Or was it perhaps practical mathematical knowledge? If so, what does 'practical mathematical knowledge' mean in this context? Finally, how does this reflect on Fine's identity as a 'mathematical practitioner'? In sections I and II of this essay I will attempt to answer such questions. In section I, I will argue that Fine's *De speculo ustorio* was a response to Bovelles's call for a practical optics, a part of practical geometry. In section II, I will show in more detail what Fine considered to be practical optics. On the basis of the dependence of Fine's work on burning mirrors on manuscripts circulating in Paris, I will argue that *De speculo ustorio* was part of a bookish culture. I will also reconstruct the history of Fine's writing of this treatise on the basis of an earlier, unfinished manuscript version of *De speculo ustorio*. The history of the writing of *De speculo ustorio* will be highly informative of Fine's understanding of practical optics.

A second significant difference between Gogava's and Fine's texts is Fine's addition of an optical context to his mathematical discussion of the parabola. The explicit presence of optics might have been self-evident in the context of a discussion of burning mirrors, but it was not for sixteenth-century mathematicians. Gogava discussed the parabolic burning mirror only as a problem of geometry, limiting his discussion to the construction and the properties of the parabola. In section III of my essay, I will connect Fine's addition of four optical postulates at the beginning of his treatise on burning mirrors to contemporary concerns about an astrological physics on the basis of a Baconian optical model. This will allow us to reflect on the importance of the perception of Roger Bacon as *magus*, and the role of burning mirrors in the shaping of this perception, in Fine's motivations to publish *De speculo ustorio*. Finally, I will briefly discuss the influence of Fine's treatise, especially in late-sixteenth-century Italy.

I. *Bovelles's Practical Geometry, Archimedes, and Fine's Optics*

Bovelles was a student of the humanist and mathematician Jacques Lefèvre d'Etaples, with whom he shared an interest in the revival of mathematics.[4] The precise nature of the relationship between Bovelles and Fine is unclear, but it has been suggested that the mutual point of contact was Antoine Loffroi, Abbot of

4 J. M. Victor, *Charles de Bovelles, 1479–1553: An Intellectual Biography* (Geneva, 1978). See also P. Brioist in this volume.

Ourscamp, to whom Bovelles dedicated his *Livre singulier et utile touchant l'art et practique de Géométrie* of 1542.[5] Loffroi was a patron of Fine, from whom he received the latter's *Arithmetica practica* of 1555. The 1542 edition of Fine's work on practical geometry, a revised version of his *Géométrie en français* (1511), was illustrated by Fine, who had a reputation as an engraver and a book illustrator, but Fine's commitment to Bovelles's book was such that he saw the work through the press.[6] Fine also reprinted other mathematical works of Bovelles, including the already mentioned *Introductio in scientiam perspectivam* as an appendix to an edition of Reisch's *Margarita Philosophica*.

In the preface to the edition of the *Géométrie pratique* of 1551, a new version enriched with mechanical 'applications' of geometry (chapter 7 contained a discussion of mills, for example), Bovelles admonished the reader 'who looked for measurements and quantities of lines and figures, and of bodies, by the art of geometry' not to forget 'the square, the rule and the compass, because the art and practice, and the profit of geometrical knowledge rely on these three [instruments]'.[7] Thus, in this preface Bovelles stressed the 'constructive' character of his *géométrie pratique*. Indeed, Bovelles was interested in developing practical means (and instruments) of construction rather than in geometrical demonstration, within the context of his project of a practical geometry, which aimed at a mathematisation of the arts.

Did Bovelles give optics a place within the field of practical geometry? In the last chapter of his *Géométrie pratique*, on the utility and excellence of geometry, Bovelles briefly mentioned optics as a science subalternate to geometry, like astronomy (also subordinated to geometry) and like music (subalternate to arithmetic). He argued that mathematical knowledge was useful for practitioners of optics or 'l'art de perspective', because 'the said perspective deals with the art of the mirror-makers, and the reflection and direction of the visible rays, which go straight or are reflected to the eye'.[8] Optics is a science of lines and angles, because – as Bovelles expressed it – one cannot know perspective (or the art of the mirror-makers), 'without knowing geometrically the nature of right and oblique angles, and of perpendicular and non-perpendicular lines'.[9] I will comment on two

5 *Ibid.*, p. 44.

6 For Fine as an illustrator of books, see A. F. Johnson, 'Oronce Fine as an Illustrator of Books', *Gütenberg-Jahrbuch*, vol. 3 (1928), pp. 107–109.

7 '[...] amy lecteur qui cherches les mesures, et quantitez des lignes et figures, et de tous corps, par art de Geometrie [...] et si n' oublie pas l' esquiere droict, la reigle et le compas: Car de ces trois despend l'art et pratique, et le profit de sçavoir Geometrique'. C. De Bovelles, *Géométrie practique composee par le noble philosophe maistre Charles de Bovelles, & nouvellement par luy reveue...* (Paris, 1551), sig. 1^{r}.

8 '[...] car ladicte perspective est comprinse sur l'art des mirouers, & sur la reverberation & directions des rais visibles che ants de droict ou soy reciprocants en l'oeil'. *Ibid.*, sig. 68^{r}.

aspects of Bovelles's understanding of optics.

First, Bovelles presented optics as a mathematical – and only mathematical, one should say – science of lines and angles, without any commitment to a physics of light. This lack of physical commitment was widespread among advocates of optics in the context of a practical geometry. These practical mathematicians considered the question whether vision comes about by extramission or intromission either irrelevant or, if they chose to respond, inclined towards extramission, and sometimes even fully endorsed this view. This reception of extramission might be surprising when one expects that, once Alhazen's intromission theory appeared on the scene (which transmitted to Roger Bacon became a corner-stone of medieval *perspectiva*), the extramission theory – marginal exceptions notwithstanding – would have lost its appeal.[10] In fact, the fate of extramission was more complicated, precisely because of Alhazen's theory of vision on the basis of which practical geometry was founded.[11] Filippo Camerota has argued that manuals on practical geometry considered measurement (of the distance between two stars, the height of a tower, etc.) as 'measurement by eye'.[12]

Since Alhazen had downplayed the contribution of refracted rays to vision, advocates of practical geometry were not compelled to adopt his intromission theory of vision, even when basing their techniques of measurement on Alhazen's optics. Consequently, Leon Battista Alberti, for example, claimed the question of the direction of the visual rays to be irrelevant for his purposes,[13] an opinion which

9 '[...] sans scavoir par Geometrie la nature des angles droicts & obliques, & des lignes perpendiculaires & non perpendiculaires'. *Idem*.

10 This suggestion is quite strong in D. C. Lindberg, *Theories of Vision: From Al-Kindi to Kepler* (Chicago and London, 1976).

11 Alhazen argued that vision was a two-stage process of 'aspectus' and 'intuitio'. The 'aspectus' is the first glance that yields only a superficial perception of an object. Next follows the 'intuitio' which gives a 'certified' impression of an object. The difference between the two stages or categories was explained in optical terms. The visual pyramid consisted only of those rays which fall perpendicularly on the eye and proceed without refraction to the glacial humor. The non-perpendicular rays contribute little to vision, because they are refracted and weakened. However, from all these rays, only the central ray, or the axis of the pyramid, falls perpendicularly on the interface between the glacial and vitreous humor, and, thus, goes unrefracted into the optic nerve. Since refraction weakens, this central ray is the strongest and perception through it is the clearest. The 'intuitio' brings points of the visible object, by successive eye movements, under the central ray. Thus perceived, it allows vision without any perceptual error, or the *certification* of sight. See Lindberg, *Theories of Vision*, pp. 84–85.

12 F. Camerota, '"Misurare per Perspectiva": Geometria pratica e prospectiva pingendi', in R. Sinisgalli (ed.), *La prospettiva: Fondamenti teorici ed esperienze figurative dall' Antichità al mondo moderno* (Florence, 1998), pp. 293–308.

13 '[...] among the ancients there was considerable dispute as to whether these rays emerge from the surface or from the eye. This truly difficult question, which is quite without value for our purposes, may here be set aside.' L. B. Alberti, *On Painting*, trans.

Daniele Barbaro in the mid-sixteenth century still shared.[14] This attitude is to be considered within the context of the sixteenth-century asymmetry in the relation between mathematics and natural philosophy.[15] This asymmetry allowed mathematicians to declare the solution to the natural philosophical question of vision or of light to be irrelevant to mathematicians, *insofar as they spoke as mathematicians*. It is important to keep this context in mind for understanding that Bovelles's definition of optics *as geometry* did not commit him to a view on the nature of vision or of light.[16]

The second interesting aspect of Bovelles's definition of optics in *Géométrie pratique* was its unusual stress on the study of reflection in connection with the art of mirror-makers – perhaps even awkward, if we take into account that it was, primarily, direct vision that was important for practical geometry. Bovelles's interest in mirror-making might have been inspired by the growing presence of makers and sellers of *cristallo* mirrors (and glass *à la façon de Venise*) in France due to the diffusion of Venetian glass-making skills over the European continent in the mid-sixteenth century. In mid-sixteenth-century Paris, the Bolognese Theseo Mutio obtained permission for making glass products *à la façon de Venise* (including mirrors), and *cristallo* products were for sale in shops of *faiseurs de miroirs de cristallin*, such as Nicolas Aufray.[17] This undeniable presence of a newly imported craft and trade of 'Venetian' mirror-making did not, however, inspire Bovelles to call for a kind of inventarisation of recipes and techniques. Instead, Bovelles foresaw the establishment of a body of practical geometry (an art of perspective or a practical optics, if one wishes) which aimed at the *mathematisation* of the art of the mirror-makers. Did Fine's work on burning mirrors match Bovelles's vision?

Let me first point out the obvious – perhaps. The audience which Bovelles envisaged for his *practical optics*, and Fine for *De speculo ustorio*, did not consist

C. Grayson, ed. M. Kemp (London, 1991), p. 40.

14 D. Barbaro, *La pratica della perspettiva*, ed. R. Fregna and G. Nanetti (Bologna, 1980), p. 6.

15 For this asymmetry, see R. S. Westman, 'The Astronomer's Role in the Sixteenth Century: A Preliminary Study', *History of Science*, vol. 18 (1980), pp. 105–147.

16 Interestingly, however, in the context of the discussion of astrological physics in section 3, extramission was still a valid choice, even if one wished to commit to a physics of light. Influenced by al-Kindi's *De radiis stellarum*, Roger Bacon allowed extramission to play a small but important part in his overall intromission synthesis. Tachau has argued that the context of Bacon's admission of visual radiation was astrological. See K. Tachau, 'Et maxime visus, cuius species venit ad stellas et ad quem species stellarum veniunt: Perspectiva and Astrologia in Late Medieval Thought', *Micrologus*, no. 5 (1997), pp. 201–224.

17 M. Philippe, *Naissance de la verrerie moderne XIIe–XVIe siècles: Aspects économiques, techniques et humains* (Turnhout, 1998), pp. 39 (for Theseo Mutio) and 183 (for Nicolas Aufray).

of the mirror-makers themselves. Even if mirror-makers would have been trained in a 'school' (and there is only one known example of such a training outside the master-apprentice relationship in France),[18] Fine's work on burning mirrors was in Latin. It did, however, fit the model of the text-book promoted at the Collège Royal, where Fine held the royal chair in mathematics, and where the whole field of mathematics, the old quadrivium and also 'new' branches, such as perspective, or optics and catoptrics, were taught.[19] The Collège Royal favored the spreading of mathematical knowledge through 'practical' approaches, and Fine published several mathematical text-books in which instruments were central. The Collège Royal was also, however, strongly attached to humanism, and, as we will see in the next section, the kind of knowledge which Fine offered in *De speculo ustorio* was 'bookish' knowledge. It was based on Fine's reading of manuscripts and books on burning mirrors rather than on visits to workshops.

In agreement with this image of mathematics, Bovelles's practical optics responded to a humanist image of Archimedes. Sixteenth-century mathematicians shaped Archimedes after their own self-image by considering him, primarily, an artificer of ingenious instruments and machines. Archimedes's legendary design of a mirror with which he burned the enemy's ships at Syracuse was a highly visible component of this image.[20] For example, in the preface of his *The pathway to knowledg* (1551), published in the same year as Fine's *De speculo ustorio*, Robert Recorde used the image of Archimedes to enoble the study of burning mirrors. 'But to retourne agayne to Archimedes', Record wrote, 'he dyd also by arte perspective (whiche is a parte of geometrie) devise such glasses within the town of Syracusae, that dyd bourne their ennemies shyppes a great way from the towne, whyche was a mervaylous politike thynge'.[21] Bovelles, also, singled out one figure to show the excellence and utility of geometry in his *Géométrie pratique*, namely the same: 'Archimedes, born in Syracuse in Sicily, [who] by means of the said art, in which he was most ingenious, and excelled, defended for a long time the city of Syracuse against the power of the Roman consul Marcus Marcellus'.[22]

18 Namely, the *verrerie* of Frizon (in the Lorraine). See *ibid.*, p. 164.

19 I. Pantin, 'Teaching Mathematics and Astronomy in France: The Collège Royal (1550–1650)', *Science & Education*, no. 15 (2006), pp. 189–207. See also Pantin in this volume and J. C. Margolin, 'L' enseignement des mathématiques en France (1540–70): Charles de Bovelles, Fine, Peletier, Ramus', in P. Sharrat (ed.), *French Renaissance Studies, 1540–70: Humanism and the Encyclopedia* (Edinburgh, 1976), pp. 109–155.

20 W. R. Laird, 'Archimedes among the Humanists', *Isis*, vol. 82, no. 4 (1991), pp. 629–638.

21 R. Recorde, *The Pathway to Knowledg, Containing the First Principles of Geometrie* (London, 1551), Preface.

22 'Archimedes natif de Syracuse en Sicile, par le moien de ladicte art, en laquelle estoit fort ingenieux & excellent, defendit longtemps ladicte ville de Syracuse contre la puissance de Marcus Marcellus Consul Romain'. Bovelles, *Géométrie practique*, sig. 69v.

II. *The Sources and the Writing of Fine's* De speculo ustorio

If Fine's work on burning mirrors was bookish knowledge, which, then, were the sources of his *De speculo ustorio*? The medieval optical tradition had not shown any special interest in the study of burning mirrors of a paraboloid shape. The parabolic burning mirror and its focal properties had, however, been discussed in antiquity, in Diocles's *On burning mirrors*, in the anonymous Bobbio fragment, and, triggered by the legend of Archimedes's burning mirrors of ships, by Anthemius of Tralles.[23] The study of the parabolic burning mirror was further developed by Arabic mathematicians, al-Kindi and his circle, Ibn Sahl and Alhazen in his *De speculis comburentibus*.[24] Of all this material, only Alhazen's *De speculis comburentibus* had been regularly available in the thirteenth century, when it served as the basis of Witelo's propositions on the burning mirror of paraboloid shape in his *Perspectiva*.[25] Interest in the study of burning mirrors increased in the fifteenth and sixteenth centuries. Indicative of this trend is that the burning mirror figured on the frontispieces of sixteenth-century editions of Witelo's *Perspectiva* in 1535 and in 1572. The frontispiece of the 1572 edition of Frederic Risner explicitly referred to the legend of Archimedes (Figs. 5.1 and 5.2).

In my introduction I mentioned that in 1548 Antonius Gogava published a treatise on burning mirrors. Gogava published it as an appendix to his edition of Ptolemy's *Tetrabiblos*.[26] The treatise included the first four propositions of Alhazen's *De speculis comburentibus* and, with the title *De sectione conica*, an edition of a treatise (which circulated in manuscript), known as *Speculi almukesi*

23 G. J. Toomer, *Diocles on Burning Mirrors* (Berlin, Heidelberg and New York, 1976), pp. 9–25; M. Cantor, 'Über das Neue Fragmentum Mathematicum Bobiense', *Hermes*, no. 16 (1881), pp. 637–642; W. R. Knorr, 'The Geometry of Burning Mirrors in Antiquity', *Isis*, vol. 74, no. 1 (1983), pp. 53–73. On the fragments on catoptrics attributed to Archimedes, compare W. R. Knorr, 'Archimedes and the Pseudo-Euclidean Catoptrics: Early Stages in the Ancient Geometric Theory of Mirrors', *Archives Internationales d' Histoire des Sciences*, vol. 35 (1985), pp. 28–104, and A. Rome, 'Notes sur les Passages des Catoptriques d' Archimède Conservés par Théon d' Alexandrie', *Annales de la Societé Scientifique de Bruxelles*, vol. 52 (1932), pp. 30–41.

24 R. Rashed, *Oeuvres philosophiques et scientifiques d'Al-Kindi* (Leiden, New York and Köln, 1997), pp. 97–102, 111–25; 'A Pioneer in Anaclastics: Ibn Sahl on Burning Mirrors and Lenses', *Isis*, vol. 81, no. 1 (1990), pp. 464–491; J. L. Heiberg and E. Wiedemann, 'Ibn Al Haithams Schrift über Parabolische Hohlspiegel', *Bibliotheca Mathematica*, no. 10 (1910), pp. 201–237.

25 Witelo, 'Perspectiva', IX. 39–44, in D. C. Lindberg (ed.), *Opticae Thesaurus Alhazeni Arabis libri septem, nuncprimùm editi. Eiusdem liber de crepusculis et nubium ascensionibus. Item Vitellionis Thuringopoloni libri X* (New York and London, 1972), pp. 398–403, partly translated in E. Grant (ed.), *A Source Book in Medieval Science* (Cambridge, MA, 1974), pp. 417–418.

26 *Cl. Ptolemaei Pelusiensis Mathematici operis quadriparti...De sectione conica, orthogona, quae parabola dicitur: Deq Speculo ustorio, libelli duo, hactenus desiderati: restituti ab Antonio Gogava Graviensi* (Louvain, 1548).

compositio. This was a thirteenth- or fourteenth-century treatise (ca. 1250–ca. 1350) of uncertain authorship, but at the time wrongly attributed to Roger Bacon (an attribution criticized by Gogava), or to an unspecified Arabic author (by Fine).[27] Gogava's edition of the *Speculi almukesi compositio* was based on a version prepared and annotated by Regiomontanus.[28] Regiomontanus became interested in optics in Italy, where he frequented the circle of students of Biagio Pelacani, to which Paolo Toscanelli, Alberti and Giovanni Fontana belonged.[29] When Regiomontanus launched his publication programme for the restoration of mathematics in Nuremberg in 1474, it included the printing of Ptolemy's *Optics* – a project which was only fulfilled by the instrument-maker Georg Hartmann, who inherited the manuscript from Regiomontanus – and, as an item on the list of Regiomontanus's own works to be published, a treatise *De speculis ustoriis atque aliis multorum generum ususque stupendi*, which most likely referred to a still-to-be-expanded version of Regiomontanus's manuscript copy of *Speculi almukesi compositio*.[30] One of the most significant changes which Regiomontanus introduced to the treatise as he had found it, consisted of leaving out the mechanical section on 'the conditions of good steel', which in the older manuscript versions of the *Speculi almukesi compositio* gathered together the *material* knowledge for the making, hammering and polishing of a parabolic mirror of steel.[31]

While Regiomontanus might not have shown any interest in *material* knowledge, the knowledge on burning mirrors which he offered was, nevertheless, *practical*. Regiomontanus's knowledge of burning mirrors belonged to the field of practical mathematics, because it was directed towards the design of a burning mirror (but not its material making). Two aspects of Regiomontanus's work on burning mirrors characterize it as practical mathematical.[32] First, the construction of the parabola was reduced to a two-dimensional problem, without reference to the three-dimensional cone, which allowed the parabola to be constructed solely on the basis of the knowledge of the focal distance. Second, the stress was on the invention of drawing instruments (compasses) which allowed the mathematician to draw a parabolic curve. We will see that this 'constructive' aspect was also strongly present in Fine's *De speculo ustorio*. This is not to claim that there is no attention to demonstration. In fact, propositions 1 to 7 are demonstrative, and, in

27 For the history of this manuscript, see M. Clagett, *Archimedes in the Middle Ages*, 4 vols. (Philadelphia, 1980), vol. 4, pp. 99–113. For Fine's attribution see below, n. 41.

28 Clagett, *Archimedes in the Middle Ages*, vol. 4, p. 101. For Gogava's edition, see *ibid.*, pp. 319–320.

29 For Regiomontanus's biography and bibliography, see E. Zinner, *Leben und Wirken des Johannes Müller von Königsberg genannt Regiomontanus* (Munich, 1938).

30 P. L. Rose, *The Italian Renaissance of Mathematics: Studies on Humanists and Mathematicians from Petrarch to Galileo* (Geneva, 1975), pp. 90–117.

31 Clagett, *Archimedes in the Middle Ages*, vol. 4, p. 178.

32 For Regiomontanus's manuscript, see *ibid.*, pp. 174–184, 357–358.

this sense, Fine's *De speculo ustorio* might differ from Bovelles's project for a practical optics. However, I will show that the finality of Fine's optics lay in the construction of a parabola, or in an instrument to draw a parabolic curve. In this sense, Fine's *De speculo ustorio* was practical mathematical, and in line with Bovelles's project.

Considering the dates of publication of Gogava's edition of Regiomontanus's version of *Speculi almukesi compositio* (1548) and of Fine's *De speculo ustorio* (1551) one would expect that Fine's work was based on Gogava's publication. Marshall Clagett has, however, convincingly shown that this is not the case. Clagett has argued that Fine's treatise was based on a now lost manuscript version of the *Speculi almukesi compositio*, at the time in the library of the Sorbonne, which also exerted its influence on the *Libellus de seccione mukesi* (presumably written shortly before 1400) of the French mathematician, Jean Fusoris.[33] It is to this work that Fine most likely refers in the preface as 'translated from the Arabic tongue in such a confused and convoluted way [...] that we could scarcely elicit any sense out of the text or discern the single diagram which would correspond to the text.'[34] Interestingly, it was the diagrams of *De speculo ustorio*, superior in terms of clarity and liveliness to Gogava's (compare Figs. 5.3 and 5.4), and for which Fine was especially gifted, which could be considered Fine's most important contribution to the study of burning mirrors. In Fine's *De speculo ustorio*, and in contrast to the manuscript tradition, text and image corresponded in unambiguous ways. Fine's diagrams, and this time in contrast to Gogava's, must also have been intelligible to those less at ease with the mathematics of conic sections. For their importance and influence, and perhaps also their quality, speaks the fact that they were borrowed (without acknowledgement) later in the century, as we will see below.

Additional evidence that Gogava's edition was not Fine's source comes from a manuscript version of the latter's *De speculo ustorio* at the Bibliothèque Nationale in Paris.[35] It is bound together with a manuscript of Fine's *De astrolabio sive planisphaerio*, prepared for publication. The manuscript version is dated to 1548, and unlike another manuscript version of *De speculo ustorio* in the Bibliothèque Nationale, which was most likely copied from the printed version, it is in Fine's own handwriting.[36] The manuscript shows that a good part of *De speculo*

33 *Ibid.*, p. 322. For Fusoris's *Libellus*, see *ibid.*, pp. 159–172. For the history of Fine's manuscript, see *ibid.*, pp. 321–329.

34 '[...] ex Arabica lingua in Latina adeo perplexe ac involute conversum [...] ut vix sensum aliquem ex ipsa potuerimus elicere litera: aut unicam conspicere figuram, quae eidem literae responderet'. Fine, 1551*d*, sig. 3^r^.

35 Bibliothèque Nationale de France, Paris, MS Lat. 7415, fols. 10–17.

36 The later manuscript, copied from the book, is Bibliothèque Nationale de France, Paris, MS Lat. 16650, fols. 2–11. These manuscripts are listed in Fine's bibliography in D. Hillard and E. Poulle, 'Oronce Fine et l'Horloge Planétaire de la Bibliothèque Sainte-Geneviève', *Bibliothèque d' Humanisme et Renaissance*, vol. 33 (1971), pp. 311–51;

ustorio – I will tell below which part – including the diagrams, identical to the ones later published, was already written in 1548, the date of publication of Gogava's book. More importantly, the manuscript is also telling of the writing history of Fine's *De speculo ustorio*. It is informative of Fine's motivitations to publish a book on burning mirrors, and of Fine's own view of his work on burning mirrors. More precisely, the writing history allows us to sustain the claim that the kind of knowledge in Fine's work was practical mathematical.

When compared with the printed book of 1551 the manuscript of three years earlier is incomplete. After the title-page, similar to the published one but without a publisher mentioned, and dated 1548, the manuscript starts at the top of folio 10 in the middle of one of the last sentences of proposition 6. The numbering of the manuscript folios is continuous until folio 17.[37] This suggests that the folios 1 to 9, containing propositions 1 to 6, were separated from the rest of the treatise at some point, but that they were already written in 1548, three years before publication. Also at the other end, the manuscript is incomplete when compared to the book. It ends in the middle of folio 17 with the first sentence of proposition 9. This means that Fine's manuscript in 1548 ends with the instrument to draw a parabolic section (in preparation for a tool with which to make a parabolic mirror) and that the section on material knowledge, not included by Regiomontanus and Gogava but present in the older manuscript versions of the *Speculi almukesi compositio* as the 'conditions of good steel', was not yet included in Fine's manuscript of 1548. Unlike the incompleteness at the beginning of the manuscript, I think that the break-off at the end reflects the state of the manuscript in 1548.

The evidence in the manuscript for this claim is corroborated by Fine's own preface in the printed book. Fine claimed that already twelve years earlier, around 1539, he had put together and demonstrated in a single book a device for describing a parabola.[38] The manuscript shows that in 1548 Fine's work on burning mirrors was still aimed at the construction and drawing of a parabolic section given the focal distance, a necessary step towards the making of a concave mirror. Thus, evidence of the manuscript shows, even more strongly than the 'completed' printed book, that Fine's knowledge of burning mirrors was practical mathematical in defining its aim as the construction of the parabola.

Therefore, the manuscript version of 1548 shows that Fine's knowledge of burning mirrors was, first, of a practical geometrical kind, and second, *bookish*, since his *De speculo ustorio* was a re-working of older manuscript material. The

---

R. P. Ross, 'Oronce Fine's Printed Works: Additions to Hillard and Poulle's Bibliography', *Bibliothèque d' Humanisme et Renaissance*, no. 36 (1974), pp. 83–85.

37 Hillard and Poulle wrongly comment that the folio numbering is interrupted, fols. 9–11 and 14–17. See Hillard and Poulle, 'Oronce Fine et l'Horloge Planétaire', p. 347.

38 Fine, 1551*d*, sig. 2^v.

additions of 1551 do not substantially alter this image of Fine's knowledge.[39] Fine added proposition 9, in which he discussed the *fabrica* of a burning mirror. However, the first part of the proposition was only a substantially shortened paraphrase of the section on the 'conditions of good steel' which he had found in the *Speculi almukesi compositio*.[40] It is indicative of the distance between Fine's *De speculo ustorio* and workshop practice that he borrowed from a thirteenth- or fourteenth-century text a section on the making of a mirror of steel, while the important technical development in mirror-making of his own time was the diffusion of techniques to make glass, or perhaps better, *cristallo* mirrors *à la façon de Venise*.

Fine also added an alternative procedure for making a burning mirror. This method consisted of casting the mirror instead of grinding it.[41] The procedure was, however, mathematised. Fine was most concerned with the preservation of the parabolic curve. His alternative procedure for making a mirror testifies more strongly to Bovelles's call for the *mathematisation* of the art of mirror-makers than to any familiarity with workshop practice. Fine also briefly mentioned techniques and products, which were used for polishing in contemporary workshops, not included in the *Speculi almukesi compositio*. The inclusion of this material knowledge assumes some familiarity with contemporary mirror-making practices, but Fine's brief notes on this hardly function as a practical guide for making mirrors (nor are they intended as such).[42] Finally, Fine also added a short note on the annular or ring mirror type, but again the addition is bookish.[43] It was based on those sections of *De speculis comburentibus* in which Alhazen had noted that the parabolic section from which the mirror is to be made may not include the vertex of

39 It is not clear why it took another three years for the book to be published. Might it be that the second longer stay in Paris of Sir John Mason played a role in this? Mason was an English diplomat and Member of Parliament, who secured a royal exhibition in the early 1530s to study in Paris. After his studies, Mason embarked on a diplomatic career that brought him back to Paris and the French court in mid-June 1550 to negotiate peace. Mason was allowed to return to England only after a marriage treaty between Edward VI and a daughter of Henry was concluded in July 1551. It was in this period of Mason's second stay in Paris that Fine dedicated *De speculo ustorio* to Mason. See P. R. N. Carter, 'Mason, Sir John (c.1503–1566)', in *Oxford Dictionary of National Biography* (Oxford, 2004 [http://www.oxforddnb.com/view/article/18278, accessed 4 May 2006].

40 Compare Fine, 1551*d*, sigs. 21^v–22^r, with the text of *Speculi almukesi compositio*, in Clagett, *Archimedes in the Middle Ages*, vol. 4, pp. 154–156.

41 Fine, 1551*d*, sigs. 21^r–23^r.

42 In the year of publication of *De speculo ustorio* King Henri II offered the Italian Theseo Mutio a sole privilige to produce glass and mirrors *à la façon de Venise* for ten years. It is not possible to tell which workshops Fine visited. For a brief discussion of the establishment of a French glass industry in the sixteenth century, see Philippe, *Naissance de la verrerie moderne*; S. Melchior-Bonnet, *The Mirror: A History* (New York and London, 2001), pp. 30–34.

43 Fine, 1551*d*, sig. 23^r.

the section.[44] Moreover, Fine's interest in the annular mirror might also have had a bookish origin. It referred to the representation of an annular mirror on the frontispiece of the 1535 edition of Witelo's *Perspectiva* (see Fig. 5.1), which in turn was possibly inspired by a paraboloidal ring mirror of considerable size – the bigger section was about 110 cm, the diameter of the smaller section about 60 cm, the total height was about 60 cm – which Regiomontanus was said to have owned.[45]

III. *Roger Bacon, Magic and Astrological Physics*

Thus far I have concentrated on the first significant difference between Gogava's and Fine's texts – a section on the material knowledge for making a mirror. I have argued that this section should not mislead us to associate Fine's work on burning mirrors too closely with craft. As I have shown, this section was a later, and even likewise bookish, addition to a treatise of which the aim was the discussion of an an instrument to draw a parabolic section. I have thus argued that Fine's *De speculo ustorio* responded to Bovelles's call for a mathematisation of the art of the mirror-makers, and that the type of knowledge which Fine offered in his book was practical mathematical and bookish rather than artisanal. In this section I will focus on the second significant difference between Gogava's and Fine's texts. Unlike the *Speculi almukesi compositio* which Fine followed most of the time (but like Alhazen's *De speculis comburentibus*), Fine included four optical postulates at the beginning of *De speculo ustorio*.[46] The first postulate defined a solar ray as a mathematical line; the second stated the law of equal angles for plane mirrors; the third expanded this law to convex and concave mirrors; and the fourth stated that a burning mirror which will reflect the incident solar rays to one point of combustion produces the quickest and most intense combustion of all burning mirrors, and that the parabolic burning mirror is such a mirror. These postulates place the otherwise strictly geometrical discussion of the parabola in an overtly optical context.

In this section I will argue that the placing of the discussion of burning mirrors in this optical context (and, in fact, the interest in burning mirrors as such) had much to do with Fine's perception of the figure of Roger Bacon. In the sixteenth century Bacon gained a reputation as a powerful magician.[47] This reputation as a magician was not so much based on the use of demonic or forbidden

44 Clagett, *Archimedes in the Middle Ages*, vol. 4, p. 330.

45 This mirror is known to have existed from a letter of Regiomontanus to Roder. See G. Rosinska, *Optyka w XV wieku miedzy nauka sredniowieczna a nowozytna Fifteenth-century Optics between Medieval and Modern Science* (Wroclaw, Warszawa, Krakow, Gdansk and Lodz, 1986), p. 187.

46 Fine, 1551*d*, sigs. 5^v–6^r.

47 G. A. Molland, 'Roger Bacon as Magician' *Traditio*, no. 30 (1974), pp. 445–460. See also G. A. Molland, 'Roger Bacon and the Hermetic Tradition', *Vivarium*, no. 31 (1993), pp. 140–160.

magic, from which Bacon had always tried to distance himself, but on his grasp of mathematics. One of the most convincing advocates of this image of Bacon was Recorde, who in the same year as the publication of Fine's *De speculo ustorio* wrote that

> [...] many thynges seme impossible to be done, whiche by arte may very well be wrought. And whan they be wrought, and the reason therof not understande, than say the vulgare people, that those thynges are done by negromancy. And hereof came it that fryer Bakon was accompted so greate a negromancier, whiche never used that arte (by any coniecture that I can fynde) but was in geometrie and other mathematicall sciences so experte, that he coulde dooe by theim suche thynges as were wonderfull in the syght of most people.[48]

The achievement of optical marvels was a highly important component of this image of Bacon as magician – not surprisingly, since Bacon had given more than enough reason for it in his own writings, most prominently in his *Epistola de secretis operibus artis et naturae et de nullitate magiae*.[49] One of these optical marvels was the construction of burning mirrors. Bacon argued that 'the use of this instrument is a task for the experimenter; the preparation of it a task for the geometer', he boasted of the skills of Petrus Peregrinus, and he mentioned that he had sent a burning mirror, together with a copy of *De multiplicatione specierum* and other works to the Pope.[50]

Among the thirteenth-century perspectivists of the Latin West, Roger Bacon was the primary advocate of the doctrine of the 'multiplication of species'.[51] It was the core of what would become the standard explanation of vision, perception and cognition. According to this theory any visible object generates or 'multiplies' species, or 'forms', 'images' or 'likenesses', of light and color in the transparant medium. The *multiplication of species* in the medium, that is, their rectilinear propagation along rays, was subjected to geometrical analysis. Moreover, the visible species was only one instance of a more general category of species. Species

48 Recorde, *The pathway to knowledg*, preface.

49 T. L. Davis, *Roger Bacon's Letter concerning the Marvelous Power of Art and of Nature and concerning the Nullity of Magic* (Easton, 1923), pp. 28–30 (optical phenomena and devices).

50 R. Bacon, *Opus tertium*, quoted in J. Hackett, 'Roger Bacon on *scientia experimentalis*', in J. Hackett (ed.), *Roger Bacon and the Sciences: Commemorative Essays* (Leiden, New York and Köln, 1997), p. 312.

51 Lindberg, *Theories of Vision*, pp. 112–114; K. H. Tachau, *Vision and Certitude in the Age of Ockham: Optics, Epistemology and the Foundation of Semantics 1250–1345* (Leiden, New York, Copenhagen and Köln, 1988), pp. 3–26; D. C. Lindberg, *Roger Bacon's Philosophy of Nature: A Critical Edition, with English Translation, Introduction, and Notes, of De Multiplicatione Specierum and De Speculis Comburentibus* (Oxford, 1983), pp. lxiii–lxxi.

denoted the effect of any agent, whether or not a percipient being was present. Bacon attributed all natural causation to the multiplication of species, and all natural causation was, thus, subjected to the same geometrical laws applied to the visible species. In *De speculis comburentibus* Bacon linked his doctrine of the multiplication of species to a mathematical model of the propagation of light, by analyzing burning mirrors.

For this doctrine of the multiplication of species Bacon was deeply influenced by a treatise of al-Kindi on astrology, known in its Latin translation as *De radiis*.[52] This connection of interests returned in the mid-sixteenth century. It was the motivation behind the publication of Gogava's publication of a treatise on burning mirrors. On the basis of the inscriptions on an astrological disc, published by Gerard Mercator in May 1551, and a text published by Mercator's son Bartholomeus, *Breves in sphaeram meditatiunculae* (1563), Steven Vanden Broecke has convincingly argued that, at that time, in Louvain a programme of astrological reform existed, which 'gave attention to rays as the physical basis of astrological causation and the relevance of optical analogy in determining the vigour of such rays'.[53] John Dee, who visited Louvain in this period, named Mercator and Gogava, a student of Gemma Frisius, as the most important mathematicians in this regard.[54] That Gogava published the treatise on burning mirrors as an appendix to a work on astrology – Ptolemy's *Tetrabiblos* – suggests that he considered the optics of the burning mirror relevant to astrological physics.

Was there a similar connection between the study of the parabolic burning mirror and optico-astrological concerns on the Baconian model of the multiplication of species in the work of Fine? The connection between celestial influences and light was certainly alive in France at the time of Fine's publication of *De speculo ustorio*. In his *De abditis rerum causis* (1548), the French physician Jean Fernel (the title-page of whose early mathematical work had been designed by Fine)[55] had drawn an analogy between celestial heat and light, a theory which was

52 Lindberg, *Roger Bacon's Philosophy of Nature*, pp. xliv–xlv.

53 S. Vanden Broecke, 'Dee, Mercator, and Louvain Instrument Making: An Undescribed Astrological Disc by Gerard Mercator (1551)', *Annals of Science*, vol. 58, no. 3 (2001), pp. 219–240, at p. 228. See also S. Vanden Broecke, *The Limits of Influence: Pico, Louvain, and the Crisis of Renaissance Astrology* (Leiden and Boston, 2003), pp. 174–181.

54 'I was [for *21. yeares ago] by certaine earnest disputations, of the Learned Gerardus Mercator, and Antonius Gogava (and other,) thereto so provoked: and (by my constant and invincible zeale to the veritie) in observations of Heavenly Influences (to the Minute of time,) than, so diligent [...].' J. Dee, *The Mathematicall Praeface to the Elements of Geometrie of Euclid of Megara (1570)*, ed. A. G. Debus (New York, 1975), sig. B.iiij^{r}.

55 Fernel's *Monalosphaerium*, with a title-page designed by Fine, was published at the Colines press. See C. Sherrington, *The Endeavour of Jean Fernel* (Cambridge, 1946), pp. 14–15, 188.

much contested at the time.[56] Just as the sun sends its light to sublunar things, Fernel argued, this celestial heat is vital for the growth of all terrestrial things. Moreover, in 1542 Fine edited a treatise of Claudius Caelestinus, *De his quae mundo mirabiliter eveniunt*, which was published together with Bacon's *De mirabili potestatis artis et naturae*. The latter work was also known under the title *Epistola de secretis operibus artis et naturae*. I have already referred to this work of Bacon in the context of my discussion of the relevance of optical marvels within the Renaissance perception of Roger Bacon as magician.

It has been shown that Caelestinus's treatise was an account and discussion of Oresme's arguments in his *Quodlibeta (De causis miribilium)*, in which Oresme adopted the Baconian doctrine of the multiplication of species.[57] Oresme was a severe critic of the judicial astrology of his own day, but he, nevertheless, allowed celestial influences, propagated according to the Baconian optical model. He limited celestial influence to motion and light propagated as rays which also produce heat, excluding all occult influences. Caelestinus and Fine included a chapter *De influentiis caelorum* to defend astrology against Oresme's attack. They drew heavily upon al-Kindi's *De radiis* for the claim that everything in the terrestrial world depended on the disposition of the stars. Moreover, they allowed occult influences from the heavens in addition to light and motion. Notwithstanding that Caelestinus and Fine made little use of specific optical arguments in their defense of astrology, given the background of the influence of al-Kindi's *De radiis*, Fine must have understood the connection between astrological causality, optical radiation and the geometry of burning mirrors.

However, we should be careful about pushing the connection between Fine's interest in and study of the parabolic burning mirror and concerns for an astrological physics too far. There is no evidence that Fine ever developed an astrological physics of the style promoted by the Louvain mathematicians, or John Dee.[58] Critical in this respect is the absence of the doctrine of the multiplication of species from *De speculo ustorio*. The optical postulates which Fine added to his treatment of the burning mirror were strictly mathematical. Fine did not commit himself to any physics of light. Rather, he kept to developing optics as a branch of

56 For Fernel's theory of celestial heat (and the analogy with light), see H. Hirai, 'Alter Galenus: Jean Fernel et son interprétation platonico-chrétienne de Galien', *Early Science and Medicine*, vol. 10, no. 1 (2005), pp. 1–35; *Le concept des semences dans les théories de la matière à la Renaissance* (Turnhout, 2005), pp. 83–103.

57 S. Caroti, 'Nicole Oresme, Claudio Celestino, Oronce Fine e i 'mirabilia naturae'', *Memorie Domenicane*, nos. 8–9 (1977–78), pp. 355–410, 357–358.

58 For the connection between optics (burning mirrors) and astrological physics in the work of John Dee, see N. H. Clulee, 'Astrology, Magic, and Optics: Facets of John Dee's Early Natural Philosophy', *Renaissance Quarterly*, vol. 30, no. 4 (1977), pp. 632–680; *John Dee's Natural Philosophy: Between Science and Religion* (London and New York, 1988), pp. 42–50. For the role in this of Dee's interest in Roger Bacon, see *ibid.*, pp. 52–57, 68–69.

practical geometry, as Bovelles envisioned the discipline. Speaking *as a mathematician* allowed Fine to do 'mathematical magic' and to mathematise Bacon's optical marvels. He avoided, however, being explicit about a connection between this mathematisation and a theory of causation.

IV. *The Influence of Fine's* De speculo ustorio *in Italy*

Fine's treatise was especially influential in Italy because of an Italian translation of *De speculo ustorio* and its adoption in the work of Giovanni Battista Della Porta. Della Porta's discussion of how to draw a parabolic section (given the focal distance), and the making of a parabolic burning mirror in his *Magia naturalis* was taken, almost verbatim, from Fine's propositions 8 and 9.[59] In fact, Della Porta borrowed the diagrams of *De speculo ustorio* as well as Fine's mistakes. In Fine's diagram (Fig. 5.5) the focal distance *ab* is doubled to produce the axis *ac* of the section. This doubled distance is taken to be the base of the right-angled cone from which the parabolic section is taken. Thus, a perpendicular *de* is constructed to *ac*, so that *da* and *ae* are each equal to *ac*. When the lines *dc* and *ce* are drawn, a right-angled (angle *dce*) isosceles triangle is constructed. This triangle is rotated about *dc* to produce the right-angled cone, with *dc* as the axis of the cone and *ce* as the radius of the cone's base. A perpendicular *fg* to *abc* through *b* is erected, with *fg* equal to *de*, and *bf* and *bg* each equal to *abc*. Also, a perpendicular *hi* to *abc* through *c* is erected, with *hc* and *ci* equal to *dc*. Thus, a line connecting *hfagi* will produce a parabola with focal distance *ab*, with vertex *a* and focus *b*. Della Porta repeated this construction in his *Magia naturalis* before describing Fine's instrument to draw a parabolic section (Compare Fig. 5.6).[60] As Clagett has shown, Della Porta also borrowed Fine's mistake to consider the radius of the right-angled cone from which the parabola is produced to be double the focal distance of the parabola.[61]

59 Clagett, *Archimedes in the Middle Ages*, vol. 4, p. 331.

60 Compare with Della Porta's construction: 'Let the distance be known how far we would have the glass to burn, namely, AB ten foot; for were it more, it could hardly be done: double the line AB, and make ABC, the whole line will be AC: from the point A, draw a right line DA, and let DA and AE be equal one to the other, and cut at right Angles by AC, but both of the must be joined to the quantitaty AC, as DCE, which in C make a right triangle, DCE. Therefore the Triangle DCE is a right angled Triangle, and equal sides: and were this turned about the Axis CD, until it come to its own place whence it parted, there would be made a right angled Cane, EDNC, whose parabolic section will be ABC: the right line DC will be the axis of the Cane, and CE will be the semidiameter of the basis of the Cane: Through the point C you must draw a line parallel to DE, and that is HI of the length of CE and CD; and by the point B draw another parallel to the said line ED, which is FBG; and let BG and BF be both of them equal to AC: so FG shall be the upright side, and HI the basis of the Parabolic section: If therefore a line will be drawn through the points HEAGI, that shall be a Parabolic section.' Quoted from the English translation in J. B. Porta, *Natural Magick* (New York, 1957), p. 372.

61 Clagett, *Archimedes in the Middle Ages*, vol. 4, p. 331.

It is worth noting that Della Porta collaborated on the construction of a parabolic burning mirror with Jacomo Contarini, the Provveditore of the Arsenal in Venice and a collector of books, manuscripts and instruments.[62] In 1580, Della Porta's patron, the Cardinal d'Este, sent him to Venice to make or obtain a parabolic burning mirror. Looking for guidance to construct a parabolic burning mirror, he turned to Jacomo Contarini, presumably not only to provide the means, but also the skills. On 29 November 1580, Della Porta wrote to his patron that Contarini had spent a day and most of a night at the Arsenal with him supervising an attempt by one of the Arsenal craftsmen to cast a parabolic mirror.[63] Contarini was also the author of a compass which allowed one to draw the different conic sections, and which, Contarini explained in a letter to the mathematician Francesco Barozzi, could be used to control the curvature of mirrors.[64] The selective borrowing of Della Porta, and Contarini's compass, show that Fine's influence was not limited to a particular geometrical construction, but that it stretched out to the development of the interest in burning mirrors as *practical mathematical* knowledge. But to whom in Italy would this knowledge have appealed?

In 1587 Fine's *De speculo ustorio* was published as an appendix to Cosimo Bartoli's translation of the *Protomathesis*, in an Italian translation by Ercole Bottrigaro, a Bolognese humanist who had also edited Ptolemy's *Geography*.[65] The network of the translator suggests that Fine's *De speculo ustorio* found its readers among patrons. Bottrigaro frequented the circle of Gian Vincenzo Pinelli (one of Europe's most important collectors of books and manuscripts at the time), which included Filippo Pigafetta and Guidobaldo del Monte.[66] Around 1597 Pigafetta discussed with the Grand Duke of Tuscany Ferdinando I the idea of a museum of military architecture, to which the Medici collection of drawing and measuring instruments would be added.[67] The project never materialized, but the

62 P. L. Rose, 'Jacomo Contarini (1536–1595), a Venetian Patron and Collector of Mathematical Instruments and Books', *Physis*, vol. 18 (1976), pp. 117–30. On Contarini's collection of instruments and his library, see F. Magani, 'Il collezionismo a Venezia al tempo del soggiorno di Galileo', and M. Zorzi, 'Le biblioteche a Venezia nell' età di Galileo', in *Galileo Galilei e la Cultura Veneziana* (Venice, 1995), pp. 137–190; See also M. Tafuri, *Venezia e il Rinascimento: Religione, Scienza, Architettura* (Turin, 1985), pp. 185–212.

63 Rose, 'Jacomo Contarini', pp. 125–126.

64 Bodleian Libary, Oxford, MS Canon. Ital. 145, pp. 12–13. See F. Camerota, *Il Compasso di Fabrizio Mordente: Per la Storia del Compasso di Proporzione* (Firenze, 2000), pp. 75–77, 245–246.

65 For Bottrigaro, see the entry in A. M. Ghisalberti (ed.), *Dizionario biografico degli italiani* (Rome, 1960–), pp. 491–495.

66 For Pinelli's library and his circle, see M. Grendler, 'A Greek Collection in Padua: The Library of Gian Vincenzo Pinelli (1535–1601)', *Renaissance Quarterly*, vol. 33, no. 3 (1980), pp. 386–416.

67 W. Prinz, 'Informazione di Filippo Pigafetta al Serenissimo di Toscana per una stanza

Stanza delle Matematiche of the Uffizi, housing a collection of mathematical instruments, was decorated around 1600 by Giulio Parigi with ancient war instruments, which were, in the original proposal, meant to be actually present in the collection. One of these instruments was Archimedes's burning mirror.[68] Thus, also in Italy, the kind of optical knowledge that Fine offered in *De speculo ustorio* primarily appealed to aristocratic circles with an interest in practical mathematical knowledge as well as in the collecting of instruments and books.

Bartoli's and Bottrigaro's Italian translation of Fine's works was dedicated to Guidobaldo del Monte. In the preface the Venetian printer Francesco Franceschi argued that mathematics 'among the other commendable and useful studies' was worthy of the 'great Lords', and thus, of aristocratic patronage.[69] It was, then, most appropriate that the book was dedicated to Guidobaldo, the aristocrat-mathematician, who himself – as Francesco Franceschi recognized in the preface – had contributed much to the study of mathematics. In Italy therefore Fine's *De speculo ustorio*, in Bottrigaro's translation, was a vehicle of the elevation of the status of mathematical knowledge. It should be made explicit though that the translation of Fine's work fed in to this larger movement. Obviously, it was not the cause of it.

The adoption of Fine's work on burning mirrors in late-sixteenth-century Italy shows that Fine contributed to the promotion of a particular sort of optics. But what kind of optics was Fine's? I have argued that it is best characterized as of the type of 'practical mathematical knowledge'. This identification grasps the 'constructive' character of Fine's optics. It was aimed at the construction of a parabolic curve through the design of an instrument. The instrument, then, that Fine offered his reader in *De speculo ustorio* was an instrument to draw a parabolic curve. That other instrument, the burning mirror, was by-and-large considered to follow directly from the construction of the parabolic curve. We have seen that Fine added a 'mechanical' section on the making of the burning mirror in the period between 1548 and 1551, but this late addition did not substantially change the fact that the finality of Fine's *De speculo ustorio* was the construction of a parabolic curve. Moreover, even in the 'mechanical' section, Fine's optics was bookish in its reliance on written knowledge in circulating manuscripts and the optical tradition.

---

da piantare lo studio di architettura militare', in P. Barocchi and G. Ragionieri (eds.), *Gli Uffizi: Quattro secoli di una galleria* (Florence, 1983), pp. 343–354; 'Dal modello al dipinto: macchine da guerra di Archimede alla fine del cinquecento', in C. Cresti, A. Fara and D. Lamberini (eds.), *Atti del convegno di studi: Architettura militare nell' Europa del XVI secolo* (Siena, 1988), pp. 409–416.

68 For a picture of Parigi's illustration of Archimedes's burning mirror in the *Stanza della Matematiche*, see S. Dupré, 'Ausonio's Mirrors and Galileo's Lenses: The Telescope and Sixteenth-century Practical Optical Knowledge', *Galilaeana: Journal of Galilean Studies*, no. 2 (2005), pp. 145–180.

69 Fine, trans. Bartoli, 1587, pp. 2–3.

It shows that Fine's practical mathematical knowledge is to be differentiated from material knowledge, which was closely associated with the craft of the mirror-maker. Fine's optics was an affair of paper, print and drawings. His investment was in the diagrams, as one would expect from someone with a reputation as an engraver.

In this sense Fine's optics answered Bovelles's call for a practical geometry (optics), or for a mathematisation of the arts (of the mirror-maker). In fact, the characterization of Fine's optics as practical mathematics, and of Fine himself as a 'mathematical practitioner' is a justified way to emphasize the difference between Fine's approach and that of other mathematicians whose optics was less directed toward construction and the design of an instrument. The fact that Fine was not explicit about the connections between the optics of the burning mirror and astrological causality in *De speculo ustorio* (although he must have understood the connection), fits the image of practical mathematical knowledge.

However, other characteristics of Fine's *De speculo ustorio* make the identification of his optics as mere practical mathematical knowledge insufficiently precise. The addition of an overtly optical context to the construction of the parabolic curve did not make Fine commit to a physics of light, but it is telling of his ambition to develop a sort of optics that was teachable at the Collège Royal alongside the quadrivium. The inclusion of demonstrations in *De speculo ustorio*, which uneasily sit together with the idenfication of Fine's optics as mere practical mathematics, also points to place his book in the category of the mixed mathematical disciplines. Fine was, then, a mathematical practitioner who took parts of the mixed mathematical field of optics into the tradition of practical mathematics, and after the transformation that resulted from this, made the outcome available as a new mélange.

We are thus justified to speak of Fine as a mathematical practitioner as long as we recognize the variety among mathematical practitioners and within the tradition of practical mathematics. Not all mathematical practitioners were interested or able to appropriate parts of the optical tradition. Notwithstanding the calls for a practical optics similar to Fine's in England, no contemporary English practitioner succesfully did what Fine accomplished.

Oronce Fine's Sundials: The Sources and Influences of De solaribus horologiis

CATHERINE EAGLETON

Among the many accomplishments noted by his contemporaries, Oronce Fine was often described as being very interested in astronomical and time-telling instruments. Writing shortly after Fine's death in 1555, his friend Antoine Mizauld explained that nobles, gentlemen and even the King of France visited Fine, and that his accomplishments included maps, books, and 'many mathematical instruments'.[1] Another biographer, André Thevet, who got his information from conversations with Fine's son Jean as well as from Mizauld, also mentions this interest in instruments. Thevet describes Fine's skill and enthusiasm for the advancement of the sciences, through lectures, demonstrations, books and the invention and making of 'sundry beautiful instruments'.[2] Evidence for the breadth of Fine's interests is given by his publications, which included a number of works about instruments: the equatorium,[3] the astrolabe-quadrant,[4] the universal quadrant,[5] the geometrical square,[6] the astronomical rings,[7] and an astronomical clock that he

1 A. Mizauld, 'Vita Orontii', in Fine, 1556*b*, sigs. v^{r}–vi^{r}: 'Possent: adhaec videre quae manu propria / Vel pinxerat, vel sculpserat, vel descripserat, / Non dico chartas, aut libros, sed mille organa / Mathematica, vel alterius artificii'. On Mizauld's biography of Fine see R. P.Ross, 'Studies on Oronce Finé (1494–1555)', unpublished PhD. diss. (Columbia University, 1971), p. 363.

2 'Que si de bouche et vive voix, ensemble par demonstrations en ses leçons il avançoit grandement ces sciences, encore plus les illustroit il par ses labeurs particuliers tant par ses escrits que par invention et fabricature de plusieurs beaux instruments, come ayant la main non moins apte et duite à fabriquer et dresser, tels organes, et les peindre, que l'esprit à les inventer'. A. Thevet, *Les vrais pourtraits et vies des hommes illustres Grecz, Latins, et Payens…* (Paris, 1584), fol. 565^{r}. On Thevet's biography of Fine see Ross, 'Studies on Oronce Finé', p. 365.

3 Fine, 1526.

4 Fine, 1527.

5 Fine, 1550.

6 Fine, 1556*a*.

7 Fine, 1557.

worked on for the Cardinal of Lorraine, adding an hour ring and astrolabe dial to it.[8] In addition, there survives an ivory sundial signed 'Opus Orontii F' and dated 1524 (Fig. 6.1a and b).[9] The crown and salamander on the central 'mast' of this instrument may link it to François Ier, King of France, suggesting that Fine was using instruments as well as publications to develop his royal connections and mathematical reputation in the years before his appointment as Royal Lecturer in Mathematics at the newly-founded Collège Royal in 1531.[10]

The year after his appointment, Fine published a mathematical book titled *Protomathesis*. In the preface, he explains his aim to show the importance of mathematics, and to place practical mathematics on a sure theoretical footing.[11] At the same time, he wanted to show the king the first fruits of his work as Royal Lecturer and, presumably, secure his position. The *Protomathesis* is a large, handsomely illustrated work, made up of four parts which would each later be republished separately. The title page explains that it consists of various works, 'no less useful than pleasing' which are for the first time to be set out clearly.[12] The four parts of the work begin with arithmetic and geometry, and then move on to astronomy and instruments.[13] Given Fine's interests in instruments, and his statements

8 In 1553, Fine published a work in Latin detailing the construction and use of the instrument. It was translated into French soon afterwards. On these and other works about the clock, see Ross 'Studies on Oronce Finé', p. 449; and D. Hillard and E. Poulle, 'Oronce Fine et L'Horloge Planétaire de la Bibliothèque Sainte-Geneviève', *Bibliothèque d'Humanisme et Renaissance*, vol. 33 (1971), pp. 311–351.

9 These ship-shaped dials are often called *navicula* since a number of medieval English manuscripts give them that name. However, the Milan dial and a number of others are made according to an altered geometry, and manuscripts describing that variant form of the instrument usually call it *organum ptolomei*, although some give the instrument no name at all. Therefore, for clarity, and the avoidance of confusion, I call the non-English and post-medieval instruments 'ship-shaped dial'. For details, including the relationship between the *navicula* and the *organum ptolomei*, see C. Eagleton, 'Medieval Sundials and Manuscript Sources', in S. Kusukawa and I. Maclean (eds) *Transmitting Knowledge: Words, Images and Instruments in Early Modern Europe* (Oxford, 2006), pp. 41–72.

10 J.-C. Margolin, 'L'Enseignement des mathématiques en France (1540–1570): Charles de Bovelles, Finé, Peletier, Ramus', in P. Sharratt (ed.), *French Renaissance Studies 1540–1570: Humanism and the Encyclopedia* (Edinburgh, 1976), pp. 109–55; and Pantin in this volume. Turner, in this volume, suggests that Fine may have presented a water clock to the king.

11 Fine, 1532*b*, sigs. AA2–[4]: 'Ad christianissimum francorum regem Franciscum, eiusce nominis primum: Orontii finei delphinatis, praefatio'.

12 *Ibid.*, title page: 'Opus varium, ac scitu non minus utile quam iucundum, nunc primum in lucem foeliciter emissum'.

13 The four parts are titled: *De arithmetica practica libri IIII*, *De geometria libri II*, *De cosmographia sive mundi sphaera libri V, propriis ipsius authoris commentaris elucidati* and *De solaribus horologiis et quadrantibus libri IIII*.

about the need to properly ground practical mathematics in theoretical understanding, it is perhaps no surprise that the fourth book, titled *De solaribus horologiis et quadrantibus libri quatuor* includes descriptions of how to make and use a vide variety of sundials. Fine explains that since there is an increasing number of types of dial, it seems to him to be important to clarify and improve on the existing sources and to add some of his own discoveries.[14] He states that sundials reflect the motions and structures of the heavens through the movement of a shadow or a plumb-line, referring the reader back to chapter 9 of the second book of the previous part of the *Protomathesis* ('De cosmographia'), where he had discussed the seasons.[15]

De solaribis horologiis is divided into four books, the first of which covers the making and common uses of sundials, including those indicating the time by thread, stylus, or other means.[16] The second book covers other types of dials, including the ring dial, cylinder dial, and universal dials that can be used at any latitude, which are all based on 'the course of the sun, or other astronomical foundations'. Added to these is a hydraulic clock that Fine himself claims to have developed.[17] The third and fourth books describe the construction and use of an astrolabe quadrant.[18] Among the dials described and illustrated in the second book is a ship-shaped dial, very similar to the one Fine had made several years previously (Fig. 6.2a and b).

This distinctive sundial was known in medieval England as the *navicula*, but more often called *organum ptolomei* by fifteenth- and sixteenth-century manuscripts, some of which are associated with Vienna. Close consideration of Fine's description and illustration of this instrument provides a way to understand the sources and influences of the work of which it is part. Firstly, I consider some of the sources used by Fine in preparing *De solaribus horologiis*, and the changes he made to his source material. The second section looks at the influence of Fine's work, both in terms of instruments made according to his instructions, and the influences his work had on other sixteenth-century sundial books. Through looking closely at just two of Oronce Fine's sundials, we see more clearly the scale of his project, and the impact it had on later writers, books, and instruments.

14 Fine, 1532*b*, 'Proloquium'.

15 On Fine and cosmography see Besse and Mosley in this volume.

16 *Liber primus, de compositione, & usu vulgarium, multiformiumque horologiorum: quibus per fili, styli, perpendiculi, alteriusque rei vmbram horae ipsae dignoscuntur.*

17 *Liber secundum, de caeteris horologiis, tum annularibus & cylindricis, tum in circulo, atque circuli quadrante descriptis: ex Solis cursu, aliisve fundamentis Astronomicis immediate pendentibus. Vbi Hydraulicum describitur horologium, ab authore recenter excogitatum.* On Fine's hydraulic clock see Turner in this volume.

18 *Liber tertius, plansphaerium, seu vulgatum Astrolabium in quadrantem vetere docet: eiusdem, vel aeque facilis cum eodem planisphaerio vsus, atque commoditatis. / Liber quartus & vltimus, ipsus quadrantis generalis vtilitatem, fructumque multiplicem, sigillatim aperteque manifestat.*

I. *Making a Book about Sundials*

Writing in the early sixteenth century, there were few printed works on astronomical or time-keeping instruments to which Oronce Fine could turn, to use as his source material. However, there was a wealth of manuscript material, including copies of many different texts about sundials and other astronomical instruments. Among these are a group of ship-shaped instruments which are usually known as *organum ptolomei* in the manuscript texts describing them. I have argued in another article that this group of texts developed from a corrupted copy of a manuscript on the English *navicula* sundial, which became available in German-speaking parts of Europe, perhaps around Vienna. The resulting instrument, renamed *organum ptolomei*, had changed more than just its name, and in the process of corruption and transmission the geometry had become simplified, and the 'mast' of the *organum ptolomei* was slightly shorter than its English ancestor.[19] Since Fine's version of the ship-shaped dial is based on the simpler geometry, and has the shorter 'mast', this means that it is possible to identify the *organum ptolomei* manuscripts, rather than the English *navicula* manuscripts, as the source for Fine's descriptions in *De solaribus horologiis* of the ship-shaped dial. Interestingly, several of the manuscripts containing texts about the *organum ptolomei* also include works by Johannes Müller von Konigsberg (more usually known by his Latin name, Regiomontanus) or Georg von Peurbach, two authors with whom Fine was very familiar, having edited their works for publication early in his career. Other distinctive instruments found in *De solaribus horologiis* are also discussed in anonymous manuscript texts found in the same manuscripts.[20]

However, the *organum ptomolei* as described and illustrated in the earliest group of manuscripts would not have worked, because of errors in, and omissions from, the description of its construction.[21] Some of the later texts on the instrument attempt to correct these errors; others do not. The author of one of the later

19 For details of the geometry of the two instruments, see C. Eagleton, *Monks, Manuscripts and Sundials: the Navicula in Medieval England* (forthcoming, 2009).

20 In 1525 a version of Peurbach's *Theoricae novae planetarum* edited by Fine was published, with a later edition in 1534, and a French translation prepared by Fine published in 1528 and 1558. See Ross, 'Studies on Oronce Finé', pp. 342–343, who discusses Fine's role in making the work of Peuerbach and Regiomontanus available in France; and R. P. Ross, 'Oronce Finé's *De sinibus libri II*: the first printed Trigonometric Treatise of the French Renaissance', *Isis*, vol. 66, no. 3 (1975), pp. 379–386, which discusses the authorities cited by Fine in his *De sinibus libri II*. An example of a distinctive instrument found in *De solarbius horologiis* that also appears in the manuscript compilations containing *organum ptolomei* texts is the universal dial appearing on pp. 159–161 of Fine, 1560.

21 There are two main groups of manuscripts about the instrument. The earlier texts are a group of five copies, with incipit 'organum ptolomei ita sit': Bayerische Staatsbibliothek, Munich, MS Lat 24105, fol. 67ᵛ; Österreichische Nationalbibliothek, Vienna, MS Lat 5303, fol. 253ʳ, and MS Lat 5418, fol. 180ʳ; Yale University Medical-Historical Library,

manuscript texts on the *organum ptolomei*, despite complaining about the confusing and inaccurate texts on the instrument that are available, and stating that his text is clearer than the others, nonetheless gives an incomplete set of instructions for making the instrument.[22] This author and others attempted to correct the errors in the *organum ptolomei* texts, as well as to clarify the construction and use of the device, in some cases redesigning the instrument to restore the missing parts and correct the geometry. This meant that by the early sixteenth century there were several related instruments, all using similar geometry, including two types of ship-shaped dial (the *navicula* and the *organum ptolomei*) and a rectangular universal rectilinear dial often known as the Regiomontanus dial (Fig. 6.3).[23]

This process of redevelopment and redesign caused confusion in the manuscript tradition: the five earliest surviving texts on the instrument all call a ship-shaped dial *organum ptolomei*; while of the four later manuscripts, one describes a ship-shaped instrument, two describe rectangular instruments, and one describes an instrument that could be either ship-shaped or rectangular. Later texts become more confusing still, with various forms of the rectangular *organum ptolomei* described (but often given other names).[24] In one manuscript, a diagram showing the rectangular form of the instrument is titled *naviculum ptolomei*, clearly showing the links back to the English tradition, but also the extent of the fluidity of this set of manuscripts.[25] It is not clear which version(s) of the texts on

---

New Haven, MS 25, fol. 92ʳ; and the now-lost copy last known in St Peter's Library, Salzburg, Inc 800, fol. 91ʳ. A later group includes versions of the text, some of which have been rewritten to try and correct the errors of the earlier texts: Yale University Medical-Historical Library, New Haven, MS 24, fol. 268ᵛ and fol. 446ʳ; Bibliotheque Royale, Brussels, MS 2962–2978, fol. 29ᵛ; Bayerische Staatsbibliothek, Munich, MS Lat 19690, fol. 79ʳ.

22 'Organum ptholomei ad multas prouincias in canone proprio vt ab ipso componitum est satis obscurum breuitate exhibetur propter quod diuersi canones feruntur a pluribus. Hic hicque sequitur plus lucide composicion eius edocet licet subscriptibus par sit tibi nec ex illo hoc ex aliis varietur instrumentum canonibus quare in sua forma et veritate permaneat. Explicit prefacio'. Yale MS 24, fol. 268ᵛ.

23 I discusses this process of corruption and revision in more detail in 'Medieval Sundials and Manuscript Sources'. The attribution to Regiomontanus is due to his inclusion of this type of dial in his *Kalendarium* (various editions, the earliest in 1474). Regiomontanus himself says that he got the design from an 'antiquus compositor'. See E. Zinner, *Deutsche und Nederlandische Astronomische Instrumente des 11.–18. Jahrhunderts* (Munich, 1979), p. 112

24 For example, Österreichische Nationalbibliothek, Vienna, MS Lat 5258, fol. 80ʳ, where the instrument is known both as *organum ptolomei* and as *quadratum horarium*, suggesting a link to Regiomontanus's printed version of the instrument, which he called *quadratum horarium*; Vienna MS Lat 5228, where it is called *horologium universale*; Yale MS 24, fol. 202ʳ, where it is referred to as *instrumentum ad horas capiendas*.

25 Lund University Library, MS47, fol. 44ᵛ. The images are available online at: http://laurentius.lub.lu.se/volumes/Mh_47/.

the ship-shaped and rectangular versions of the *organum ptolomei* Fine used when compiling his *De solaribus horologiis*, but it is clear that he took what was probably an incomplete and inaccurate text on the ship-shaped dial and, like others before him, rewrote it in order to clarify, correct and improve the description of its construction, including it along with the rectangular form of the instrument, which he called simply *horologium universale*.

The manuscript compilations in which Fine would have read these texts on astronomical instruments tend to be anonymous collections of short texts. Beyond the inclusion of texts on the most important instruments, including astrolabe, quadrant and cylinder dial, a typical fifteenth-century manuscript compilation including works on scientific instruments has no standard contents, and no particular organization of the texts within the manuscript volume. The texts are presented as separate works collected together within the book's covers, with no linking passages between texts, and no introductory section setting out what the book contains, nor the rationale behind its ordering or the selection of works included. For example, a late-fifteenth-century manuscript formerly from Melk Abbey, near Vienna, now in the Yale Medical-Historical Library, includes several works by Peurbach and Regiomontanus along with anonymous works on various types of sundials, on the signs of the zodiac, and on mathematical topics including the squaring of the circle.[26] About half of the works relate to instruments, with

[26] Yale MS 24, described in C. U. Faye and W. H. Bond, *Supplement to the Census of Medieval and Renaissance Manuscripts in the United States and Canada* (New York 1962), pp. 57–58. A full list of contents is: fols. 1^{r}–64^{r}: Regiomontanus, *Calendarium* (printed work, edition of 1474, Nuremberg); fols. 65^{r}–131^{r}: Peurbach, *Theoricae novae planetarium*; fols. 133^{r}–134^{r}: Treatise on chronology, incipit: *Mensium autem lunarium*; fols. 135^{r}–137^{r}: Astrological treatise, inc.: *Vilescit fortisal sciencia signorum*; fols. 137^{r}–144^{r}: Treatise on the four seasons, inc.: *Animalium et hominum corpora*; fols. 144^{r}–154^{r}: Peurbach, *De septem planetis*; fols. 155^{r}–172^{r}: Astesanus Astensis, *Summa* (fragment); fols. 173^{r}–201^{r}: Peurbach, *Quaestio disputa*; fols. 202^{r}–215^{r}: *Instrumentum ad horas capiendas ab elevacione poli*; fols. 215^{r}–220^{r}: *Canon horologii nocturnalis*; fols. 221^{r}–230^{r}: *Cum scientia astronomica non completur sine debitis instrumentis*; fols. 230^{r}–235^{r}: *Tractatus kilindri accurtatis*; fols. 236^{r}–237^{r}: Treatise on sundials, inc.: *Horum dum vide velis*; fols. 238^{r}–245^{r}: Peurbach, *Compositio citharae horariae*; fols. 246^{r}–252^{r}: *Compositio horologii in anulo*; fols. 252^{r}–254^{r}: *De usu huius anuli*; fols. 255^{r}–258^{r}: Regiomontanus, *Compositio quadrantis*; fols. 258^{r}–267^{r}: *Hiis itaque compositis*; fols. 267^{r}–272^{r}: *Composicio organi Ptolelmei*; fols. 272^{r}–274^{r}: *De compositione instrument incentionum*; fols. 275^{r}–310^{r}: *De dorso astrolabii*; fols. 311^{r}–317^{r}: Treatise on the signs of the zodiac; fols. 318^{r}–347^{r}: Johannes von Gmunden *De instrumento aequatorio eiusque usu*; fols. 349^{r}–369^{r}: Astronomical and astrological tables; fols. 371^{r}–392^{r}: Prosdocimus de Beldomandis, *Compositio astrolabii*; fols. 393^{r}–407^{r}: Peurbach, *Algorithmus*; fols. 409^{r}–415^{r}: Peurbach, *Fabrica et usus instrumenti*; fols. 417^{r}–428^{r}: *De compositione et usu torquati*; fols. 429^{r}–436^{r}: Johannes Simonis de Zelandia, *Speculum planetarum*; fols. 437^{r}–445^{r}: *De sinu*; fols. 446^{r}–448^{r}: *Organum Ptolomaei*; fols. 449^{r}–453^{r}: Nicholas of Cusa, *Quadratura circuli*; fols. 453^{r}–

others on astronomical, mathematical and calendrical subjects also copied. No fewer than three different texts on the *organum ptolomei* are copied, but no link made between the three, either by the scribe or by later annotators.[27] In short, this manuscript is a compendium rather than a connected book. Therefore, in his compilation of texts from a variety of sources, including the *organum ptolomei* manuscripts, Fine did more than simply correct the inaccurate texts, he compiled them into a work that went beyond the medieval manuscripts that were his sources. Looking closely at the ordering of material in *De solaribuis horologiis*, we see that significant changes were made by Fine to his source material, so that the book brought some geometrical order to the enormous variety of sundials available. Unlike the medieval compilations, Fine's book has a linking rationale, and proceeds from simple geometrical principles to more advanced instruments, in line with his aim to give practical mathematics a solid theoretical foundation. Significantly, the sections of the books are called *propositio*, borrowing geometrical language to divide a practical work.

The first book of *De solaribus horologiis* covers dials that tell the time by measuring the direction in which the sun is observed, and begins with a diagram and description of the basic geometry underpinning a sundial for the latitude of Paris.[28] Fine then explains how this geometry can be applied to make a variety of dials, including horizontal dials (proposition 2) and vertical dials (proposition 3). Later in Book 1, Fine moves on to talk about dials that are portable, and can be used at more than one latitude, and gives examples of dials for various latitudes, aligned in various ways, and constructed on various surfaces and shapes. The first book ends with instructions for how to make and use a nocturnal, for telling the time by the stars, a moon dial, and a dial for hours counted from dawn or dusk rather than from midnight or midday, based on the same geometry as the earlier, simpler examples in the book. Book 2 deals mainly with dials that tell the time from the altitude at which the sun is observed, including quadrants as well as the cylinder dial, ring dials, a universal astrolabe and universal ring dial. At the end of the second book of *De solaribus horologiis*, just before the description of the hydraulic clock, Fine includes the universal rectilinear dial (*Propositio XV*) and the ship-shaped dial (*Propositio XVI*, but in the 1560 edition wrongly headed *Propositio XV*). In the text, he explicitly links the two similar instruments, and the ship-shaped dial is described as an alternative version of the preceding instrument, although Fine also points out that this version is more useful than the universal

---

454ʳ: Nicholas of Cusa, *De sinibus et cordis*; fols. 455ʳ–459ʳ: *De virga visoria*; fol. 460ʳ: *Baculus Iacob instrumentum quo*.

27 At fols. 202ʳ, 268ʳ and 446ʳ.

28 'Descriptio Prototypi generalis fabricandorum horologiorum, ad Parisiensem latitudinem, seu poli borealis eleuationem 48 graduum, in aliorum exemplari figurata'. Fine, 1560, p. 5.

rectilinear dial.[29]

Looking more closely at the descriptions of the two instruments, Fine's aim to underpin the practical aspects of sundial-making with the necessary theoretical understanding, and to give structure to his work, can clearly be seen. As is the case in his manuscript sources, in many places, Fine refers the lines on the dial to the heavenly and astronomical lines to which they relate: the equator, the tropics, or the ecliptic. However, he goes further than his sources in explicitly linking parts of the description of the construction of a particular sundial to the astronomical explanations of the *Cosmographia*, which preceded *De solaribus horologiis* in the *Protomathesis*. For example, after describing the construction of the parallel lines representing the latitudes at which the universal rectilinear dial can be used, Fine adds a note referring back to the second chapter of the fifth book of his *Cosmographia*, where the climates are discussed in theoretical terms.[30] In the description of the ship-shaped dial, Fine refers the reader back to the second book of his *Geometria* for an account of how to use the shadow square on the back of the instrument for measuring the heights and depths of things, and in describing the unequal hours diagram he refers back to the description of the same diagram on a quadrant in proposition 8 of the second book of *De solaribus horologiis*.[31]

The book does not have a wholly theoretical focus, however, and Fine includes instructions that give precise information about the physical form of the instrument, or practical tips relating to the construction of an actual dial. For example, in the description of the universal rectilinear dial, he explains that the hour lines can be subdivided, and different coloured lines used to separate the parallel lines,[32] and in the description of the ship-shaped dial, Fine explains that the mast should be fitted inside a hollowed-out area inside the instrument, and that the mast should not be too thick.[33] There is also an intriguing reference to his own instruments, when in the section on the universal rectilinear dial he says 'just as we do in our instruments for sale'.[34]

In *De solaribus horologiis*, Fine has grouped similar instruments together, and made links between them, sometimes directly referring to another, comparable instrument. His book proceeds from the principles underlying a simple dial for

[29] The universal rectilinear dial is headed: 'Aliud insuper horologium vniversale, rectilineum, super quadrangulo plano delineare' (*ibid.*, p. 175) while the ship-shaped dial is headed: 'Idem quod antecedens tradidit horologium, in formam navis, amplioris quidem vtilitatis, conuertere' (p. 183). The opening line of the instructions for making the ship-shaped dial refer to the universal rectilinear dial as the 'antecedentis horarii generalis prototypus'.

[30] Fine, 1560, p. 177.

[31] *Ibid.*, pp. 186 and 189.

[32] *Ibid.*, p. 178.

[33] *Ibid.*, p. 185.

[34] *Ibid.*, p. 177: 'quemadmodum in venalibus instrumentis nostris fecimus'.

the latitude of Paris, to the construction of more complex dials and quadrants, some of which can be used at any latitude. It links back to the astronomical theory of the previous part of the *Protomathesis* in places, and geometrical explanations are given in order to demonstrate how the dial is constructed, and how the various lines relate to the heavenly lines. This shift from a compilation of texts to a single-authored work, and from an unordered to an ordered collection of writings about astronomical instruments, is not unique to Oronce Fine, however, since it is clear that around the same time other authors were engaged in similar activity. Georg Hartmann, in a manuscript work compiled in 1527, set out the geometrical basis for the construction of sundials before he went on to describe and illustrate a range of different types of dial.[35] Around the same time, in 1531, Sebastian Münster published the first edition of what was to become a very influential work: his *Horologiographia*, which took a similar, ordered approach to Fine's books. Indeed, the 1533 revised edition of Münster's book included material from Fine's *De solaribus horologiis*, along with new material taken from works published by other authors.[36] From Fine, Münster added both text and diagrams, including geometrical explanations and new instruments, and in some cases reproducing the diagrams with the same decorative flourishes that they had had in Fine's book, although without any reference to Fine.[37] It is clear that both authors were engaged in a similar project – to create a sundial book that progressed from the basic to the more advanced, and was a connected whole, complete with detailed diagrams.

Throughout the sixteenth century there was a rapid increase in publications on practical geometry, aimed not necessarily at practitioners, but also probably printed to meet the evolving upper classes' interest in mathematical subjects.[38] The purpose of these was as much to teach geometry and astronomy as to teach the construction and use of sundials, and this helps us to make sense of the effort made by authors like Fine and Münster to present their works on sundials as

35 G. Hartmann, 'Compositiones horologiorum et aliorum instrumentorum', in Herzogin Anna Amalia Library, Weimar, MS Fol. Max 29, fols. 13r–72v. This work includes a *navis* dial which, from its geometry clearly derives from the *organum ptolomei*, but whose design also links it to the medieval English *navicula*. The manuscript does contain an instrument called *organum ptolomei*, but this is not a ship-shaped dial, and is instead an instrument for finding solar altitudes.

36 S. Münster, *Horologiographia* (Basel, 1533). Sections on each zodiac sign (pp. 178 to 201) are taken from Hyginus, and in a section on using your hand as a sundial Jacob Koebel is named as the source (chapter xlix).

37 Zinner, *Deutsche und Nederlandische Astronomische*, p. 456. Sections taken from Fine include chapters iii, xlvi and xlviii.

38 The contested nature of mathematical expertise, and the increasing interest in mathematical subjects, is discussed in e.g. K. Hill, '"Juglers or Schollers?": Negotiating the Role of a Mathematical Practitioner', *The British Journal for the History of Science*, vol. 31, no. 3 (1998), pp. 253–74.

organised works, proceeding from first principles to the construction of dials. These sorts of works would appeal to a student readership whose interests and geometrical skills were at a basic level, but also to educated readers looking for a reference work.[39] Fine's concern with improving the teaching of mathematical subjects, and his stated aim to provide a solid theoretical grounding for practical mathematics, underpinned this organisation of material from a range of sources including some of the many manuscript compilations of texts about instruments. Through the compilation of *De solaribus horologiis*, Fine helped to create a new genre: the sundial books that later became so popular but which did not exist before 1500.

II. *Making Instruments and Making Mistakes*

At the beginning of the posthumous re-edition of the sundialling sections of *Protomathesis*, published in 1560, there is a message from the printer to the reader, complaining about the difficulties and costs involved in printing mathematical books, explaining that this is because the costs and efforts involved are so large, and the demand for accuracy so great.[40] A book like *De solaribus horologiis* would, therefore, have been troublesome and costly for the printer, since it is illustrated throughout with diagrams, which had been re-cut for the posthumous edition. Fine took pride in the provision of clear and accurate diagrams for his works, and, despite their smaller size, the diagrams in the 1560 edition retain all of the detail that was in the 1532 edition.[41] Fine's diagrams tend to show the whole instrument, overlaid with its construction lines, so that the illustration as a whole could have two functions: to understand the geometrical and astronomical basis on which the instrument was constructed, and to understand the parts of the instrument and how to use it. Although the construction lines are not needed when using the instrument, most of them are shown so as not to obscure important parts of the scales, indicating perhaps that as well as clarifying the text on the construction of the instrument, the diagrams were intended to be copied.[42]

39 See Cifoletti in this volume.

40 'In quibus vel notula vna peccare, lineola aut apiculo aberrare, summum & intolerandum habetur piaculum'. Fine, 1560, Typographus lectori. See I. Pantin, 'Les problèmes de l'édition des livres scientifiques: l'exemple de Guillaume Cavellat', in P. Aquilon and H.-J. Martin with F. Dupuigrenet-Desrousilles (eds), *Le livre dans l'Europe de la Renaissance* (Paris, 1988), pp. 240–251.

41 A. F. Johnson, 'Oronce Finé as an Illustrator of Books', *Gutenberg-Jahrbuch*, vol. 3 (1928), pp. 107–109, argues that Fine did his own woodcuts, while Ross, 'Studies on Oronce Finé', pp. 32–58, argues that Fine prepared the drawings but someone else cut the blocks.

42 O. Gingerich, 'Astronomical Instruments with Moving Parts', in R. G. W. Anderson, J. Bennett and W. F. Ryan (eds), *Making Instruments Count*, (Aldershot, 1993), pp. 63–74, argues in the case of books on equatoria that some of the diagrams of this instrument of planetary motion were intended to be copied rather than cut out. He points

An object kept at the Whipple Museum of the History of Science gives evidence that the diagrams were indeed copied in order to produce an instrument (Fig. 6.4). It is an exact copy of the illustration in Fine's book, of the same size and with the same decorative features as the diagrams in the 1560 edition of Fine's work. There are only two small differences: the mast on the actual instrument is slightly wider (10mm) than that on the instrument illustrated in the text (8mm), and the unequal hours diagram on the back is numbered differently. It is signed 'SF' and has an inscribed date of 1620. Although it has been suggested that 'SF' might be Samuel Foster of Emmanuel College, Cambridge, and the Royal Society, London, the evidence is extremely thin: the initials 'SF' are not unusual enough that this suggestion should be regarded as anything more than speculation.[43] Whoever its owner, this instrument provides clear evidence that someone used Fine's diagrams of the ship-shaped dial as the basis for making, or commissioning, their own instrument.[44]

Extending this point a little, if it is possible to construct a working instrument by copying the diagrams in Fine's book onto a piece of brass, then it would also be possible to use the same printed diagrams, perhaps mounted on a piece of wood or ivory, to make functioning instruments. Indeed, in sixteenth-century Europe there was a growing trade in single-sheet woodcuts, many of which showed portraits, religious images, coats of arms or pictures of strange celestial or terrestrial phenomena. Sometimes a single image would appear as both a book illustration and as a single-leaf,[45] and among the surviving single-leaf prints are images of instruments, including a sundial dated 1551 and designed by famous instrument maker Georg Hartmann, and a set of scales for a cylinder dial by Apian, printed in 1567, that could be wrapped around a cylindrical block and fitted with a gnomon in order to make a working instrument.[46] From studies of Italy, France and England, it is clear that there, too, printed instruments were gaining popularity during the sixteenth-century.[47] Sundial books also provided diagrams that

out that copies of Peter Apian's *Instrument Buch* included a separate set of the diagrams on heavy paper to allow their assembly, and that around quarter of surviving copies have these plates present.

43 D. J. Bryden, *Catalogue 6: Sundials and Related Instruments* (Cambridge, 1988), suggests that 'SF' might be Samuel Foster, on whom see J. Venn and J. A. Venn, *Alumni Cantabrigensis to 1751*, 4 vols, (Cambridge, 1922–7), vol. 1, pt. 2, p. 163.

44 The Adler Planetarium, Chicago, has in its collections a silver polyhedral dial copied from one of the diagrams in Fine's book (M324), and the Websters' Instruments Makers database (at http://www.adlerplanetarium.org/research/collections/websters/index.shtml) notes the existence of an identical instrument in wood.

45 W. L. Strauss, *The German Single-leaf Woodcut, 1550–1600* (New York, 1975), p. 2.

46 Strauss, *German Single-leaf Woodcut*, pp. 1, 82 and 258.

47 See D. J. Bryden, 'The Instrument-Maker and the Printer: Paper Instruments Made in Seventeenth Century London,' *Bulletin of the Scientific Instrument Society*, vol. 55 (1997), pp. 3–15; and A. J. Turner, 'Paper, Print, and Mathematics: Philippe Danfrie and

could be pasted onto board, and used as an instrument. Sebastian Münster's *compositio horologiorum*[48] provides illustrations of quadrants to be copied, and a plumbline added; scales for a cylinder dial that can be traced and wrapped round an appropriate piece of wood; and two illustrations accompany his description of the universal rectilinear dial, one showing the construction of the instrument, and the other showing its finished form, without the geometrical lines.[49] In *De solaribus horologiis*, several instruments are presented in this way, with their composite pieces separated and printed so that an instrument could be made from them.[50] These printed instruments would have been relatively cheap to produce, once a woodblock had been cut, and would have enabled some of the costs about which Fine's printer complained to be recouped.

Soon after Fine's death in 1555 and the republication of the sundial sections of *Protomathesis* in 1560, his sundial books, along with those of Sebastian Münster, provided the source material for another book on dialling, this time in French. Jean Bullant's *Horlogiographie*, published in 1564,[51] aimed, as he stated in the preface, to present the practicalities of sundials for 'men of the compass'.[52] This book – written in French – includes many of the instruments from *De solaribus horologiis*, and his debt to Fine is clear in both the content and the structure of the book. Among the instruments Bullant discusses are the universal rectilinear dial and the ship-shaped dial which, following Fine, he explicitly links together, saying that the two instruments are the same as each other, except that

---

the Making of Mathematical Instruments in Late 16th-Century Paris,' in C. Blondel et al. (eds), *Studies in the History of Scientific Instruments* (London, 1989), pp. 22–42. A discussion of Georg Hartmann's role in creating a market for these printed instruments is found in Chapter 3 of S. Karr-Schmidt, 'Art – a User's Guide: Interactive and Sculptural Printmaking in the Renaissance', unpublished PhD. diss. (Yale University, 2006).

48 S. Münster, *Compositio horologiorum* (Basel, 1531). A second edition, with the title *Horologiographia* was published in 1533.

49 Münster, *Compositio horologiorum*, pp. 29 and 30.

50 Fine, 1560, pp. 87–89: diagrams of the composite parts of the nocturnal.

51 J. Bullant, *Horologiographie* (Paris, 1564). The dedication is dated 1561, indicating that the work was prepared soon after Fine's death in 1555 and the republication of *De solaribus horologiis* in 1560.

52 'Ie laisse à cercher les grands secrets desdits quadrans, à ceux qui sont plus curieux, pourautant que ie n'ay la theorique:mais seulement ay curieusement cerché par la pratique du compas, plusieurs diuerses sortes & manieres d'horloges [...] Pour auquels donner commencement, entrée, & intelligence, ay assemblé ce petit traicté & recueil, tiré par la pratique du compas des autheurs qui par cy deuant en ont escrit, comme Sebastien Muster, & le Tresexcellent, & Tresdocte Mathematicien Oronce Finé. Et apres auior de long temps fait les espreuues d'iceux quadrans & horloges, ay bien osé mettre & reduire en nostre vulgaire ce petit traicté, pour le proffit & commodité des artizans & gens de compas [...] Ce qu'est vne delectation & industrie par ce que nuls par cy deuant n'en one escrit en nostre vulgairs'. Bullant, *Horologiographie*, sig. Aiij^{r-v}.

'in this description the dial is made in two pieces'.[53] In many places, Bullant gives more detail than Fine on the practical aspects of the dial, including how to make and use it, while reducing some of the sections that linked to astronomical theory. For example, the description of the universal rectilinear dial opens with a sentence about which areas of the world the dial is useful in, before the description of its construction begins.[54] There is more information about the ordering of the zodiac signs, with a reference to the sixteenth chapter of Bullant's book, but the reference to Fine's *Cosmographia* has been removed.[55] In describing the folding arm that is to be attached to the instrument, Fine says only that it should be made from a strong material, where Bullant goes into details, saying that it should be made from copper, brass, or some other metal.[56] There are two illustrations of the instrument, of which Bullant says one is to show the construction method, and the other to show the finished instrument.[57]

In the description of the ship-shaped dial, Bullant gives a longer explanation of the parts of the dial than does Fine, before beginning the description of how to construct it.[58] This may be related to the changes that Bullant has made to the design of the instrument which, although still ship-shaped, has been altered. The instrument is illustrated by three diagrams, one showing the construction method, and two showing the parts of the finished instrument (Fig. 6.5a, b and c).[59]

From these, it is clear that Bullant has changed his source material, and changed the design of the ship-shaped sundial – changes that allow it to be copied and made into a paper instrument more easily than can Fine's version, with its slim, fragile, mast. Many features of the ship-shaped dial as presented by Fine are preserved – the shadow square and unequal hours diagram have been brought onto the front, making the instrument one-sided, and more suitable for printing. In place of a fragile narrow mast, he has made the instrument out of two rotating

53 'La description de l'horloge suiuante est conforme à la deuant dite, sinon qu'en ceste description l'horloge ou instrument est fait de deux pieces, en l'vne desquelles pieces, assauoir la plus grande (qui est quasi comme la mere de l'astrolabe) est descrit seulement l'eschelle de hauteur, & la ligne de l'horizon. Et l'autre piece set met & applique sur la premiere, & est attachee sur le centre d'icelle, en sorte qu'elle puisse estre demenee & tournee d'vne part & d'autre, en laquelle sont descrites les heures tant de deuant que d'apres midi, & aussi le zodiac meridien, à la dextre d'icelle table ioignant la ligne de 12 heures du midi'. *Ibid.*, p. 102. Diagrams of the universal rectilinear dial are on pp. 99 and 101; diagrams of the ship-shaped dial on pp. 106–107.

54 *Ibid.*, p. 102, and Fine, 1560, p. 94.

55 Bullant, *Horologiographie*, p. 96, and Fine, 1560, p. 177.

56 Bullant, *Horologiographie*, p. 100, and Fine, 1560, p. 180.

57 Bullant, *Horologiographie*, p. 98.

58 *Ibid.*, p. 102, and Fine, 1560, p. 183.

59 Bullant, *Horologiographie*, pp. 106 (the composition diagram) and 107 (the composite parts).

plates, and instead of a slider, Bullant suggests using a folding metal arm like that on the universal rectilinear dial. The changes Bullant has made are meant, therefore, to improve the practical usefulness of Fine's sundials, and to enable 'men of the compass' to use and appreciate them. He took Fine's practical work grounded in theory and book learning, and reduced the importance of the theory in favour of giving more information about the practicalities of actually making the dials described.

Bullant's book was one of a number of publications on sundials that made use of *De solaribus horogiis*. Parts of the book were reworked and translated, appearing in works on sundials published across Europe. In France, Claude de Boissiere published a work which summarised material from *De solaribus horologiis*.[60] In Italy, Niccolò Tartaglia praised Fine's instruments in his *Nova Scientia*,[61] and the Italian translation of the *Protomathesis*, published in Venice in 1587, influenced Giovanni Paolo Gallucci's *Della fabrica et uso di diversi stromenti*, published in 1587.[62] Scholars in England (including Robert Recorde and John Dee) acknowledge Fine as an expert on dialling.[63] However, very few authors picked up Fine's ship-shaped dial, despite the fact that almost all of the increasing number of sundial books published included the universal rectilinear dial among the instruments they described, and Fine had explained how the ship-shaped variant was more useful because of the scales on the back. The ship-shaped dial, perhaps because of its more complex form – requiring two pieces rather than one, as noted by Bullant – had lost out to the simpler and more common universal rectilinear dial.

It did not disappear completely, though, and occasional references to the ship-shaped dial may be found. One of these is an anonymous manuscript text about sundials compiled in English in 1653, which draws heavily on Fine's and Bullant's books.[64] Around the same time, Athanasius Kircher included the ship-shaped dial in his 1646 *Ars magna lucis et umbrae*. Here, the dial is no longer a ship, but has become the *columba*, shaped like a dove (Fig. 6.6), the imagery

60 C. de Boissiere, *De la proprieté & usages des Quadrans de l'invention d'Oronce Finé Dauphinois, Lecteur & Professeur du Roy és sciences Mathematiques, & de Pierre Apian excellent Mathematicien* (Paris, 1565).

61 See S. Drake and I. Drabkin, *Mechanics in Sixteenth Century Italy* (Madison, 1969), p. 67.

62 On the Italian translation of the *Protomathesis* see Dupré in this volume.

63 For information about Fine's influence on sixteenth-century English mathematicians, see S. K. Heninger 'Oronce Finé and English Textbooks for the Mathematical Sciences', in D. B. J. Randall and G. W. Williams (eds.), *Studies in the Continental Background of Renaissance English Literature: Essays Presented to John L. Lievsay* (Durham, 1977).

64 British Library, London, MS Sloane 2102, fols. 21^{r} to 64^{v}, titled: 'The First Booke of the Principles of Geometrie seruing for the making of Sonne Dialles, the Measuring of things, diuers Astronomical Instruments, & for all Architecture'. The description of dials draws on Münster and Fine, as well as Thomas Fale and others.

strongly linking it to the Pope. The link to the ship-shaped dial is clear, however, since Kircher acknowledges his source for this instrument to be Fine.[65]

Kircher's use of Fine's *De solaribus horologiis* as the source for his ship-shaped dial is not uncritical, since Kircher at the same time mentions an error with the construction of the dial: with the 'mast' straight up, the thread would be suspended from the right point on the latitude scale, but with the 'mast' fully tilted, the thread would be suspended from a point slightly lower than it should be. To correct for this, Kircher proposes a modified way of using the instrument in order to compensate for the error, explaining that the difference is barely two degrees.[66] Interestingly, this same error, and the modification to correct for it, were also of concern to one of the owners of the Whipple Museum's ship-shaped dial, copied directly from the diagrams in the 1560 edition of *De solaribus horologiis*. This instrument has an alternative scale on the back of the mast, constructed in order to give the corrected latitude scale for use when the mast is fully tilted.[67] It has been rather roughly scratched on, suggesting that this was a modification made by a later user of the instrument, sometime after it had been made in 1620. Whether the modification was made as a result of the publication of Kircher's suggestion, or independently, we cannot know.

This seventeenth-century reference to errors in Fines construction of the ship-shaped dial, and the modification of an instrument, are not the earliest evidence of concerns about inaccuracies in Fine's book. One of the most important and influential books on sundials, Christoph Clavius's *Gnomonices* (1581), describes fixed and portable dials, including a number of the instruments found in *De solaribus horologiis*, among them the universal rectilinear dial (Fig. 6.7).[68] In the text, Clavius names Fine as his source of his information on the universal rectilinear dial, saying that 'almost everyone follows him,' but warning the reader that there are mistakes in the construction of the zodiac scale on the instrument, referring perhaps to the alternative method given by Fine for the construction of this scale.[69]

65 A. Kircher, *Ars magnis lucis et umbrae* (Rome, 1646), p. 507. On the *columba* dial, see C. Eagleton, J. Downes, K. Harloe, B. Jardine, A. Mosley and N. Jardine, *Instruments of Translation* (Cambridge, 2003); abridged version in L. Taub and F. Willmoth (eds) *Instruments and Interpretations: A Festschrift for the Whipple Museum of the History of Science* (Cambridge, 2006), 255–282.

66 Kircher, *Ars magnis lucis et umbrae*, p. 506. In his description of the *columba*, Kircher refers to the 'mast' as the 'pedum'.

67 The latitude position at the equinox should be $l' = l/(\cos d)$, where l is the latitude position at the solstice, and d the maximum solar declination. Measuring the position of the latitude marks on the Whipple instrument confirms that they are in the expected places for this 'correction' of the scale: the mark for 40 degrees latitude is 26mm from the pivot (it should be at 25.2mm), 50 degrees is at 36.5mm (36.7mm) and 60 degrees is at 51mm (52.0mm).

68 C. Clavius, *Gnomonices libri octo* (Rome, 1581), p. 637.

69 Clavius, *Gnomonices*, p. 637: 'In constructione porrò posterioris huius Zodiaci hallu-

For Clavius, it is nonetheless clear that Fine's *De solaribus horologiis* was an important source, albeit in places flawed. Other sixteenth-century mathematicians were not so kind, and in 1546, mathematician Pedro Nunes published a harsh criticism of Fine's work, which was reprinted in 1573.[70] This short publication lists errors made by Fine, including *Reprehensio XV* and *XVI*, which criticise *De solaribus horologiis*. Nunes claims that Fine gave incorrect descriptions of both horizontal and vertical dials, and proceeds to give a lengthy demonstration of the errors in the first proposition of the first book. He then goes on to criticise proposition seven of the first book, where Fine discusses the basic geometry of dials that can be used at more than one latitude, concluding that 'whatever other dials Oronce makes in this way on this foundation, are false' and not worth examining any further.[71] Nunes, therefore, attacked the foundations of Fine's book – the sections on which other parts of the book were based.[72]

Criticisms of Fine by Nunes, Clavius and Kircher, among others, show us more of his influence than we might otherwise have seen. Many authors used material from Fine's works without naming him, but Clavius and Kircher, perhaps motivated by Nunes and his criticisms, explained the source of their material at the same time as warning their readers about potential problems with it. The use of parts of Fine's *De solaribus horologiis* in the works of Münster and Clavius, as well as other books on sundials throughout the sixteenth and seventeenth centuries, shows the influence of Fine as an authority on this subject in the mid-sixteenth century. His diagrams could have been, and in one case are known to have been, copied to make working instruments, and many later sundial books include very similar diagrams to those found in *De solaribus horologiis*.[73] A major source for anyone writing on sundials in this period, Fine helped to shape the genre of sundial books, and his work was reworked, revised, translated, plagiarised, and criticised.

---

cinatus est Orontius, quem ferè omnes sequuntur'. Fine, 1560, p. 180 describes 'eandem zodiacum generalem aliter delineare', involving transferring the divisions from one part of the instrument to another, rather than constructing the scale as previously described on p. 179.

70 P. Nunes, *De erratis Orontii Finei* (Coimbra, 1546), with another edition in 1573. On Nunes and his criticisms, see Leitão in this volume.

71 'Et falsa sunt igitur quaecumque alia horologia per huiusmodi Orontii fundamenta conficiuntur. Reliqua autem inclinata, & pendula solaria horologia ab eo tradita, examinandi otium non est'. Nunes, *De erratis*, p. 54.

72 This was, perhaps, a little unfair considering that many of the 'errors' were not Fine's own, but present in the source material with which he was working. However, Fine did claim that his book was placing practical mathematics on strong theoretical foundations, so for there to be errors in those foundations was problematic.

73 Compare, for example, the diagrams of the universal rectilinear dial found in Fine and Clavius with other sundial books of the period.

This paper has tracked two closely-related instruments – the ship-shaped dial and the universal rectilinear dial – to better understand the sources and influences of Oronce Fine's *De solaribus horologiis*. Rooted in the manuscript tradition, Fine probably used a wide range of manuscript sources in compiling his works, some of which may have been available to him thanks to François Ier's support for scholarship and the collecting of manuscripts.[74] A particular group of manuscripts, some of which were copied in and around Vienna, and are linked to the university, formed the basis for a number of the descriptions of instruments in *De solaribus horologiis*, but Fine nonetheless deserves significant credit for moulding this dispersed, inconsistent and sometimes erroneous source material into a coherent and organised whole. From the confused and inaccurate manuscript tradition of the *organum ptolomei*, Fine published a corrected and unambiguous text about this instrument. His construction of a ship-shaped dial with corrected geometry in 1524 shows that by this date he had already worked on the *organum ptolomei* manuscript descriptions, eight years before the descriptions of the instrument were published in 1532 as part of the *Protomathesis*. In this work, a range of types of sundial are presented – fixed and portable – within a framework progressing from the simple to the complex. The detailed diagrams in the fourth book, *De solaribus horologiis*, probably supervised personally by Fine, combine the theoretical and practical functions of the book. From the illustrations, one could understand how a particular instrument is constructed, or make and use one's own.

The influence of *De solaribus horologiis* was wide, both directly, and indirectly through the use of Fine's descriptions of sundials by important writers including Sebastian Münster and Christoph Clavius, whose books became the standard works on sundials across Europe. However, perhaps in part because of the errors noted by Nunes, but also because of the passage of time, by the end of the sixteenth century Fine had been eclipsed by others. More importantly, though, *De solaribus horologiis* helped to create the genre of sundial books that lasted throughout the following centuries. Fine was one of a number of scholars who in the first decades of the sixteenth century pioneered the presentation of material about astronomical instruments in a new way, taking the contents of medieval manuscripts and working them into something with wider appeal, with a more structured approach and a stronger theoretical grounding. Returning to Mizauld and Thevet, with whom this paper began, we should remember Oronce Fine not so much, perhaps, for the inventing of new instruments, but for the re-invention of old ones.

74 For the history of the Bibliothèque Nationale and its constituent collections, see L. Delisle, *Le cabinet des manuscrits de la Bibliotheque imperiale (puis nationale)*, 3 vols. (Paris, 1868–81), vol. 1.

Cosmography and Geography in the Sixteenth Century: the Position of Oronce Fine between Mathematics and History

JEAN-MARC BESSE *

In *The Self-Made Map*, Tom Conley devoted detailed study to the different ways in which Oronce Fine chose to have himself represented in certain frontispieces of his works, whether directly in portraits or indirectly through signatures.[1] Conley has shown, in particular, the constituent role of Ptolemy (pictured as engaged in taking measurements and contemplating the cosmos) in defining the identity, both scholastic and personal, of this scholar who came from the French Dauphiné province. By choosing to represent Ptolemy at the start of his books, Fine is in fact, according to Conley, symbolically representing himself, or rather the lineage to which he lays claim, in the exercise of his calling and the assertion of his knowledge (Figs. 7.1 and 7.2).

In fact, Fine is merely subscribing to an iconographic tradition: he draws inspiration from a Venetian edition of Regiomontanus's *Epitome of the Almagest* (*Epytoma in almagestum ptolemei*, 1496) in which the author can be seen seated with Ptolemy under a celestial sphere. For sixteenth-century geographers, Ptolemy was *the* cosmographer, as in the very title given to Ptolemy's work by his first translator, Jacopo d'Angelo da Scarperia.[2] This translator called the work by Ptolemy *Cosmographia*, and it is under this title that the work was first known and spread through Europe. In the dedication to Pope Alexander V, Jacopo d'Angelo justifies his choice as follows:

> Our author [Ptolemy] calls the whole work in Greek a Geography, or description of the earth. This title has not been altered by Manuel Constantinopolitanus, the wisest man in our century [...]. But we choose to translate it as Cosmography [...]. Indeed, if Pliny and the other Latin authors who have described this place that is the earth, have called their works Cosmographies, and their authors cosmographers, I do not see why we should not use the

* The present essay was translated from the French by Laura Walters.

1 T. Conley, *The Self-Made Map. Cartographic Writing in Early Modern France* (Minneapolis and London, 1996). See especially chapter 3: 'Oronce Finé: A Well-Rounded Signature', pp. 88–134.

same term to refer to this work by Ptolemy, whose subject is the same [...]. *Cosmos* in Greek – in Latin *mundus* – which means the heavens and the earth, is the basis of things throughout the work.[3]

The images on which Conley comments contain more than a mere personal utterance, they express more than an iconographic representation of the cosmographer's ego: they also emblematically represent the instruments, the methods and the orientations of cosmographic knowledge itself. The presence of Ptolemy in the frontispieces of Fine's works (as in those of other cosmographers of the same period) reflects a positioning and a choice with respect to the type of scholarly orientation that Fine aimed to promote in the field of cosmography. In other words, the picture of Ptolemy is the figuration of a theoretical choice, and of an interpretation of what cosmographic knowledge is, or should be, and the place of geography within it. In fact, at more or less the same time, Rheticus (in the letter of dedication to Duke Albert of Prussia that accompanied his *Chorographia* [Wittenberg, 1541]) also states that 'only mathematicians following in the footsteps of Ptolemy can reform geography', adding, 'mathematicians have initiated ways of studying the earth in relation to the heavenly space'.[4]

Yet if geographers inherited the Ptolemaic system (from which they derived most of their concepts), Ptolemy also handed down a frame of thought which was not always well suited to their specific objectives. Indeed, their resort to authors such as Strabo, Pliny or Pomponius Mela can be explained by the inadequacy of the Ptolemaic conceptual system to lend itself to the descriptive ambitions of certain geographers of the time. These authors offer examples, or even models, of a more fruitful approach, giving history (generally understood as natural as well as

2 Here, reference can be made to Robert Mahew, albeit on a different plane (iconography) and in a different period (the sixteenth century), for his reflections on the importance that historians should give to what he calls the 'textual tradition' of geography. The geographers of this period not only take care, via text and illustration, to define the nature, the aim and the limitations of their knowledge, they also seek to clarify the type of intellectual and perceptive stance underpinning that knowledge. See R. Mayhew, 'The Character of English Geography c. 1660–1800: a Textual Approach', *Journal of Historical Geography*, vol. 24, no. 4 (1998), pp. 385–412.

3 '... Ceterum Geographiam, hoc est Terrae descriptionem, auctor hic noster omne opus graece nuncupat. Quam appellationem vir saeculi nostri eruditissimus Manuel Constantinopolitanus [...] non mutavit. Sed nos in cosmographiam id vertimus [...] Nam si Plinius ceterique latini, qui terrae situm descripserunt, opus suum cosmographicum appellant et auctores ipsi cosmographi dicuntur, nescio cur Ptolomei opus, quod idem tractat, eodem vocabulo apud nos appellari non debeat. [...] Cosmos nomen graeci mundus latine, qui terram caelumque ipsum, quod per totum hoc opus tanquam rei fondamentum adducitur [...].' Ptolemy, *Cosmographia*, trans. Jacopo d'Angelo (Vincenza, 1475), unpaginated.

4 Quoted by J. Babicz, 'La résurgence de Ptolémée', in M. Watelet (ed.), *Gérard Mercator cosmographe: le temps et l'espace* (Antwerp, 1994), pp. 50–69.

civil history) its place. Thus, at the time when Fine was producing his writings, other choices were being made elsewhere, and other definitions of cosmography and geography were becoming available, for instance in Joachim Vadian, Peter Apian or Sebastian Münster.[5] As will be shown below, cosmography and geography were envisaged not only through the perspective of astronomy and mathematics, but also through the perspective of history (or indeed of theology).[6] We need therefore to place Fine's stance in relation to the general context of theoretical considerations on cosmography and geography in the first half of the sixteenth century.[7] A particular focus here will be the way in which the relations between cosmography and geography were envisaged, and especially what degree of independence from cosmography was given to geography.

I. *Definitions*

What are cosmography and geography for Oronce Fine? How does he define these forms of knowledge and their relationships in the different texts on the subject? On this particular point, Fine's stances are not new, positioned as they are in the reiterative tradition of treatises on the sphere inherited from Sacrobosco, which Fine himself edited as early as 1521:[8] cosmography is the 'contemplation' of the sphere, or the 'machine of the world'.[9] Since the world comprises two parts, cosmography is also made up of two fields of knowledge, between which this contemplation, or doctrine of the world, is divided: 'the description [of the world] is properly named Cosmography: covering the first part, Astronomy, and also Geography, that is to say the fabrication and the ratiocination of both heavens and earth.'[10]

5 For which see Mosley in this volume.

6 For the status of geography in Germany, see G. Strauss, 'Topographical-Historical Method in Sixteenth-Century German Scolarship', *Studies in the Renaissance*, vol. 5 (1958), pp. 87–101; *Sixteenth-Century Germany. Its Topography and Topographers* (Madison, 1959). For England, see. L. B. Cormack, *Charting an Empire. Geography at the English Universities, 1580–1620* (Chicago and London, 1997), and S. J. G. Mendyk, *'Speculum Britanniae': Regional Study, Antiquarianism, and Science in Britain to 1700* (Toronto, 1989). For Italy, see D. Defilippis, *La rinascita della corografia tra scienza e eruditone* (Bari, 2001).

7 N. Broc, *La Géographie de la Renaissance (1420–1620)* (Paris, 1980); F. Lestringant, *L'Atelier du cosmographe, ou l'image du monde à la Renaissance* (Paris, 1991); J.-M. Besse, *Les grandeurs de la Terre. Aspects du savoir géographique à la Renaissance* (Lyon, 2003).

8 See e.g. Sacrobosco, ed. Fine, 1538.

9 Fine, 1552, p. 6r.

10 *Ibid.*, p. 1r. The Latin version reads: 'Unde Cosmos à Graecis dicitur: & quae de Mundo traditur disciplina, Cosmographia (de qua praesentis tractare est instituti) respondenter vocitatur. Est enim Cosmographia, mundanae structurae generalis ac non iniucunda descriptio: primam Astronomiae partem, atque Geographiam, hoc est, caeli

According to Fine, cosmography is therefore the association, or rather the succession, of two bodies of knowledge: that concerning the heavens, and that concerning the earth. The order is not indifferent, since it is one of the principles propounded by Fine that knowledge of the earth (i.e. geography) depends on knowledge of the heavens (i.e. astronomy). Cosmography is thus, in a sense, a way of surveying the world as a whole, heavens and earth, but from the perspective of the heavens. The reference to Ptolemy and to what Thomas Kuhn has called the 'mathematical tradition' is essential here.[11] It is common to Fine and all his contemporaries, being perfectly summed up by the representation of cosmography found in Apian's *Cosmographia* (Fig. 7.3). What does this image tell us? That cosmography surveys the heavens and the earth with the same gaze, and that this viewpoint is also a mathematical construction, or rather a projection. In other words, geography should be apprehended through the perspective of astronomy, and this perspective is that of mathematics (or, more particularly, of geometry).[12]

This subordination of geographical discourse to the cosmographic viewpoint is derived from Ptolemy's *Geography* itself: geographical measurement is based on the teachings of astronomy and geometry. The Ptolemaic 'tradition' is not found solely in the *Geography,* it was already present well before in the *Almagest* – a treatise that does not contain just a theory of celestial bodies, their positions and their movements. Indeed, Book 2, where Ptolemy in fact announces the *Geography* as an independent work, is entirely devoted to the study of the terrestrial globe, and its inhabited areas in particular. The main principle of the study, set out at the start of the treatise, may be seen as one of the foundations of the cosmographic approach: 'To succeed in the synthesis intended, we must begin by presenting the overall relationship between the heavens as a whole and the earth as whole. In the treatise which we propose, then, the first order of business is to grasp the relationship of the earth taken as a whole to the heavens taken as a whole.'[13]

What, then, is cosmography? What is it to be a cosmographer? To answer these questions, we need to return to the definition of the cosmos (or *mundus* in

---

terraeque rationem comprehendens.' Fine, 1555, book 1, chapter 1.1, no pagination. The wording of this text is reminiscent of that used by Jacopo d'Angelo in his edition of Ptolemy's *Cosmographia* (see above, n. 3).

11 T. Kuhn, 'Mathematical versus Experimental Tradition in the Development of Physical Science', in *The Essential Tension: Selected Studies in Scientific Tradition and Change* (Chicago and London, 1977), pp. 31–65.

12 See here how the *Cosmographiae Introductio* relates cosmography to astronomy and astronomy to geometry: 'Cum cosmographiae notitia sine praevia quadam astronomiae cognitione et ipsa etiam astronomia sine Geometriae principijs plaene haberi nequat [...].' M. Waldseemüller, *Cosmographiae introductio cum quibusdam geometriae ac astronomiae principiis ad eam rem necessariis...* (Saint-Dié-les-Vosges, 1507), sig. A ii^{v}.

13 Ptolemy, *the Almagest*, trans. G. J. Toomer (Princeton, NJ, 1998), p. 37.

Latin) that was given by Jacopo d'Angelo: the heavens and the earth taken together, as one. Cosmography is the description of this whole, but it is also, and perhaps above all, in metonymic mode, the description of the earth in relation to the structure of this whole. Cosmography is not astronomy in its entirety. In the many, reiterative definitions suggested by geographers in the sixteenth century at the start of their treatises, cosmography is described in specific manner as a description of the earth 'in relation to the space of the heavens', as noted earlier by Rheticus, or as a description using the 'circles of the heavens' as Apian writes.[14] It is probably Apian who provides the most explicit definition:

> This science considers firstly the Circles, of which we imagine the supreme heavenly Sphere to be composed. Thereafter, according to the distinction and the distribution of the said Circles, it states the situation of the lands [that are beneath them], their measure and their proportion; furthermore it shows the proportion of climates, and the diversity of days and night; it also specifies the four principal points of the world, that some call the four corners of the earth, east, west, north and south. It studies the movement of fixed stars, and of erratic bodies, also called the planets, their rising and setting, and over which nations they move. And likewise all things that belong to the consideration of the heavens, like elevations of the Pole, Parallels and Meridians, all of which diverse matters are clearly demonstrated by Mathematics in the realm of Cosmography.[15]

This description can be viewed as an inventory of the themes that a treatise on cosmography should broach, through the sixteenth century and beyond. Inventories of this sort are to be found in the prefaces of a large number of works: in Martin Waldseemüller's *Cosmographiae introductio* (1507), in Henricus Glareanus's *De geographia liber unus* (1527), in Fine's *De mundi sphaera*, in the cosmography lectures given by Gerard Mercator and edited by his son Bartholomew in 1563, in the *Commentarius in sphaeram Ioannis De Sacro Bosco* by Christoph Clavius

14 John Dee writes along the same lines in his 'Mathematicall Praeface' to the first English edition of Euclid's *Elements*: 'Cosmographie, is the whole and perfect description of the heavenly, and also elementall parte of the world, and their homologall application, and mutall collation necessarie. This Art requireth *Astronomie*, *Geographie*, *Hydrographie* and *Musike*. [...] This matcheth Heaven, and the Earth, and aptly applieth parts Correspondent [...]'. John Dee, *The Mathematicall Praeface to the Elements of Geometry of Euclid of Megara (1570)*, ed. A. G. Debus (New York, 1975), fol. biii^r.

15 P. Apian, *Cosmographie, ou description des quatre parties du Monde...* (Antwerp, 1581), p. 5. The Latin version is: 'Imprimis enim contemplatur Circulos, ex quibus illa supercoelestis Sphaera componi intelligitur. Deinde ex ipsorum distinctione, terrarum illis subiectarum situs, et locorum symmetriam seu commensurationem, Rationem insuper Climatum, Dierum, Noctiumque diversitates, Quatuor mundi Cardines, Stellarum quoque fixarum necnon errantium Motus, Ortus et Occasus, et quibus Verticales moventur, et quaecunque as coeli rationem pertinent, ut Poli elevationes, Paralelos, et Meridianos circulos, et caetera iuxta Mathematicas ostentiones demonstrat.' P. Apian, *Cosmographia ... per Gemmam Frisium...* (Paris, 1551), p. 1^r.

(first edition 1570), thus contributing to the establishment of a paradigm for a mathematical brand of geography, echoes of which are to be found in the *Introduction à la géographie* by Guillaume Sanson (1681).[16] This conception of cosmography places geographical study in a dual dependency, dependency on Ptolemaic astronomy on the one hand, and on geometry on the other.

Geography, as defined by Fine, is a mathematical knowledge of locations and measures of distance.[17] It studies 'the description and situation of terrestrial and maritime places, their longitude, latitude, distances one from the other, distinctions in climate and in winds.' Its main objective is cartography: the geographer's task is to 'reduce the whole earth, or a part of it, to a suitable flat form, corresponding to the layout of the sphere.' Finally, it is based mainly on projection:

> This is performed by way of circles applied to the said terrestrial globe, which is in the middle of the whole world, as the centre of it; the outside surface is round and circular throughout. For in this manner, using imaginary circles in the celestial sphere, we can gain knowledge of the stars and their movements; we can likewise, using the corresponding circles on the terrestrial globe, have a representation of all that has been said above.[18]

It can also be observed that this practice of projection at the same time implies a theory of 'correspondence' or 'similarity' between the terrestrial and celestial spheres.[19] Geography developed a sort of dependency on, and formal analogy to, astronomy. It would not be unreasonable to consider that in the theory of the cosmographer Fine, geography is a form of astronomy applied to the study of the surface of the earth.

How, though, could astronomy and geography be distinguished, if in either case it was the same body of knowledge that was implemented? One observation may be made in relation to this question: rather than by way of a particular content and method, it is by way of a certain orientation of viewpoint and thought that,

16 Mercator's lectures were published by his son as B. Mercator, *Breves in sphaeram meditatiunculae, includentes methodum et isagogen in universam Cosmographiam, Hoc est Geographiae pariter atque Astronomiae initia ac rudimenta suggerenetes* (Cologne, 1563).

17 Hydrography is set aside here, for it is described by Fine in the same terms as geography.

18 Fine, 1552, p. 44^{v}. On Fine's projection techniques, see J.-J. Brioist in this volume.

19 'Hinc fit, ut coelestium cum terrestribus mutua quaedam circulorum videatur esse respondentia: adeo ut quemadmodum per circulos in Coelo prudenter imaginatos, syderum venantur habitudines, ita per respondentes in globo terrestri, locorum positiones, atque distantias, & quae utrique, Coelo videlicet & Terrae sunt communia, consequenter obtineamus. Non sunt tamen omnes circuli, quos coelesti sphaerae deputamus, ad Geographicam contemplationem necessarii; neque singuli qui ad ipsum geographicum videntur spectare negocium, ipsi Coelo coaptandi.' Fine, 1532*b*, pp. 141^{v}–142^{r}.

for Fine and his contemporaries, geography and astronomy could be distinguished. It might be useful here to refer back to the iconography: images (both visual and literary) are used by theoreticians of cosmography in the sixteenth century not only to represent themselves, but also to characterise the meaning and the metaphysical orientation of their knowledge. Hence it is through the representation of an alternation between two mental orientations that cosmography came to be defined: on the one hand astronomy, metonymically represented by the figure of Ptolemy, is shown to be looking towards the sky (Fig. 7.4); on the other, geography is pictured as a return to the earth. It is this return that Fine, at the start of chapter 5 in the *Sphère du monde*, signals when he broaches the question of geography: 'Returning finally from the machine of the heavens to the terrestrial globe [...].'[20]

The work of the cosmographer, from this point of view, consisted in articulating the heavens and the earth in a panoptic, all-embracing manner. This unifying view initially alternated, first turning towards the heavens and then towards the earth, to determine the real measure of its extent. It is probably Henricus Hondius who gave the clearest description the moral and intellectual sphere covered by cosmography, a sphere unifying the earth and the heavens, taking form within the system of alternating contemplation of the two:

> God has convened us on earth, beneath the heavens, so that sometimes, looking downwards, we see the earth, and sometimes, raising our eyes, we contemplate the heavens. When looking upwards, and carefully considering this admirable machine of the celestial spheres, we become astronomers. When lowering out eyes towards the earth, and busying ourselves measuring its extent, we are geographers. By these two sciences man becomes worthy to inhabit the world.[21]

This process of alternation is again expressed in the frontispiece of Mercator's edition of Ptolemy's *Geography* (Fig. 7.5).[22] The illustration, which Mercator also used on other occasions, is especially significant; indeed, let us consider the gestures proffered by Ptolemy, the prince of geographers, as represented by the scholar who came be known as the 'modern Ptolemy'. The left hand is pointing downwards to the terrestrial globe by means of the compass used to measure it. The first finger of the right hand is raised and pointing to the celestial globe (with representations of the constellations), which is the basis for terrestrial measurement. The two arms symbolise two viewpoints: the cosmographer is he who links the two globes, in the manner of an intercessor, in the space formed by his two open arms. The cosmographer's body itself, in this picture, is the place where, for the

20 Fine, 1552, p. 44^{v}.

21 H. Hondius, *Nouveau Théâtre du Monde* (Amsterdam, 1639), preface to the reader.

22 G. Mercator, *Tabulae Geographicae Cl. Ptolemaei ad mentem auctoris restituae et emendatae...* (Cologne, 1578).

future reader, the perceptive and intellectual space of the cosmographer is imprinted.

These directional illustrations need to be taken seriously. In iconographic form, they express the paradigm that provided a framework for cosmographic knowledge. This iconography shows the symbolic system underpinning cosmography, its position, and the mental or perceptive attitudes that are the source from which cosmographic knowledge proceeds. The pictures do not tell us about the earth directly, but about how and whence the earth must be viewed so as to capture its cosmographic reality, and this reality – this 'truth' – is the earth considered in its entirety, in its unity and in its structure, from the viewpoint of the heavens.

These alternating cosmographic viewpoints, which are a way of travelling visually across the world, are no doubt connected with the traditions that had coalesced around the 'flight of the soul', inherited from Plato and stoicism.[23] It should be noted that a possible origin of this dual visual representation is the well-known picture of the winged chariot found in Plato's *Phaedra* (246a and after): cosmography, like the human soul, is a sort of two-horsed chariot, representing a dual attraction, part upward-looking (astronomy) and part downward-looking (geography).[24] But in cosmography, what ensures communication and unity between these two gazes (or viewpoints) is the use of the methods and tools of mathematics, the practice of measurement and projection. In cosmography, it is mathematics that plays the part that belongs to dialectics in Platonic philosophy.[25]

II. *Geography*

Strabo developed a line of thought on geography that resembles the viewpoint described above. Indeed, when he drew up the list of kinds of knowledge required for the study of geography, he recalled, with a reference to Hipparchus, that it is impossible to capture geographical information 'without prior analysis of celestial phenomena and geometry'(I,1,13).[26] Knowledge of celestial phenomena thus con-

[23] See for example, R. Miller Jones, 'Posidonius and the Flight of the Mind through the Universe', *Classical Philology*, vol. 21, no. 2 (1926), pp. 97–113; P. Hadot, *Exercices spirituels et philosophie antique* (Paris, 1993 [3rd ed.]) and *Qu'est-ce que la philosophie antique?* (Paris, 1995).

[24] Plato uses the image of a harnessed chariot and a driver, borne up by wings, to describe the soul. For the human soul, which (unlike those of the gods) is mixed, one of the harnessed horses is pulling upwards towards the stars and the procession of the gods, while the other is pulling downwards. The driver has the task of guiding the harnessed pair in a balanced manner.

[25] See I. Hadot, *Arts libéraux et philosophie dans la philosophie antique* (Paris, 1984), esp. chapter 4: 'La naissance du cycle des sept arts libéraux dans le néoplatonisme et les conditions de sa réception par le Moyen Âge chrétien.' On the relationship of Fine's work to Platonism, see Axworthy in this volume.

[26] Strabo, trans. H. L. Jones, 8 vols. (London and New York, 1917), vol. 1, pp. 27–29.

stitutes the basis for any expounding of regional characteristics. However, he adds:

> Once our thoughts have thus been raised, we should nevertheless not neglect the terrestrial globe. It would indeed be ironic, in our eagerness to clearly picture the inhabited world, to presume to embark upon the study of celestial phenomena, and use this for teaching, having no concern for the terrestrial globe of which the inhabited world is a part, for its dimensions, its characteristics, and its place in the whole universe [...]. It therefore appears that geography specifically concerns both the study of cosmography and the study of geometry, uniting that which lives on the earth with that which moves in the heavens, bringing them to unity as if they were close, and not separated as are the heavens and the earth. But to so diverse knowledge we must also add description of those things that live on the earth, animate beings, plants, and in general of all that the land and the sea produce, whether useful or harmful.[27]

Thus, beyond the mere consideration of astronomy and geometry, geography implied a manner of conversion in the way people looked at things and in the way they thought – previously eyes and thought were directed towards the heavens, and they should now turn towards the terrestrial globe. In a first phase, geography identified with cosmography, according to the principle of unity of earth and the heavens as we have just seen. But geography in Strabo's view – and this is probably his distinguishing feature in relation to the nature of knowledge – does not aim solely to study the relationship between the earth and the heavens; it also envisages objects the nature of which is specifically *terrestrial* (living beings, products of the earth or the sea), the very existence of which attests to the presence of a new order of reality, to a certain degree independent from the heavens, that needs to be accounted for using appropriate discourse. This discourse is neither geometrical nor astronomical. It is the discourse of description, or *historia*, and here geography, in the true meaning of the term, found its *raison d'être*.

Alongside this first type of mental construction or representation – the 'vertical' type seen above – certain cosmographers among Fine's contemporaries developed another, a 'horizontal' construction or representation, in which the relationship between cosmography and geography are envisaged in differing modes. For instance, while the contents of Apian's *Cosmographia* are close to those of Fine's treatise on the sphere, Apian distinguishes, if confusedly, cosmography from geography on the basis of their respective objects and methods. Geography, he writes, differs from cosmography because 'it distinguishes the earth by mountains, rivers, streams, seas, and other more renowned things, without regard for the circles of the Sphere', while cosmography 'determines and divides out the earth by the circles of the heavens.'[28] For Apian, geography, which aims to represent the whole of the *orbis terrarum*, takes on a sort of autonomy with regard to

27 *Idem*.

28 P. Apian/R. Gemma Frisius, *Cosmographia* (Paris, 1553), fols. 3 and 4.

cosmography and astronomy. This point is driven home by his position according to which the function of geography is to assist in the understanding and the perusal of histories, and to facilitate the exercise of memory: geography indeed provides memory with 'the order and situation of localities and places'.[29] In other words, geography is positioned on the side of rhetoric, while cosmography (as is shown in the rest of Apian's treatise) is an application of mathematics.

The origin of this distinction between cosmography and geography can be found in the works of Joachim Vadian, whose *Geographiam Catechesi* and the comments he provides for his edition of Pomponius Mela (1518 and 1522 respectively) had wide repercussions in the German-speaking world.[30] The definitions provided by Vadian draw away from the subordination of geography to cosmology, and highlight the autonomy of this field of investigation and the intellectual approaches that are specific to geography:

> Geography, if one follows the Etymology of the word, is the description of the position of lands in relation to the ocean and the Mediterranean Sea, with the enumeration of all the places found therein [...]. Geographer is the name given to those who have sought to describe the earth in particular, its extent, or the most important regions, such as Pliny, Dionysius, Strabo [...], and Mela except for the first book where he mentions the principles of the world [...]. It is said that cosmography gives an account of the earth and the heavens taken together, endeavouring above all to show how the main parts of the earth extend according to latitude and longitude. Ordinarily the geographer, to the enumeration of place, adds history, indicating the origins of cities, of races, of nations, of peoples, and also whence names of things are derived, as well as the prodigious works of nature all over the earth. The cosmographer also enumerates regions, cities, rivers, seas and mountains, but only for the purpose of defining their boundaries, stating where each starts and ends, or else for the purpose of showing under which parts of the heavens they are situated.[31]

Thus, for Vadian, if the cosmographer entertains the concerns of a geometer, the intellectual orientation of a geographer is different. Geography describes the sur-

29 'Et est grandement proufittable à ceulx qui désirent parfaictement sçavoir les histoires et gestes des Princes et autres fables : car la paincture ou limitation de paincture facilement maine a memoire l'ordre et situation des places et lieux, et par ainsi la consummation et fin de la Geographie est constituée au regard de toute la rondeur de la terre [...].' *Ibid.*, fol. 4r.

30 P. Mela (ed. J. Vadian), *Libri de situ orbis tres, adiectis Ioachimi Vadiani Helvetii in eosdem scholiis: Addita quosque in Geographiam Catechesi: & Epistola Vadiani ad Agricolam digna lectu* (Vienna, 1518), on which see W. Näf, *Vadian und seine Stadt St Gallen*, 2 vols. (St. Gallen, 1944 and 1957).

31 'Geographia tamen, si Etymon vocabuli sequimur, propriè ea est quae terrae situm, ut extra intraque ad oceanum nostrumque mare se habet, cum locorum passim iacentium enumeratione describit [...] Ab illa Geographi dicuntur, quibus terrae describendae per singulas, ut expensa est, aut celebres magis regiones cura fuit. Tales Plinius, Diony-

face of the earth without making use of the 'projective' relationship with the heavens that characterises cosmography. Its object is the distribution of terrestrial and maritime, continental and oceanic masses on the surface of the globe. To the mathematical-astronomical dividing-up of terrestrial space, the geographer adds the natural divisions. But, in addition, geography widens the scope of its objects of study: added to the indication of the position of a place is that of its 'properties', i.e. its origin, its natural features, the men who have followed one another therein, the beliefs and customs, animals, plants, and other remarkable things found there, and so forth. The references for this enterprise are Strabo, Pliny, Pomponius Mela and Dionysius of Alexandria, authors who exemplify a particular approach to the construction of geographical discourse: *periegesis* and collection.

The models for writing (or discourse) in cosmography and geography are consequently basically different, even if they can be articulated, according to Vadian. The cosmographer, says Vadian (and paraphrasing Strabo), studies the longitude and latitude of place, the length of day, the position and the movements of the stars, intervals and climates, and the nature of the air. Here the frequently reiterated layout of sixteenth century treatises on cosmography can be recognised. The cosmographer's tools are first and foremost instruments of measurement:

> The study of the instruments intended to conduct measures of this sort and to represent the earth according to meridians and parallels is the concern of the Cosmographer. The greatest of which [...] is Ptolemy. [...] I dare therefore say that the Geographer is closer to the poet and the historian, in that their processes of description are alike. This is why, as we can see with due attention, Strabo, Pliny and others came to the assistance of Poets and Histo-

---

sius, Strabo fere, et Mela, nisi quod et hic quan quam paucissimis, a primi libri principio, de toto mentionem facit [...] Cosmographia vero quanquam utriusque, coeli inquam et terrae rationem habet, magis tamen id curat, ut quibus terrae extantis partibus, quod coeli secundum longitudinem latitudinemque spatium adsit, ostendat. Cumque Geographus praeter locorum enumerationem, et historiam addat, et plerunque quae civitatum, quae gentium, nationum, populorum origo fuerit, atque unde data rebus nomina, tum et illustria nonnunquam naturae sive miranda opera indicans, in terrae situ multo esse uberior soleat: Cosmographus non alio sine regiones, oppida, amnes, maria, montesque enumerat, quam ut vel terminos statuat regionum, quaeque exordia, et qui fines sint enumeret, vel sub qua coelestis superficiei parte illa universa sint demonstret [...].' Mela (ed. Vadian), *Libri de situ orbis tres*, sig. aiii^{v} (I quote here the Basel edition, published by Heinrich Petri in 1557). The preface to the 1522 (Basel) edition develops the same distinctions: '[...] locorum terrae notitiam, coeli ratione adhibita, quam Cosmographiam appellant [...]', '[...] terrarum situm, marium flexus, amnium decursus, montium tractus, civitatum, populorum, gentium ritus, mores, ingenia, denique & animalium effigies, naturam, discrimina, duce Geographia, quae haec fere universa secum trahit [...].' Mela (ed. Vadian), *Libri de situ orbis tres*, unpaginated.

> rians in their descriptions of places upon the earth [...]. The Cosmographer, on the contrary, is turned towards Geometry and Astronomy.[32]

The geographer, then, shares the tools of his descriptive approach with the historian and the poet. Geography entails first and foremost the acts of reading (it rests mainly on the perusal of various authors, according to Vadian), of writing and of representation, while the cosmographic mode rests on mathematics (and is thus more precise, says Vadian, based on axioms that are ascertained). Thus, Ptolemy and Strabo, mathematics and history, are the two poles between which 'cosmography' and 'geography' are separated out in the sixteenth century. As Antonio Possevino states in his *Bibliotheca selecta*: 'If it is manifest that Ptolemy carries off the prize in the implementation of mathematical disciplines, Strabo is the clear victor in historical methods, which are what enlighten geography.'[33]

Sebastian Münster, in his *Appendix geographica* (added to the edition of Ptolemy's *Geography* that he produced in 1540), explicitly attempted to characterise the distinctions between cosmography and geography. The second chapter in this *Appendix* sets out a distinction that he relates to Strabo, which is close to that elaborated by Vadian:

> Geography has two descriptions, one that is common and one that is specific. Common Geography is the description of the earth, region by region, and of the things inside and outside of it, of the seas, the mountains, the rivers, and then of the histories, and of all things wherever they be found that are admirable examples of nature and worthy of our notice. It is in this manner that Strabo comprehends it in Book I [of his *Geography*] [...]. It is the same definition that is followed by Pomponius Mela, Dionysius the African, Pliny, Solin, and Herodotus who show the surface of the earth by dividing it into parts, rarely using the calculations of astronomy. In this way [geography] is

32 'Addo quod instrumentorum ad illiusmodi dimensiones attinentium, et picturarum similiter terrae cum debita meridianorum parallelorumque descriptione, tota ratio ad Cosmographiam attinet, quam primus propre omnium, post multiplicem veterum diligentiam, absoluisse existimatur Cl. Ptolemaeus, Aegyptius natione, qui Antonini temporibus floruit, Mathematices universae citra controversiam princeps. Constat autem Geographiae cum poëtica, cumque historia convenire, vel ea causa quod non dissimili in reddendis locis describendi modo utitur, unde magna Straboni, Plinio, caeterisque, si diligentius paulo perpenderimus,ex Poëtis et Historicis in locorum terrae descriptione, auxilia fuerunt, ut Plinius li. 1 Strabo ubique (nam eius opus Homeri doctrina illustre factum est) indicat. Cosmographus ad Geometricam Astronomicamque inclinans, aliorum, praeter ea quae retulimus, propè securus est.' Mela (ed. Vadian), *Libri de situ orbis tres*, sig. aiii^{v}.

33 A. Possevino, *Bibliotheca Selecta qua agitur De Ratione Studiorum* (Venice, 1603), p. 295. Possevino is quoting Martin Velser, *Rerum Augustanarum vindelicarum libri octo* (Venice, 1594), book 1, p. 6. On Possevino and geography, see J.-M. Besse, 'Quelle géographie pour le prince chrétien? Remarques sur Possevino', *Laboratoire italien. Politique et société* (2008), pp. 122–143.

> differentiated from cosmography, which, when it pictures the world, embracing everything, encompassing the full magnitude of the earth [...] observes quality and quantity according to place, longitude, climate latitude, the reasons for the length or shortness of day and night, the nature of the stars that are placed above us, to the west and to the east, with the distances between places, the nature and position of elements, and so forth. This other [specific] definition of geography does not differ from cosmography, such as Strabo, and Ptolemy the prince in this art, have established it.[34]

There is no doubt that Münster sets his enterprise under the word *cosmographia*. But what Münster actually *does*, while naming it cosmography, in fact corresponds to what he himself, with Vadian, calls *geography*, in the Strabonian rather than the Ptolemaic use of the term. Once again with Münster, there is a hinging of the perspectives of cosmography and geography, but with a slide, so to speak, towards geography. In other words, to be a cosmographer in the sixteenth century was not merely to place the study of the surface of the earth in a Ptolemaic perspective, solely dependent upon astronomy. It was also, in the lineage of Strabo's *Geography*, to look towards the collection and the inventory of information of all sorts relating to the terrestrial realities that are to be described. Girolamo Ruscelli, in the preface to his edition of Ptolemy's *Geography* (1561) at once ratifies this semantic duality and condemns it:

> And others hold cosmography to be [the discipline] that, without taking account of particular magnitudes or the measures of distance between places, seeks to describe and narrate the nature and the properties of countries, and of the things that are found in them, the customs, the peoples, and the remarkable events that have occurred over time.[35]

34 'Geographia duplicem habet descriptionem, unam communem, alteram propriam. Geographia communis est terrae, eorumque quae intra eam et extra sunt, marium, montium, fluviorumque per regiones descriptio, tum historiarum et earum rerum quae à natura in singulis locis admirandae relictae sunt et annotatione dignae. Hoc modo eam Strabo accipit lib. I. [...] Et hacq uidem ratione terram terrenaque sunt ferè persecuti Mela Pomponius, Dionysius Afer, Plinius, Herodotus, paucique alii, qui terrae situm per partes aperverunt, coeli ratione raro admodum adhibita, per quod à Cosmographia differt quod haec orbem designans, totum amplecti solet, et terrae amplitudine, in quam omnia vergant, supposita, quantitatem et qualitatem observare secundum situm, longitudinem, latitudinem climatum, rationem dierum, noctiumque longitudinem et brevitatem, syderum verticalium orientum atque occidentium naturam, cum locorum distantia, natura et positione elementorum etc. Haec in qua maltera geographiae definitio à cosmographia non differt, quae admodum Strabo et huius artis princeps Ptolemaeus scripsit de situ orbis opere instituto.' S. Münster, *Appendix Geographica* to his edition of Ptolemy, *Geographia universalis* (Basel, 1540), 'Geographia quomodo à Cosmographia differat', fol. 158^{r}. On Münster's cosmography see M. McLean, *The 'Cosmographia' of Sebastian Münster: Describing the World in the Reformation* (Aldershot, 2007).

35 'Et altri [...] vogliono, che Cosmografia sia quella, che senza curarsi della particolar

Cosmography is still, in this case, motivated by the 'totalising' ambition that has been mentioned earlier, but here it is on another register, no longer that of cartographic or geometrical representation, but on a textual register, based on writing and description. This is a new organisational pattern for geographical knowledge, a spatial organisation of knowledge that is not adjusted by geometry: a science, but not a science based on mathematics.

Thus, ultimately, it appears that the definition of cosmography elaborated by Oronce Fine bears witness to the presence of a mathematical tradition within the history of geographical thought. At the same time, set against this mathematical tradition, we must also acknowledge the existence of a rhetorical tradition – or more accurately a historical tradition – alongside it, which presents itself, if not as an alternative, at least as another manner of practising geography, and another manner of looking at things. This is the 'horizontal' view of the traveller (or more strictly the historian or the enquiring researcher), not solely the 'vertical' view of the astronomer or the metaphysician.

To be precise, a third approach to cosmography should be added to these: the approach that is illustrated by Mercator, who does not move along quite the same lines. If Mercator's position is fairly close to that of Fine in his exaltation of the role of mathematics in cosmography, there is something new with the Flemish scholar: a *cosmopoïesis*, or a discourse on the world that is articulated with the narrative of the book of Genesis, which Mercator postulates as forming the real foundations of cosmography: 'Before all else', he writes in the legend accompanying the world map in his *Atlas*, 'he who studies geography must consider the creation of the world [...].'[36] The same approach is to be found in Münster in the preface to his *Cosmographia.* Cosmography for Mercator and Münster was not independent from discourse on the creation of the world. Reference should be made, in particular, to meditations on the Sphere to be found both in Mercator's *Atlas*, and in his lessons on the Sphere to which I referred above: here we can observe that cosmography finds its place in the prolongation of a hexameral discourse (with reference to the biblical narrative of the creation) in the tradition of the spiritual exercises inherited from Antiquity and the Middle Ages. Finally, it seems that it is in an area mapped out by relationships and tensions between mathematics, history and Christian theology, that cosmography developed in the sixteenth century.

---

quantità ò misura delle lontananza de' luoghi, attenda à descrivere & narrar le nature & proprietà de' paesi, & delle cose, che in esse sono, i costumi, i popoli, le cose notabili accadute di tempo in tempo [...].' G. Ruscelli, *La Geografia di Claudio Tolomeo Alessandrino, novamente tradotta di Greco in Italiano, da Girolamo Ruscelli...* (Venice, 1561), p. 4.

36 G, Mercator, *Atlas sive Cosmographicae Meditationes de Fabrica Mundi et Fabricati Figura* (Duisburg, 1595), sig. Aii^r.

Early Modern Cosmography: Fine's Sphaera Mundi *in Content and Context*

ADAM MOSLEY

Oronce Fine first published *De cosmographia, sive sphaera mundi* as the third part of his *Protomathesis* of 1532. Variant forms of the text were issued on several subsequent occasions, in both Latin and French, up to and including the year of his death (1555), and in 1587 it appeared, along with other works from the *Protomathesis*, in Italian translation. It was also excerpted in his own *Canonum astronomicorum libri II* of 1553.[1] In this essay, I propose to explore two related issues raised by this text: the character and significance of Fine's cosmographic work, and the nature of the Early Modern genre of which it was such an apparently successful example. Fine's cosmography was not, I shall show, the blend of descriptive geography and proto-ethnography that the discipline has sometimes been taken to be. While this may disappoint some, it does mean that the study of the work promises to add to our understanding of the mathematical culture (or cultures) of the sixteenth century.

I. *Locating Cosmography: Time, Place and Role*

As a genre and a discipline, cosmography should probably be considered a child of the Renaissance. Although there was a prior history of usage of the terms 'cosmography' and 'cosmographer', it does not seem to have been particularly extensive: the words are not well-attested in classical Latin or Greek,[2] and despite a

1 For a bibliography of Fine's works, see D. Hillard and E. Poulle, 'Oronce Fine et L'Horloge Planétaire de la Bibliothèque Sainte-Geneviève', *Bibliothèque d'Humanisme et Renaissance*, vol. 33 (1971), pp. 311–351, at pp. 335–349; R. P. Ross, 'Oronce Fine's Printed Works. Additions to Hillard and Poulle's Bibliography, *Bibliothèque d'Humanisme et Renaissance*, vol. 36 (1974), pp. 83–85.

2 H. G. Liddell and R. Scott, *A Greek-English Lexicon* (Oxford, 1940), ascribe *kosmographos* to Joannes Gazaeus, a sixth-century poet and grammarian, and lists *kosmographia* as the title of a work by Democritus, according to Diogenes Laertius (third century AD). Aside from the use of *cosmographia* by Cassiodorus, C. Lewis and C. Short, *A Latin Dictionary* (Oxford, 1879), note only two uses of the word *cosmographus*; perhaps because of their obscurity, the abbreviated references to the texts in question are not expanded in their apparatus.

prominent late-Latin usage in the *Divine institutions* of the sixth-century monk Cassiodorus, in which Ptolemy's *Geography* was referred to as one of several key works for learning cosmography, seem not to have enjoyed much currency through the Middle Ages.[3] Thus the *Cosmographia* of the twelfth-century poet Bernardus Silvestris, perhaps the most significant of the medieval texts to have been accorded this name, was not so-called by its author, and must have received that designation at some later juncture.[4] When cosmography was utilised in medieval texts, it appears to have been applied with almost equal fervour to ancient writers on geography, to the Pentateuch (described as the *cosmographia Divinae Historiae*) and to the travel account of Aethicus Ister, a supposed translation from the Greek falsely attributed to St. Jerome.[5] A decisive moment in the history of cosmography as a subject, therefore, was when the Italian scholar Jacopo d'Angelo chose to produce a Latin translation of Ptolemy's geographical work under this title in 1409. Apparently he preferred *cosmographia* to *geographia* as the former Graecism was already known to him from other texts, while the latter was not.[6]

Selecting cosmography as the Latin name for Ptolemy's *Geography* did not, however, entirely settle the issue of what cosmography was, or how it should be understood in relation to other forms of knowledge.[7] For one thing, Early Modern scholars soon revived 'geography' as the proper name for the enterprise embodied by Ptolemy's work, with the result that the two terms subsequently enjoyed a simultaneous currency. For another, Early Modern scholars had access to other ancient texts which made it clear that Ptolemy's *Geography* did not represent the totality of the discipline. Strabo, for example, had carefully distinguished between the enterprise of descriptive geography, his main concern, and mathematical geography, which played only a minor part in his work.[8] Thus, while the works of some ancient authors other than Ptolemy, such as the *De situ orbis* (or *Chorographia*) of Pomponius Mela, also appeared under the heading 'cosmography', a writer such as Strabo might be labelled an authority in the subject without the attri-

3 *Cassiodorus Senatoris Institutiones*, ed. R. A. B. Mynors (Oxford, 1961), I.XXV, p. 66.

4 B. Silvestris, *Cosmographia*, ed. P. Dronke (Leiden, 1978). The work is divided into two sections, entitled *Megacosmus* and *Microcosmus* respectively, and so-named in the *summa operis* that precedes them in many of the manuscripts.

5 This, at least, is the impression I have acquired from study of the texts in the *Patrologia Latina* (where the terms occur more frequently in the apparatus than elsewhere), and from L. Thorndike and P. Kibre, *A Catalogue of Incipits of Medieval Scientific Writings in Latin* (Cambridge, MA, 1963).

6 M. Milanesi, 'Geography and Cosmography in Italy from XV to XVII century', *Memorie della Societa Astronomica Italiana*, vol. 65 (1995), pp. 433–468, at p. 443.

7 See, in addition to my comments below, Besse in this volume.

8 *The Geography of Strabo*, trans. H. L. Jones, 8 vols. (London, 1917–1936), vol. 1, I.I.12–21, pp. 22–47; D. Dueck, *Strabo of Amasia. A Greek Man of Letters in Augustan Rome* (London and New York, 2000), pp. 53–62.

bution to him of a specific example of the genre.[9]

As a result of this uncertainty it was up to Early Modern writers on cosmography to define the boundaries of their field. Since, broadly speaking, cosmography could be conceived of as either the *sum* or the *intersection* of subjects such as astronomy, geography and hydrography, authors attempted to give accounts of cosmography that situated it with respect to these disciplines and to related fields such as topography, chorography and navigation, but that would not result in any terminological redundancy. The results, and the emphases, varied from author to author, and from context to context. Martin Waldseemüller, for example, signalled in both the title of his *Cosmographiae introductio* of 1507, and in an opening passage setting out the order in which the topics would be covered in this work, that knowledge of cosmography had to be grounded in principles derived from astronomy and geometry.[10] Yet he presented his initial astronomical discussion as very much preliminary to the treatment of cosmography proper, by which he seemed to mean knowledge of the lands and waters of the earth described in writing, projected onto a plane, or depicted on a globe.[11] Similarly, Gemma Frisius published a text in 1530, *De principiis astronomiae et cosmographiae deque usu globi*, whose title differentiated between, and thereby implicitly contrasted, astronomy and cosmography. The work itself, however, made no evident distinction between them. Peter Apian, in his *Astronomicum caesareum* of 1540, visually distinguished between four vocations, including astronomer and cosmographer, by distributing them between the four cardinal directions and depicting them with clothing and accoutrements characteristic of their work: the astronomer, at south, equipped with a celestial globe and universal astrolabe, was positioned opposite the cosmographer, at north, with terrestrial globe and dividers, while a priest, with a chasuble and chalice, and a poet, furnished with a book, appeared at west and east.[12] However, by beginning his earlier *Cosmographicus Liber* of 1524 with a chapter headed, 'What is Cosmography, and how does it differ from Geography

9 See F. E. Romer, *Pomponius Mela's Description of the World* (Ann Arbor, 1998), esp. pp. 5–6.

10 'Cum Cosmographiae noticiae sine praevia quadam astronomiae cognitione et ipsa etiam astronomia sine Geometriae principiis plaene haberi nequeat [...].' M. Waldseemüller, *Cosmographiae introductio cum quibusdam geometriae ac astronomiae principiis ad eam rem necessariis* (St. Dié, 1507), sig. Aii^{v}.

11 'CAPUT .IX. DE QUIBUSDAM COSMOGRAPHIAE RUDIMENTIS Propositum est hoc libello quandam Cosmographiae introductionem scribere: quam nos tam in solido quam plano depinximus.' Waldseemüller, *Cosmographiae introductio*, sigs. aiiii^{v}–b^{v}; fold-out facing sig. av^{r}. On the map and globe, see the introduction by J. Fischer and F. von Wieser to the facsimile edition by C. G. Herbermann, *The Cosmographiae Introductio of Martin Waldseemüller* (Freeport, NY, 1969), pp. 1–30, at pp. 15–30.

12 For a reproduction of this image, see M. Watelet (ed.) *Gerard Mercator, Cosmographe* (Brussels, 1994), p. 13.

and Topography', Apian also strove to distinguish cosmography from other disciplines concerned with the description of parts of the earth.[13] Rembert Dodoens published a *Cosmographica in astronomiam et geographiam isagoge* in which, notwithstanding this title, he claimed to have, 'gathered into one place those things which are properly cosmographic, having cut out all those things which pertain to geographical precepts, and reserved them for another book, concerning the elements of geography, which we have decided to publish.'[14] But John Dee, in his *Mathematicall Praeface* to Henry Billingsley's English edition of Euclid's *Elements* (1570), declared that cosmography, as the 'whole and perfect description of the heavenly, and also elementall parte of the worlde [...] requireth *Astronomie, Geographie*, *Hydrographie*, and *Musike*.'[15] Francesco Barozzi, in his *Cosmographia* of 1585, defined cosmography with the aid of a branching diagram: cosmography, he indicated, combined with judicial astrology to comprise 'astrology, or astronomy'; it was itself, however, a composite of astrology 'or astronomy, properly called', and geography.[16] Yet according to the Jesuit Antonio Possevino, who with the aid of Christoph Clavius surveyed the mathematical sciences as part of his *Bibliotheca selecta* of 1593:

> The treatment of cosmography follows astronomy; since it describes the world, consisting of the four elements and heaven, it considers first the circles from which the supercelestial sphere is understood to be composed [...] But subject to Cosmography is Geography, which shows the method of describing the terrestrial world and placing it before our eyes.[17]

13 'Quid sit Cosmographia & quo differat a Geographia et Topographia'. P. Apian, *Cosmographicus Liber* (Landshut, 1524), p. 1ʳ. See also U. Lindgren, 'Was verstand Peter Apian unter "Geographie"' and 'Was verstand Peter Apian unter "Kosmographie"', in K. Röttel (ed.), *Peter Apian. Astronomie, Kosmographie und Mathematik am beginn der Neuzeit* (Eichstätt, 1995), pp. 153–157, 158–160.

14 '[...] quae proprie Cosmographica sunt, in unum congesserimus, omnibus quae ad Geographica precepta spectant, resectis, & ad alium libellum, quem de Geographicis elementis in publicum dare decrevimus, reservatis.' R. Dodoens, *Cosmographica in astronomiam et geographiam isagoge* (Antwerp, 1548), sig. Aiiiiʳ.

15 *The elements of geometrie of the most auncient philosopher Euclide of Megara*, trans. H. Billingsley (London, 1570), sig. biiiʳ.

16 'Astrologia, sive Astronomia alia est / Cosmographia, quae totam Mundi Machinam, seu Sphaeram, & eius principia, partes, & passiones considerat. / Iudiciaria, seu Divinatrix, quae ex motibus, & Aspectibus stellarum futuros eventus in hisce inferioribus praecognoscere docet. / Astrologia, vel Astronomia, proprio nomine dicta, quae caelorum, & stellarum tum principia partes, & passiones considerat. / Geographia, quae totum terrae, & aquae globum quoad universialiores eius partes describit.' F. Barozzi, *Cosmographia* (Venice, 1585), sig. b2ᵛ.

17 'Astronomiam, tractatio sequitur Cosmographiae quae cum describat, Mundum e quattuor elementis, ac Caelo constantem, circulos in primis contemplatur, e quibus supercaelestis sphaera componi intelligitur. [...] Cosmographiae autem Geographia subditur, quae rationem Orbis terrarum describendi, atque sub oculos ponendi prae-

There was, in short, no clear consensus about how to define cosmography vis-à-vis astronomy and geography throughout the sixteenth century.

If cosmography was susceptible to a variety of characterisations, it also experienced differing levels of institutionalisation and support. In Spain and Portugal there were cosmographic institutions and licensing systems that entitled cosmographers, so identified, to prepare navigational charts and construct mathematical instruments. Iberian cosmographers also participated in the training and examination of ship's pilots and masters.[18] Outside of the Iberian territories, however, few individuals attained the *title* of cosmographer, and those who did held it as an honour as much as an office. Egnatio Danti, for example, was made cosmographer to the Grand Duke of Tuscany, and in that role tutored members of the ducal family and other members of the court in geography and mathematics, wrote treatises on a number of mathematical topics, designed and used astronomical instruments, and designed and installed the decorative maps of the *Sala di Geografia* in the Palazzo Vecchio.[19] André Thevet served four French kings as royal cosmographer, without specific duties, apparently on the strength of his publications in the field.[20] At other times and places it was entirely possible for individuals to pursue the same range of activities, and with similar expertise, without receiving the recognition of this name. Accuracy compels us to say, therefore, that Oronce Fine, while an author on cosmography, and on occasion a cartographer, was not himself a cosmographer.[21]

II. *Oronce Fine's Cosmography: Context and Content*

Although *De sphaera* was the only cosmographic work that Fine authored himself, it was not the only one that he helped to produce. In conjunction with the

bet.' A. Possevino, *Bibliotheca selecta* (Rome, 1593), p. 213.

18 See U. Lamb, 'Introduction', in *A Navigator's Universe. The* Libro de Cosmographiá *of 1538* (Chicago and London, 1972); U. Lamb, 'Cosmographers of Seville: Nautical Science and Social Experience', in F. Chiapelli (ed.), *First Images of America: The Impact of the New World on the Old* (Berkeley, 1986), pp. 675–686; V. Brotóns, 'Astronomy and Cosmography 1561–1625. Different Aspects of the Activities of Spanish and Portuguese Mathematicians and Cosmographers', in L. Saraiva and H. Leitão (eds.), *The Practice of Mathematics in Portugal* (Coimbra, 2004), pp. 225–274.

19 See F. Fiorani, *The Marvel of Maps: Art, Cartography and Politics in Renaissance Italy* (New Haven and London, 2005), especially pp. 41–54; T. Settle, 'Egnazio Danti and Mathematical Education in Late Sixteenth-Century Florence', in J. Henry and S. Hutton (eds.), *New Perspectives on Renaissance Thought: Essays in the History of Science, Education and Philosophy in Memory of Charles B. Schmitt* (London, 1990), pp. 24–37.

20 F. Lestringant, *Mapping the Renaissance World: The Geographical Imagination in the Age of Discovery* (Berkeley and Los Angeles, 1994), especially p. 10.

21 On Fine as a cartographer, see L. Gallois, *De Orontio Finaeo Gallico Geographo* (Paris, 1890), and J.-J. Brioist in this volume.

Petri press of Basel, he prepared an edition of Gregor Reisch's *Margarita philosophica* that was published in 1535, but on which he may have worked prior to the appearance of the *Protomathesis*.[22] As well as containing a chapter on astronomy, 'both that which is truly mathematical and cosmographical, and that which is called judicial astrology', the volume included Jodocus Honterus's *Rudimentorum cosmographiae* as one of several appendices on mathematical topics. Later, in 1551, the printer Guillaume Cavellat, with whom Fine collaborated on several occasions, issued Martin Borrhaus's *Elementale cosmographicum* together with Peter Apian's *Cosmographiae introductio*. According to Cavellat's preface to the reader, having been passed Borrhaus's text for his opinion, Fine had returned it to the publisher corrected and augmented with notes and illustrations for publication, even though he did not wholly approve of it.[23] Fine was therefore implicated in the printing of four treatments of cosmography other than his own.

One characteristic shared by these different texts on cosmography was their emphasis on the astronomical component of the subject. While Reisch subsumed cosmography under the heading of astronomy, Honterus was one of those who defined it in such a way as to imply that it encompassed aspects of both astronomy and geography. 'Cosmography', he wrote, 'is the description of the whole world; that is the description of heaven and earth and those things contained within them. Astronomy is that which examines the rising, setting, and motion of the stars [...] Geography is that which describes the situation of the earth with an enumeration of its places.'[24] Borrhaus's text was subtitled, 'in which the rudiments of the whole of both astronomy and geography are taught, with very brief and very certain demonstrations'; it was divided into two sections, *Astronomica* and *Geographica*, the former almost twice as long as the latter. And Apian, although his text began with the distinction between geography and chorography, went on to outline the nature of cosmography, its indebtedness to astronomy, and its intimate relationship to geography, in a passage revealing for its complexity as well as

22 R. P. Ross, 'Studies on Oronce Finé', unpublished PhD. diss. (Columbia University, 1971), p. 397, suggests that Fine's edition of Reisch was prepared in the early 1520s.

23 'Orontii Finaei, Regii Mathematicarum professoris, ad publicam utilitatem foelici quodam sydere natur fidissimum de eo sentire iudicium. Is igitur, libellum ipsum nedum approbavit: sed pro sua humanitate, qua studiosos omnes iuvare non cessat, emendatum, figurisque & annotationibus, (quae antea in illo desiderabantur) illustratum nobis reddidit.' M. Borrhaus, *Elementale cosmographicum quo totius & Astronomiae & Geographiae rudimenta, certissimis brevissimisque docentur apodixibus* (Paris, 1551), p. 2^{r-v}.

24 'COSMOGRAPHIA est totius mundi, id est, coeli & terrae, & eorum quae in iis continentur, descriptio. Astronomia, est quae scrutatur, ortus, obitus, motusque syderum [...] Geographiae, est quae terrae situm cum locorum enumeratione describit.' G. Reisch, *Margarita philosophica* (Basel, 1535), p. 1440. The elided text contains a definition of astrology; the passage concludes with a definition of chorography.

its clarity:

> Here it should be noted in passing that by the designation cosmography is understood the method [*ratio*] of geography; though cosmography tends rather to geometry and astronomy. For shapes, climates, magnitudes, and other things of this kind belonging to it cannot be understood without knowledge of those. However, the two subjects combine, and are so connected to one another, that for those causing one to be taught in whatever way without the other, there could easily be a danger of making it lame. But cosmography would seem to be more exact, because since it uses certain axioms it remains consistent with itself. Geography, both through the passing of centuries, and in the variable teachings of illustrious writers, changes with time, as all mortal works do, and is not always consistent with itself and not certain. Since besides the enumeration and history of places it also adds very many things about what the origin of cities, races, nations, and peoples was and thence the names given to things, then sometimes also indicates something of nature or amazing works, it tends to be much broader regarding the description of the Earth. But cosmography, although it has regard for each side, that is for heaven and earth, and combines the earthly with the celestial into one, cares more to show what the measure of heaven is, according to longitude and latitude, for those parts which are of the earth. And it does not enumerate regions, towns, rivers, seas, and mountains, other than to do so as to establish regional boundaries, and enumerate what the beginnings and the limits are, or to show that all those things are beneath part of the heavenly surface: in this exercise caring almost solely that it be understood that places are located beneath this heaven, what the latitude is, what the longitude, and in addition what the difference of the gnomon and shadows is for different places (whence arises the theory of the length and shortness of days).[25]

25 'Hic obiter notandum quod Cosmographiae ratio Geographiae vocabulo intelligatur: Cosmographia potius ad Geometriam Astronomiamque sese inclinat. Nam figurae climata, magnitudines, & alia huiusmodi propria, sine illarum scientia deprehendi nequeunt, coeunt tamen, suntque adeo interse connexae, ut alteram sine altera utcunque traditam, mancam esse facile constare possit periculum facientibus: sed acutior videtur esse ipsa Cosmographiae, quod cum certis quibusdam axiomatibus utatur, certa sibique constans manet. Geographia cum saeculorum decursu, tum vero multifaria illustrium scriptorum traditione, ut omnia mortalium opera, tempus mutat, nec semper consona sibi, nec certa est: quum praeter locorum enumerationem, & historiam, addat & plerunque quae civitatum, quae gentium, nationum, populorum origi fuerit, atque unde data rebus nomina, tum & illustria nonnunquam naturae sive miranda opera indicans, in terrae situ multo unberior esse solet: Cosmographia vero quamquam utriusque, coeli inquam & terrae rationem habet, & terrena cum coelestibus in unum coniungit, magis tamen id curat, ut quibus terrae extantis partibus, quod caeli secundum longitudinem latitudinemque spacium adsit, ostendat. Nec aliter regiones, oppida, amnes, maria, montesque enumerat, quam ut vel terminos statuat regionum: quaeque exordia, & qui fines sunt, enumeret, vel sub coelestis superficiei parte illa universa sint demonstret: hoc prope unice curans, ut sub quo coelo sita loca, quae latitudo, quae longitudo, praeterea quae gnomonis umbrarumque pro locorum latitudine differentium (unde brevitatis longitudinisque dierum ratio manat) diversitas

Apian's account takes us to the core of cosmography as it was, in practice, conceived in the period. For despite their failure to agree on how its relationship to astronomy and geography ought to be expressed, most authors shared an understanding of the subject as concerned with the description and delineation of the surface of the earth in terms related to the heavens. And as Apian pointed out, while particular geographical features, and man's knowledge of them, might change over time, the principles of locating those features with respect to the heavens would always be the same. Accordingly, the basis for all treatments of cosmography, howsoever far they eventually extended into the realms of descriptive geography, was the doctrine of the Sphere. Fine signalled this as clearly as any by giving his *Cosmographia* the alternative title of *De sphaera mundi*. That name connected the third part of the *Protomathesis* with the long tradition of works on spherical astronomy of which Sacrobosco's *De sphaera* is the best known example. Subsequent editions of his text made the point even more emphatically. By 1541, it had become *De mundi sphaera, sive Cosmographia, Primave Astronomiae parte*, and from 1551 onwards, it was issued as *De mundi sphaera, sive Cosmographia* [...] *in qua tum prima astronomiae pars, tum geographiae ac hydrographiae rudimenta pertractantur*. The 'first part of astronomy' must be understood here to mean spherical astronomy, as opposed to the planetary astronomy traditionally taught through the medium of the *Theoricae planetarum*. In Fine's *Sphaera mundi*, fully four fifths of the text were devoted to it rather than to the surface of the earth. Thus, while the first book treated the general structure of the universe, the second introduced the divisions of the celestial sphere, the third covered the rising and setting of the stars, and the fourth dealt with solar astronomy and gnomonics, it was not until the fifth book that Apian provided the promised geography and hydrography. The subtitle in the 1555 edition gives a fair idea of this fifth chapter's contents: 'wherein is treated geographical, chorographical, hydrographical foundations, that is parallels, climata, longitudes and latitudes of places, and the travel distances of these, plus the exact delineation of charts, both particular and universal.'[26] But everything prior to that was basic astronomy.

In some quarters cosmography has attracted scholarly attention because of a perceived association with the expansion of European horizons, the intellectual challenges posed by the New World discoveries, and the emergence of new forms of ethnographic and natural historical writing.[27] There is certainly an element of

---

sit, intelligatur.' P. Apian, *Cosmographiae introductio* (Paris, 1551), pp. $3^{r-v}/3^{v}$– 4^{r} (the edition exists in two variant states).

26 'Ubi de geographicis, chorographicis, & hydrographicis tractatur institutis, utpote, de parallelis, climatibus, longitudinibus atque latitudinibus locorum, ac illorum viatoria distantia, de chartarum insuper tam particularium quam universalium exacta descriptione.' Fine, 1555, p. 44^{v}.

27 See, for example, M. B. Campbell, *Wonder and Science: Imagining Worlds in Early Modern Europe* (Ithaca, NY, 1999).

truth in this view. As we have already seen, the office of cosmographer was closely associated with the practical business of transoceanic travel in the Iberian peninsula, and Martin Waldseemüller's *Cosmographiae introductio* is chiefly known for championing the name of 'America'.[28] Gemma Frisius justified the appearance of his *De principiis astronomiae et cosmographiae* to readers, despite the prior existence of works by the likes Ptolemy, Strabo, Stephanus, Mela and Pliny, on the grounds of the existence of parts of the world not mentioned by these authors.[29] Both his work, and Apian's *Cosmographiae introductio* included a brief geographical survey that featured New World cannibals as well as the traditional old world monsters.[30] Yet because cosmography could be considered, as Apian suggested, the *method* of describing the world, rather than the description itself – because the term implied a universal treatment, without necessarily aspiring to exhaustiveness – there was no obligation on cosmographies to cover the New World discoveries. Fine's *Sphaera mundi* was among those that that did not. The examples of the genre that have excited most attention, the universal cosmographies of Münster and Thevet, were actually anomalous in their pursuit of encyclopaedic and descriptive accounts of the world on the model of authors such as Pliny and Strabo.[31] Even these tautologously titled texts, however, retained some place for the mathematics of the sphere. Recognition of this fact imparts a new importance to Fine's *Sphaera mundi*, because it was among the longest of all the works on cosmography to remain focused on this, the discipline's astronomical core.

III. *Heaven and Earth: Natural Philosophy, Astrology and Religion in Fine's* Sphaera mundi

Fine's *Sphaera mundi* was not, however, exclusively mathematical. Historians of astronomy have long recognised that the distinction between mathematical and natural philosophical approaches to the heavens, widely discussed in literature of and on the Early Modern period, was not only contested by sixteenth-century mathematicians, but also difficult to maintain in presentations of the subject – especially so, perhaps, in elementary textbooks.[32] And natural philosophy was, in

28 Herbermann, *Cosmographiae Introductio*, pp. v, 70.

29 G. Frisius, *De principiis astronomiae et cosmographiae deque usu globi* (Antwerp, 1530), sig. A^v.

30 Frisius, *De principiis astronomiae*, sigs. E^v–K4^v, especially K3^v; Apian, *Cosmographiae introductio*, pp. 36^r/44^v.

31 On Thevet, see Lestringant, *Mapping the Renaissance World*; Campbell, *Wonder and Science*. For Münster, see G. Strauss, 'A Sixteenth-century Encyclopedia: Sebastian Münster's Cosmography and its Editions', in C. H. Carter (ed.), *From the Renaissance to the Counter-Reformation: Essays in Honor of Garrett Mattingly* (New York, 1965), pp. 145–164.

32 N. Jardine, *The Birth of History and Philosophy of Science. Kepler's* A Defence of Tycho against Ursus *with Essays on its Provenance and Significance* (Cambridge, 1988), pp.

turn, a matter of faith; concerned as it was with study of the universe created by God, it was perceived and promoted as a pious activity. A sixteenth-century reader would not have been surprised to discover that both natural philosophy and the divine made an appearance in Fine's cosmography. These were important aspects of his enterprise, and to omit consideration of them here would give a rather distorted picture of what the work was about.

Like other astronomical treatises, Fine's *Sphaera mundi* introduced natural philosophical considerations at the outset, when discussing the structure of the universe. Unlike many such works, however, Fine emphasised not only the shape of the universe, and the principles of celestial motion, but also the idea of celestial causation. Thus in Book 1, Chapter 1, Fine introduced the distinction between the elemental region of the cosmos, liable to generation and corruption, and the celestial region, entirely free of change, and stressed the importance of the relationship between them. The elements, he indicated, combined in various proportions to provide the *material* cause of inferior things, but the 'celestial machine' constituted the *formal* and *specific* cause; thus sublunary things took nourishment from the elements, but life from the heavens.[33] This was natural (or physical) astrology of a fairly commonplace sort, doctrinally and terminologically dependent upon Aristotelian philosophy, and it cropped up at intervals elsewhere in the text.[34] In the chapter of Book 1 devoted to the number and order of the celestial orbs, Fine supplied a table of the celestial bodies listing, in addition to their names and symbols, the complexion of each (hot or cold, moist or dry), whether they were considered benevolent or maleficient, and with which of the seven metals they were supposed to correspond. And in the chapter of Book 2 concerned with the signs of the zodiac, he explained the names of these with reference to the qualities shared by the sun in the sign and the creature or thing after which it was named, providing thereby an account of seasonal change. Thus:

> [...] the fifth sign, on account of the intense heat, with dryness introduced, takes the name of the lion. For the lion is the strongest animal, and is of a hot and dry complexion. And the sixth sign is ascribed to the virgin, inasmuch as she is sterile and very weakly quickening: for then the heat is diminished, and the introduced dryness predominates. Whence the increase of things ceases,

---

243–244.

33 'Quasi elementa diversimode commixta, atque invicem proportionat, sint causa materialis: caelum vero sua virtute, & actione continua, quae mediante lumine atque motu diffunditur, formalis & specifica causa omnium rerum quae in his generantur inferioribus, & vitam a caelo, alimentum vero ab ipsis capiunt elementis.' Fine, 1555, p. 1^{r-v}.

34 See R. Lemay, 'The True Place of Astrology in Medieval Science and Philosophy: Towards a Definition', in P. Curry (ed.), *Astrology, Science and Society: Historical Essays* (Woodbridge, 1987), pp. 57–73; E. Grant, *Planets, Stars, and Orbs: The Medieval Cosmos, 1200–1687* (Cambridge, 1994), pp. 569–617.

> and all things become sterile, perhaps with the extreme decoction of some things removed.[35]

Judicial astrology also made an appearance: in the last chapter of Book 2, Fine discussed the division of heaven into the twelve astrological houses. This was a contested topic, and one that Fine treated in greater depth in his *De duodecim caeli domiciliis* of 1553. As he stated in the preface to that work, he had limited himself in the *De sphaera mundi* to summarising, 'the division of the celestial houses that is rational and superior to all others, and wholly necessary for those who practise judicial astrology', without touching on the 'various lesser opinions'.[36]

Astrology was therefore a significant minor component of Fine's *De sphaera mundi*, even if it was not a subject that he pursued very deeply or very far in the pages of the book. One issue that he touched upon was the physical basis of celestial change in the sublunary world, a much debated topic in scholastic treatments of the heavens, and one that was attracting increasing attention from scholars interested in accomplishing a mathematical reform of astrology. Although he did not address the controversy directly, scattered remarks allow the reader to infer something of his views. Thus, he initially stated that heaven acted on the elemental realm through the diffusion of its 'virtue and ceaseless activity' via the mediation of both light and motion. In the chapter on the zodiac signs, he spoke in quite general terms of the initiation of changes in the qualities of the air, but went on to specify that such changes depended on just the projection of the rays of the sun and the prior disposition of the sublunary realm.[37] And in the discussion of astrological houses, he cashed out the idea of 'light and motion' in terms of planetary irradiation of the sublunary realm, varied by both the individual motion proper to each of the planets and the diurnal rotation of the outermost sphere in which all of them shared.[38] Thus, the light emanating from the various heavenly bodies was

35 'Quintum signum, ob intensam caliditatem, cum introducta siccitate, Leonis nomenclaturam accepit: est enim Leo animal fortissimum, calidae atque siccae complexionis. Sextum porro signum, Virgini adscriptum est, utpote sterili & admodum debili animanti: minuitur enim tunc calor, & introducta dominatur siccitas: unde rerum augmentatio cessat, fiuntque omnia sterilia, dempta forsitan nunnullorum extrema concoctione.' Fine, 1555, p. 9^{r}.

36 '[...] duodecimo capite secundi libri nostrae Cosmographiae, seu Mundanae sphaerae rationalem, ac caeteris omnibus praestantem caelestium domiciliorum partitionem, iis qui iudiciariam profitentur Astrologiam admodum necessariam quantum videlicet ad eum locum spectare videbatur, summatim perstrinxerimus: Cum tamen hac de re, variae tandem subortae sint opiniones [...].' Fine, 1553, p. 2^{r}. The controversy over house division is treated in J. North, *Horoscopes and History* (London, 1986).

37 'De iis tantum intelligas velim qualitatibus, quae ex sola proiectione radiorum ipsius Solis, & praevia dispositione horum inferiorum pendere videntur.' Fine, 1555, p. 10^{r}.

38 'Quemadmodum enim sydera (potissimum errantia) propria & intrinseca latione singulas Zodiaci peragrando partes, pro varia suorum radiorum proiectione, propriae virtutis seu naturae potestatem rebus inferioribus multifariam imprimunt: haud

the actual causal agent in Fine's conception, but the motion of these bodies was critical to the generation of mutiple effects in the sublunary realm.

Overall, the attitude towards astrology exposed in *De sphaera mundi* seems consonant with the view expounded in a fifteenth-century text that Fine edited for publication in 1542, the *De his quae mundo mirabiliter eveniunt* of Claudius Caelestinus. This work drew heavily upon the criticisms levelled by Nicolas Oresme at astrological theory, including his rejection of any kind of 'influence' of the heavens on the earth other than that comprised by celestial motion and light, but stepped back from Oresme's wholesale denial of the value of judicial astrology.[39] Thus Fine's conception of astrology seems to have been underpinned by a coherent philosophy, even if it was not one that was original to him.

As R. P. Ross has noted, Fine tended not to make ambitious programmatic statements in the body of his texts, reserving such comments for his dedicatory prefaces.[40] Most notably, in dedicating the *Protomathesis* to his monarch and patron, Fine sought to promote himself and his art by claiming a place for mathematics as the proper basis for all forms of scholarly study.[41] This is hardly unusual for the period, but the terms in which he did so merit our attention. Mathematics, he asserted, possessed a superior status to other disciplines, both in virtue of holding a place between the study of sensible matter (physics) and intelligible immaterials (metaphysics), and because it enjoyed the highest degree of stability and certainty. Thus Plato had reputedly accepted only students with some knowledge of mathematics, and had employed it in successfully inquiring into 'difficulties concerning God'.[42] Mathematics, therefore, was a route to knowledge of the divine.

In subsequent prefaces to the *De sphaera mundi*, Fine elaborated on this theme using arguments that cohered across the various editions. In dedicating the 1542 edition to the French chancellor, Guillaume Poyet, Fine declared that three things contributed to the cultivation of 'true philosophy' – contempt for external goods, insatiable desire for external happiness, and incessant illumination or enlightenment (*illustratio*) of the good and pious mind – and that the third of these, which paved the way for the other two, was best accomplished through con-

---

dissimiliter ad ipsum primum & universalem motum, veluti partes ipsius Universi dietim circunducta, pro diversa eorundem syderum irradiatione, quam ascendendo descendendoque repectu horizontis, ac ipsum praeterlabendo meridianum singulis contrahunt revolutionibus, horum inferiorum qualitates rursum immutare videntur.' *Ibid.*, p. 18^{v}.

39 See L. Thorndike, 'Coelestinus's Summary of Nicolas Oresme on Marvels. A Fifteenth Century Work Printed in the Sixteenth Century', *Osiris*, vol. 1 (1936), pp. 629–635. I am grateful to Stuart Clark for bringing this article to my attention.

40 Ross, 'Studies on Oronce Finé', p. 344.

41 On which see Axworthy in this volume.

42 'Plato insuper multa scrutatu difficilia de Deo, Mathematicarum praesidio plus caeteris Philosophis dogmata [...].' Fine, 1532*b*, sig. AA2^{r}.

templation of the divine works, especially those parts of Creation that were both *visible* and *constant*.[43] As Fine stressed in this preface, in his dedication of the Latin edition of 1551 to England's Edward VI, and in the 1555 dedication to Antoine Olivares, Bishop of Lombez, there were traces of divinity to be found in the nature of every thing, no matter how small. Through God's visible works, he pointed out, alluding to St. Paul's *Letter to the Romans* I.20, the invisible attributes of God, his eternal power and divinity, could therefore be perceived. And yet, each of his prefaces suggested, if study of the terrestrial part of Creation was a worthy enterprise, study of the celestial was worthier still.[44] The celestial realm constituted the eter-

43 'TRIA SUNT, GRAVISSIME Ac integerrime Cancellarie, in tam varia & admiranda rerum pulchritudine, quibus verae philosophiae, hoc est, divinae munificentiae donum potissimum excolitur: contemptus videlicet externorum bonorum, perpetuae felicitatis inexplebile desiderium, & assidua piae ac bonae mentis illustratio [...] ad utriusque facilem adsequutionem, tertium viam parat, & sua facilem reddit opera [...] Atqui mentem ipsam tunc maxime videmur illustrare, cum rerum naturam mira foecunditate refertam perdiscere: ac Deum ipsum per ea quae visibilia sunt, & semper eodem modo se habent, agnoscere conamur.' ['There are three things, most eminent and virtuous Chancellor, amongst such varied and admirable excellence of things, by which the gift of true philosophy, that is of divine munificence, is best cultivated: namely contempt of external goods, insatiable desire for eternal happiness, and assiduous enlightenment of the pious and good mind. Of which the first is, it seems, the most honourable of all, just as nothing is more remarkable than the second: but for easy pursuit of each of these, the third paves the way, and renders its labour easy. For what could be more pleasant and joyful, what more charming in this course of human life, than that the mind and intellect furnish itself with pious, and good, seeds of the virtue and disciplines which render God closer to us, and make us ultimately participants in his ineffable and eternal happiness. And also we seem to most greatly illuminate the mind when we attempt to investigate the nature of things, filled with amazing fecundity, and recognise God himself through those things which are visible, and always hold themselves in the same way.'] Fine, 1542, sig. *ii^{r}.

44 'Si per ea quae visibilia sunt [...] invisibilia Dei opera, cuiusmodi est eius sempiterna potentia, atque divinitas, clarissime (ut inquit Paulus apostolus) perspiciuntur: nullum studium censeri debet utilius, iucundius, ac dignius Christiano quovis homine, nedum antistite, una divinorum operum contemplatione. In quibus primas tenet, caelestis & mundanae sphaerae structura [...].' ['If through those works of God which are visible (...) invisible things such as his eternal power and divinity are most clearly perceived (as the apostle Paul said), no study ought to be judged more useful, more joyful, and more worthy of any Christian man, not to mention a priest, than a contemplation of the divine works. Amongst which the structure of the celestial and worldly sphere holds first place (...).'] Fine, 1555, sig. * ii^{r}.

45 'INTER admiranda naturae sive Dei miracula, duo sunt, Edoarde rex inclyte, quae omnium miraculorum superare videntur admirationem: Mundus scilicet, & homo. Quorum partes insigniores sunt rursum duae: utpote, immortalis vel aeterna, & ea quae corruptioni, atque mutationi semper obnoxia est. Mundi namque pars aeterna, est ipsum caelum [...].' ['Among the marvelous things of nature or miracles of God, there are two, glorious king Edward, which seem to transcend the admiration of all miracles: namely the World, and man. The chief parts of which are also two: namely the

nal and immortal part of the world, like the soul of man.[45] It showed the divine majesty and power, blazing forth, and was the eternal resting place of all blessed spirits.[46] And it was studied through mathematics, which by means of its certain principles and demonstrations, unwavering order, and pure and forever inviolable essence, readily reconciled the divine with the human, and transformed the human mind into a celestial intelligence.[47] Man, and only man, had been *made* to contemplate the works of God.[48] The implication, therefore, was that to do so was both a duty and a privilege.

In asserting that the objects of mathematical study occupied a place between the realms of the sensible-corporeal-corruptible and the intelligible-immaterial-eternal, Fine was drawing on commonplaces of Platonic and Neoplatonic thought.[49] His claim that study of the eternal and unchanging heavens was wor-

immortal or eternal, and those which are always liable to corruption and alteration. For the eternal part of the world is heaven (...).' Fine, 1551*b*, sig. aii^{r}.

46 '[...] certe divinam maiestatem, ac aeternam Dei potentiam undiquaque relucescere perspicit [...] & non modo eorum, quae in his generantur inferioribus, formalem ac specificam esse causam, sed omnium beatorum spirituum aeternum habitaculum.' ['Certainly, he will perceive that the divine majesty and external power of God shine out everywhere (...) and that it [heaven] is not only the formal and specific cause of those things which are generated in these inferior realms, but also the eternal dwelling place of all blessed spirits.'] Fine, 1555, sig. *ii^{r-v}.

47 'Ad cognitionem porro ipsius divini & semper admirandi opificii, pulchre & ex omni parte suum referentis opificem: divinae illae ac fidissimae artes, quae solae Mathematicae, hoc est, disciplinae meruerunt adpellari, non utiles tantum, sed omnibus modis videntur esse necessariae. Utpote, quae tum principiorum & demonstrationum certitudine, ordine stabili, tum pura & inviolabili semper essentia: divina humanis (inter quae medium obtinuere locum) facile conciliant, & mentem ipsam humanam in caelestem transformant intelligentiam.' ['But for knowledge of the divine and always marvellous creation, beautiful and in every respect referring back to its creator, those divine and most faithful arts (as the Mathematical disciplines alone deserve to be called) seem to be not merely useful, but in every way necessary. Inasmuch as those, through both the certainty, the established order, of their principles and demonstrations, and an always pure and imperishable essence, readily reconcile the divine with the human (between which they hold the middle place) and transform the human mind into a celestial intelligence.'] Fine, 1542, sig. *ii^{r}.

48 'Homo itaque sic efformatus est, ut utranque suam originem aeternam videlicet, & corruptibilem recognoscere possit & debeat: hoc est, incolere atque gubernare terrena, & simul intelligere & admirari quae caelestia sunt. Nempe cui soli inter animantia, portio mentis ab ipso Deo, caeli & animae, ac omnium eorum quae Mundus comprehendit opifice atque rectore, concessa est.' ['Man, therefore, is so formed that he can and should understand both his eternal and corruptible beginning: that is, cultivate and govern earthly things and at the same time understand and wonder at those which are celestial. For to him alone among all the living things was granted a portion of mind by God himself, creator and ruler of heaven, and soul, and all the things which the world contains.'] Fine, 1551*b*, sig. aaii^{r}.

49 S. Cuomo, *Ancient Mathematics* (London, 2000), pp. 24–25, 180–186, 234–241.

thier than study of the corruptible terrestrial realm could have been separately sourced: Ptolemy, for example, in the first book of his *Syntaxis*, claimed a higher status for mathematics than physics on precisely these grounds; Aristotle had said something rather similar in *De Partibus Animalium*, even as he pointed out that knowledge of celestial things was harder to obtain.[50] And in one his prefaces, Fine remembered to cite Ovid's *Fasti* in support of the claim that those who first took the trouble to 'ascend' to the celestial regions were 'happy souls'.[51] In other respects, however, the language of Fine's prefaces evokes a particular natural philosophical tradition temporally closer to home. In emphasising *contemplation* and *illumination* as the means by which the human mind approaches the divine, and in using *Romans* I.20 as the scriptural underpinning for his notion of *how* consideration of the natural world was a pious activity, Fine's account strongly echoes the thirteenth-century Franciscan approach to nature identified and described by Roger French and Andrew Cunningham.[52] This was also informed by Neoplatonism, but it was a Christian Neoplatonism adapted to the particular needs of the friars minor, the remit of the order, and their time.

Whether Fine was consciously appropriating the Franciscan tradition, and whether he drew on specific material in order to do so, is difficult to say. It is likely that the once characteristic phrases and attitudes were widely dispersed by the sixteenth century, and could therefore be found in a number of places. Caelestinus was not a source of such material for Fine; as late as he was, he must, in any case, have been a member of the Benedictine Celestines, rather than one of the short-lived Celestine branch of the Franciscan Order suppressed in 1317.[53] But Fine published his *De his quae mundo mirabiliter eveniunt* with the *De Mirabili potestate artis & naturae* attributed to Roger Bacon, and it is possible that he studied other works by the Franciscan *Doctor Mirabilis*. Medieval optical writings are a likely point of contact; Fine is known to have studied Witelo's *Perspectiva* prior to preparing his *De speculo ustorio*, may have been acquainted with Bacon's work on burning mirrors, and could have read other products of what, according to French and Cunningham, was a particularly Franciscan obsession with the behaviour of light.[54]

Whatever the sources of Fine's pious defense of cosmography, something needs to be said about its significance. Because dedicatory prefaces praising the subject of a book, and commending its study to the dedicatee, were so conven-

50 See G. Toomer (ed. and trans.), *Ptolemy's Almagest* (London, 1984), I. 1, pp. 35–37; R. McKeon (ed.), *The Works of Aristotle* (New York, 1941), I. 5, pp. 656–658.

51 'Felices igitur animae, (ait Ovidius) quibus haec cognoscere primum, Inque domos superas scandere cura fuit.' Fine, 1551, sig. aaii^{v}. The source is the *Fasti*, I. 297–298.

52 R. French and A. Cunningham, *Before Science: The Invention of the Friars' Natural Philosophy* (Aldershot, 1996).

53 P. Day, *A Dictionary of Religious Orders* (London and New York, 2001), p. 77.

54 Ross, 'Studies on Oronce Finé', pp. 302, 323, and Dupré in this volume.

tional in the period that the absence rather than the presence of such a text is to be considered remarkable, it is tempting to ignore the precise terms in which such defenses are put. Conversely, the assumption that Fine was wholly sincere in making such claims might lead us to overlook the possibility that they might have served his interests in other ways as well. Yet Fine would not be the first human being who sincerely believed what it was most convenient to believe, if indeed that were to turn out to be the case. And promoting the study of mathematics in general, and cosmography in particular, on grounds such as these, would certainly have been useful to Fine in early sixteenth-century Paris. It would have distinguished his work from that of the scholastic natural philosophers – scholars whom he criticised for causing a decline in the study of mathematics at the University – even as he continued to draw on certain Aristotelian doctrines, particularly celestial causality, helpful to his case.[55] But it would also have offered an alternative approach to mathematics to the one sponsored by scholars such as Jacques Lefèvre d'Etaples and Symphorien Champier, promoters of a speculative or mystical mathematics affiliated with 'Pythagorean' numerology and Christianised Kabbalah.[56] In both of these ways it could have served him well as a defense of the teaching and study of what was, in truth, often quite basic mathematical material, without ceding the ground claimed by his competitors for students and readers in sixteenth-century France, that their approaches (and not his) were rooted in true Christian philosophy. It was, in other words, a way of promoting mathematical study highly consonant with what might be termed Fine's theoretical and academic approach to practical mathematics.

IV. *Theoretical Practical Mathematics and Fine's* Sphaera mundi

The notion of *theoretical* practical mathematics might seem to be oxymoronic, but it is a useful analytic tool given current understandings of sixteenth- and seventeenth-century mathematical culture. Historiographically, and semantically, *practical* mathematics has become so bound up with the terms mathematical *practice*, and mathematical *practitioner*, that to speak of any one of these three is, it might be assumed, to speak of the others. This situation owes something to the work of a pioneer in the field, E. G. R. Taylor, whose *The Mathematical Practitioners of Tudor and Stuart England* could be said to have been inclusive to a fault. Taylor's adoption of the term 'mathematical practitioner' was presumably informed by its contemporary usage; so, at least, her later statement, that this was a name that 'they had bestowed on themselves in the days of Henry VIII', would seem to imply.[57] Yet Taylor made no attempt to identify period instances of the

55 Ross, 'Studies on Oronce Finé', p. 357. See also Pantin and Cifoletti in this volume.

56 See B. P. Copenhaver, 'Lefèvre d'Etaples, Symphorien Champier, and the Secret Names of God', *Journal of the Warburg and Courtauld Institutes*, vol. 40 (1977), pp. 189–211.

57 E. G. R. Taylor, *The Mathematical Practitioners of Hanoverian England, 1714–1840*

label, to reveal its connotations as it was used on particular occasions, or to justify its application to a set of individuals who differed from one another quite considerably in terms of social status. The identity of the mathematical practitioner was instead established by Taylor in two other ways: specifically, yet ambiguously, in the form of a biographical register that frequently distinguished individuals as instrument-makers, surveyors, mathematical practitioners, hydrographers, etc., and yet grouped them altogether as 'The Practitioners'; and more generally, and descriptively, as users and promoters of mathematical instruments. For Taylor, the essence of mathematical practice was 'the use of geometrical instruments for precise measurement' or 'the application of geometry to everyday affairs and skills'; thus, 'the handling of instruments was the very badge of the mathematical practitioner.'[58] It is apparent that, for Taylor, the *unity* of mathematical practice was one of its most important attributes, and that the mathematical practitioner disappeared, as a category, when industrialisation and professionalisation imposed new distinctions between instrument-makers and instrument-users.[59]

Subsequent studies have challenged some of Taylor's assumptions, even as they have built on her account. Stephen Johnston has done the most to refine our picture of the mathematical practitioner by studying the individuals who self-consciously developed this role for themselves in sixteenth-century England.[60] In the process, he has characterised the mathematical practice of the English mathematical practitioners as predominantly commercial, urban and vernacular, occurring outside both the academy and the court. The unity of mathematical practice, which was so important to Taylor, features in Johnston's account as part of the rhetorical armamentarium of individuals who often, in fact, sought to use their superior command of mathematics to distinguish themselves from other workers in their fields – an armamentarium in which claims of mathematics' utility was perhaps the principal weapon. Johnston has also acknowledged the diversity of European mathematical culture as a whole. Other scholars, however, have exported the term 'practitioner' from England to the Continent, sometimes intending to connote by it the full range of individuals with mathematical expertise, sometimes applying it just to those with an interest in mathematics' practical applications.[61]

(Cambridge, 1966), p. 3.

58 E. G. R. Taylor, *The Mathematical Practitioners of Tudor and Stuart England* (Cambridge, 1954), pp. 7, 10; Taylor, *Mathematical Practitioners of Hanoverian England*, pp. 3–4.

59 Taylor, *Mathematical Practitioners of Hanoverian England*, pp. 3–4, 105.

60 S. Johnston, 'Making Mathematical Practice: Gentlemen, Practitioners and Artisans in Elizabethan England', unpublished PhD. diss. (University of Cambridge, 1994); 'The Identity of the Mathematical Practitioner in 16th-Century England', in I. Hantsche (ed.), *Der 'Mathematicus": Zur Entwicklung und Bedeutung einer neuen Berufsgruppe in der Zeit Gerhard Mercators* (Bochum, 1996), pp. 93–120.

To the extent that the latter becomes the norm, there is some danger of attributing the characteristics of English mathematical practitioners to all individuals active in the field. To do so would be to extend Taylor's notion of the unity of mathematical practice *qua* practical mathematics across the whole of Europe's rich cultural landscape.

Practical mathematics – or at least, practical geometry – is clearly the right term to use of the mathematical tradition focused on instruments and the solving of problems in mensuration, surveying, time-finding, navigation and similar fields. It was, after all, used in that way in the Latin West from at least the twelfth century onwards, following the appearance of Hugh of St. Victor's *Practica geometriae*.[62] Needless to say, prior to the sixteenth century this tradition, as represented to us through the available evidence, was not predominantly commercial, urban and vernacular, but was most strongly affiliated with the Latinate settings of the university and monastery. Whether it is helpful to write of practical geometry as an aspect of mathematical *practice* is open to question. Both *practice* and *practitioner* carry connotations of activity from which it is difficult to escape, yet even the extensive treatment accorded to the English mathematical practitioner has not entirely settled whether the *actual application* of geometry to the solution of real-world problems, or the *promotion*, through teaching and authorship, of such an application, constituted the activity in which they most characteristically engaged. The medieval tradition of practical geometry purported to be *about* the solving of problems in the field, but it was at least partially a tradition of the study and scriptorium. The teaching and learning of this kind of mathematics, even if manipulation of instruments was involved – and instruments were often possessed by the medieval institutions that owned practical mathematical and astronomical texts – did not necessarily involve any attempt to apply the knowledge outside of the classroom.[63] Hence the utility of *theoretical* practical mathematics as a description of a certain approach to the subject.

In its earliest form, Fine's *Sphaera mundi* was intimately associated with practical mathematics of more than one sort. As part of the *Protomathesis*, Fine's cosmography was initially sandwiched between his *De arithmetica* and *De geome-*

---

61 See, for example, J. Bennett, 'Geometry in Context in the Sixteenth Century: The View from the Museum', *Early Science and Medicine*, vol. 7 (2002), pp. 214–230; M. Biagioli, 'The Social Status of Italian Mathematicians, 1450–1600', *History of Science*, vol. 27 (1989), pp. 41–95; M. Mahoney, *The Mathematical Career of Pierre de Fermat* (Princeton, NJ, 1973), pp. 1–25.

62 H. L'Huillier, 'Practical Geometry in the Middle Ages and the Renaissance', in I. Grattan-Guinness (ed.), *Companion Encyclopedia of the History and Philosophy of the Mathematical Sciences*, 2 vols. (London, 1994), vol. 1, pp. 185–191. See also P. Brioist in this volume.

63 See J. Viellard, 'Instruments d'astronomie conservés à la bibliothèque du college de Sorbonne au XVe et XVIe siècles', *Bibliothèque de l'École des Chartes*, vol. 131 (1973),

tria, and his *De solaribus horologiis*. The first of these was actually a treatise on *practical* arithmetic, while the second comprised two books, one an introduction to the basic universal principles of geometry, the other concerned with the application of geometry to the solving of traditional problems of mensuration, and with mathematical instruments such as the geometrical quadrant and the cross-staff. The status and the shortcomings of the third text, as an example of a work on the practical mathematical topic of dialling, are discussed elsewhere in this collection. Not surprisingly, in this context, the distinction between theoretical and practical branches of the mathematical disciplines was one that exercised Fine himself. As he indicated in the preface to the second book of *De geometria*, 'practical geometry is of use to all men, while theoretical geometry is of special interest to university students and other lovers of philosophy'.[64] But of course, this second more select group was precisely the audience for Fine's teaching in the Collège Royal, and the one that this and other of his publications seemingly intended to engage. He repeatedly indicated that his auditors were among those desirous or needful of a copy of his works.[65]

Even when divorced from the other texts with which it had originally appeared, Fine's *Sphaera mundi* was itself problem- and instrument-oriented to a limited degree. Unlike some authors on spherical astronomy and cosmography, Fine did not stress the cognitive or epistemological significance of the construction or study of an armillary sphere, perhaps because the illustrations of Book 2 were trusted to convey to the reader the characteristics of the celestial circles which they needed to know.[66] He did, however, introduce his discussion of certain circles employed in conjunction with the horizontal coordinate system as worthwhile because, 'from them a good part of astronomy, and the whole construction and theory of the instruments which they call astrolabes seems to depend.'[67] Another chapter of this book concerned 'hour circles, and the theories

587–593; K. Eagleton, 'Instruments in Context: Telling the Time in England, 1350–1500', unpublished PhD. diss. (University of Cambridge, 2004).

64 See Ross, 'Studies on Oronce Finé', p. 164.

65 E.g. '[…] libros cum ob distributorum exemplariorum raritatem, ab innumeris (nostris potissimum auditoribus) desiderari saepius audieremus […].' Fine, 1542, sig. *ii^v; '[…] conscripsimus: partium ut auditoribus nostris, vero caeteris rerum caelestium amatoribus […].' Fine, 1555, sig. ii^v.

66 See, on this, A. Mosley, 'Spheres and Texts on Spheres: The Book-Instrument Relationship and an Armillary Sphere in the Whipple Museum of History of Science', in L. Taub and F. Willmoth (eds.), *The Whipple Museum of the History of Science: Instruments and Interpretations* (Cambridge, 2006), pp. 301–318; A. Mosley, 'Objects of Knowledge: Mathematics and Models in Sixteenth-Century Cosmology and Astronomy', in S. Kusukawa and I. Maclean (eds.), *Transmitting Knowledge: Words, Images, and Instruments in Early Modern Europe* (Oxford, 2006), pp. 193–216.

67 '[…] aliorum inter immobiles circulorum subsequitur contemplatio: quos hoc loco diffinire commodissimum existimavimus, utpote, a quibus bona pars astronomiae, ac

of sundials depending upon them.'[68] This was not a detailed discussion of the construction and theory of dials, for which Fine referred the reader to his *De solaribus horologiis*, but contained a description and depiction of five categories of dial – equatorial (equinoctial), horizontal, vertical (and meridional), hanging (polar), and lateral (vertical and oriental or occidental) – and an explanation of their basic characteristics with reference to the obliquity of the horizon. He pointed out that the hour-lines on an equinoctial dial occurred at equal intervals, and that the delineation of these lines was invariant with latitude, claiming that neither was true of the other types of dial that he mentioned.[69] Later on, in a chapter of Book 4 devoted to the variation of solar altitude with time, he referred to the design features of the other main kind of geometrical time-finding instrument, the altitude-dial, 'of such a kind as many old-style quadrants are'.[70] The following chapter treated the shadows cast by the Sun, explained the distinction between *umbra versa* and *umbra recta*, and introduced the *quadratum geometricum* or shadow-square, 'which is drawn on quadrants and on the back of astrolabes', and using which, from shadow-lengths or from sightings, the 'lengths, heights, and depths of things could be measured.'[71] It also noted that geographers used only meridional *umbrae rectae*, in order to establish the distance from the equator of particular places. But in making remarks of this kind, Fine was evidently more interested in expounding principles than in describing procedures. His text assumed a realm of practical mathematics, in which instruments were deployed and problems were solved, but *doing* took second place to *knowing* and *understanding* in Fine's presentation. Although we cannot be certain about just how such material would have been taught face-to-face, it seems clear that, to the extent that this work was a work of practical mathematics, it was, if anything was, practical mathematics of the theoretical sort.

Having said that, there are places in the text in which Fine chose to give precise instructions about how to perform a particular operation. Through Books 1

universa instrumentorum (quae vocant astrolabia) tum fabrica, tum ratiocinatio pendere videtur.' Fine, 1555, p. 16^{r}.

68 'De circulis horariis, & pendentibus ab illis solarium horolgiorum rationibus.' *Ibid.*, p. 16^{v}.

69 Fine is evidently presupposing – somewhat unfairly given that the equinoctial dial must be adjusted for latitude – that the 'lateral' dial is not drawn on a surface that can subsequently be mounted on another vertical surface at the latitude-appropriate angle, and that the polar dial is not drawn on a surface of adjustable inclination.

70 '[...] insolaribus horariis, adminiculo predictarum altitudinum fabricatis (cuiusmodi sunt quadrantes plerique veteres) horarum intervalla sive lineamenta, tam antemeridianis, quam pomeridianis horis indifferenter adcommodentur.' Fine, 1555, p. 42^{r}.

71 '[...] quadratum illud geometricum, quod tum in quadrantibus, tum in Astrolabiorum dorsum figuratur. Quo videlicet earundem umbrarum, aut visualium radiorum adminiculo, rerum longitudines, & altitudines, atque profunditates dimetiuntur.' *Ibid.*, p. 43^{r}.

to 4, the operations in question are principally the extraction of information from one of the tables provided. In the fifth book, however, Fine explained, without entering into details, how to determine terrestrial longitude by observation of lunar eclipses or the lunar meridian altitude, how to construct maps that display an eighth, a quarter, or half of the terrestrial surface, and how to draw and colour wind-roses. Only the latter operation is at all unusual in the context of a tradition that took its starting point from Ptolemy's *Geography* and Sacrobosco's *De sphaera*.

V. *The Fortunes of Cosmography: Fine's* Sphaera mundi *at Home and Abroad*

Evaluations of Oronce Fine often consist of a statement to the effect that, though neither at the cutting edge of his discipline, nor especially original, he did much to promote the study and teaching of mathematics in sixteenth-century France.[72] The apologetic tone of such remarks reflects historians of science and mathematics continuing thrall to narratives of progress and stories of revolutionary change. We ought rather to embrace the oportunity to study, through Fine and his works, such apparently unrevolutionary topics as the nature of sixteenth-century mathematical education, the breadth and depth of the international mathematical community, and the extent to which particular tools and approaches were deployed and discussed within and without the Early Modern academy. Fine was much more representative of the mathematical culture of the period than the extraordinary figures who have tended to monopolise historians' attention. Arguably, indeed, he did much more to shape the nature of that culture than many of those who are now more famous than him.

Historians often use, as a measure of the popularity or influence of a particular book, the frequency and speed at which it was reprinted. It is an inadequate proxy for a detailed history of publication and reception, and one frequently compromised by lack of knowledge of, or attention to, the print-runs involved, the success or failure of the commercial enterprises involved, the means of distribution, the possibility that successive editions were in competition with another, and the distinction between perceived and actual demand. In most cases, for example, we lack something akin to Fine's testimony in the preface to the 1542 edition of the *De sphaera mundi*, that the previous edition (the *Protomathesis*) had completely sold out.[73] Using this measure, however, as crude as it is, it would seem that the most successful cosmography of the century was the one composed by Peter Apian and augmented by Gemma Frisius in 1529. But Fine, with Latin, French and Italian editions of his work, probably comes second in the list of sixteenth-century writers of cosmography as judged by the frequency with which he was published.

[72] E.g. Ross, 'Studies on Oronce Finé', pp. 375–385.

[73] See n. 65, above.

What would be desirable, as a means of qualifying this view, would be a systematic attempt to accumulate information about the owners and readers of works such as Fine's *Sphaera mundi*: a collaborative effort akin to the *Reading Experience Database* being developed at the Open University, but devoted specifically to mathematical and astronomical texts of the Early Modern period.[74] In the absence of such a resource, observations collected on an *ad hoc* basis by individuals may serve the same purpose in a much more limited way. On this evidence it would seem that Fine's *Sphaera mundi* was indeed known and studied, both at home and abroad. At home, André Thevet appropriated elements of Fine's pious defense of cosmography for his own *Cosmographie universelle*.[75] Outside of France, Fine's text was owned by such Scandinavian and German notables as Tycho Brahe, Johann Scheubel, Landgrave Wilhelm IV of Hessen-Kassel, and Elector August of Saxony (the latter two both in the form of the *Protomathesis*).[76] A copy of the 1551 Latin edition found its way into the possession of the aspotate bishop, Andreas Dudith, who lived in modern-day Breslau.[77] In England, it was possessed by several men of property, and by scholars at Cambridge University, and was certainly read and referred to by Robert Recorde.[78] It was cited in the Low Countries by the Scots Catholic James Cheyne, a professor at Douai, in one of his

74 See http://www.open.ac.uk/Arts/RED/.

75 A. Thevet, *La cosmographie universelle* (Paris, 1575), sigs. Aiii^v– av^v. It was also owned, along with several other of Fine's works, by the sixteenth-century lawyer, cartographer and mathematician, Jean I du Temps of Blois. See A. Marr, 'A Renaissance Library Rediscovered: the 'Repertorium librorum Mathematica' of Jean I du Temps', *The Library*, vol. 9, no. 4 (2008), pp. 428–470.

76 W. Norlind, *Tycho Brahe: En levnadsteckning med nya bidrag belysande hans liv och* (Lund, 1970), p. 345; B. Hughes, 'The Private Library of Johann Scheubel, Sixteenth-century Mathematician', *Viator*, vol. 3, (1972), p. 422; J. Leopold, *Astronomen, Sterne, Geräte: Landgraf Wilhelm IV. und seine sich selbst bewegenden Globen* (Lucerne, 1986), p. 214; H. Watanabe O'Kelly, *Court Culture in Early Modern Dresden* (Basingstoke, 2002), p. 82.

77 J. Jankovics and I. Monok, *András Dudith's Library: A Partial Reconstruction* (Szeged, 1993), p. 68, no. 114.

78 R. J. Fehrenbach, 'An Inventory of Books in the Possession of Sir Roger Townshend, ca. 1625', in R. J. Ferhrenbach and E. Leedham-Green (eds.), *Private Libraries in Renaissance England: A Collection and Catalogue of Tudor and Early Stuart Book-Lists*, 3 vols. (Marlborough, 1992), vol. 1, pp. 79–135, at p. 98; N. Krivatsy and L. Yeandle, 'Books of Sir Edward Dering of Kent (1598–1644)', in *idem*, pp. 137–269, on p. 220; E. Leedham-Green, *Books in Cambridge Inventories: Book-Lists from Vice-Chancellor's Court Probate Inventories in the Tudor and Stuart periods*, 2 vols. (Cambridge, 1992), vol. 1, pp. 51, 504, 524; R. Recorde, *The castle of knowledge* (London, 1556), p. 2. See also S. K. Heninger, 'Oronce Finé and English Textbooks for the Mathematical Sciences', in D. B. Randall and G. W. Williams (eds.), *Studies in the Continental Background of Renaissance English Literature: Essays Presented to John L. Lievsay* (Durham, NC, 1977), pp. 171–185.

own university text books.[79] In Italy, Barozzi criticised Fine as he criticised other contemporary writers on cosmography; these comments should not be taken too seriously, however, since Barozzi considered Ptolemy's *Syntaxis*, not the *Geography*, as the subject's foundation, and strongly deplored what he took to be the excessive reliance of cosmographic authors on Sacrobosco's *De sphaera*.[80] Possevino, in contrast, guided by Clavius, rated Fine's *Sphaera mundi* rather highly; tellingly, however, he listed it among the astronomical texts and not the geographies.[81] Institutionally, the text seems also to have enjoyed a certain measure of success. A copy of the *Protomathesis* was held by Wittenberg university library in 1536; whether that is very significant is rather difficult to tell, although it is interesting to note that it was listed under *philosophi latini recentiores*.[82] Of more obvious importance is that the *Sphaera mundi* was used, according to Charles Schmitt, as a textbook in Pisa in the later part of the century, and the New Oxford statutes of *ca*. 1564 listed it as an acceptable alternative to Sacrobosco for students of the arts course.[83] Perhaps the English appeal of the book was enhanced by the 1551 dedication to Edward VI.[84]

Cosmography was a rich and varied tradition in sixteenth-century Europe. Until it has been studied more comprehensively, it is difficult to say precisely how influential Fine's work on it was. What is evident, however, is that to study Early Modern cosmography is, at least in part, to study the problem- and instrument-oriented field of practical mathematics. The case of Fine's *De sphaera mundi* reminds us that this branch of mathematical culture straddled the divide between the academy and other sites of mathematical expertise in the period, and could be promoted on grounds other than the insistent appeal to utility that has often been identified as characteristic of its advocates – grounds that situated it with respect to, and placed it in dialogue with, natural philosophy.

79 J. Cheyne, *De geographia libri duo* (Douai, 1576), sig. B5ᵛ.

80 Barozzi, *Cosmographia*, sigs. b3ᵛ, b5ᵛ– b6ʳ, c4ᵛ.

81 A. Possevino, *Bibliotheca selecta*, p. 201; J.-C. Margolin, 'L'enseignement des mathématiques en France (1540–70): Charles de Bovelles, Fine, Peletier, Ramus', in P. Sharratt (ed.), *French Renaissance Studies, 1540–1570: Humanism and the Encyclopedia* (Edinburgh, 1976), pp. 109–155, at pp. 124–125.

82 S. Kusukawa, *A Wittenberg Library Catalogue of 1536* (Cambridge, 1995), p. 117.

83 C. Schmitt, 'The Faculty of Arts at Pisa at the Time of Galileo', *Physis*, vol. 14 (1972), pp. 243–272, at p. 260; G. Strickland, *Statuta Antiqua Universitatis Oxoniensis* (Oxford, 1931), pp. xciv, 390.

84 See also the manuscript of a sixteenth-century English translation noted by P. Brioist in this volume.

Oronce Fine and Cartographical Methods

JEAN-JACQUES BRIOIST

Questioning the means by which Fine drew his 1536 world map, Monique Pelletier reflects on his many careers and concludes: 'Fine's printed works reveal that this mathematician's interests were indeed manifold and his cartographic works certainly rank among his most original works.'[1] Fine's cartographic craft is revealed in two types of document: his treatises on cosmography on the one hand, where mapmaking is arguably devoted to but a tenth of the pages, and his beautiful engraved plates on the other, which still foster his fame.[2] Among Fine's engraved maps we find *Nova totius Galliae descriptio*, a map of France, Switzerland and the Low Countries published in 1525 by Simon de Colines, and the heart-shape ('cordiform') world map bearing the title *Recens et integra orbis descriptio* (Fig. 9.1).[3] Such maps, engraved as they were on large plates for printing as standalone artifacts, are the masterworks of a skilled engraver. In the leasing of the plates to several printers, these maps were used to illustrate a number of books, among which are Vadian's 1530 edition of Pomponius Mela's *Chorographia* or Grynaeus's

1 M. Pelletier (ed.), *Cartographie de la France et du monde de la Renaissance au siècle des Lumières* (Paris, 2001), p. 11. See also M. Pelletier, 'Die herzförmingen Weltkarten von Oronce Fine', *Cartographia Helvetica*, vol. 12 (1995), pp. 27–37; F. Lestringant and M. Pelletier, 'Maps and Descriptions of the World in Sixteenth-Century France', in D. Woodward (ed.), *The History of Cartography Volume 3: Cartography in the European Renaissance*, 2 vols. (Chicago and London, 2007), vol. 2, pp. 1463–1479, esp. pp. 1464–1467: 'Oronce Fine and the Ptolemaic Tradition'; M. Pelletier, 'National and Regional Mapping in France to about 1650', in Woodward, *History of Cartography*, vol. 2, pp. 1480–1503, esp. pp. 1480–1483, and the bibliography of studies of Fine's maps provided at p. 1482, n. 7.

2 Fine's treatises dealing with cosmography are the *Protomathesis* (1532) and the several editions of *De mundi sphaera* (first edition 1541). On the *Protomathesis* see Axworthy and P. Brioist in this volume. On *De mundi sphaera* see Besse and Mosley in this volume.

3 *Nova totius Galliae descriptio. Orontius F. Delphinas faciebat 1525 Parisii*. Reprinted in 1553 by Jérôme de Gourmont. Wood engraving of 4 sheets bound together; 69 × 94.5 cm. Copy consulted: Bibliothèque Nationale de France, Paris, Département des Cartes et Plans, ref. Réserve Ge B 1475. *Orontii Finaei Recens et integra orbis descriptio* (*ca.* 1534). Wood engraving of 2 sheets bound together; 52 × 59.5 cm. Copy consulted: Bibliothèque Nationale de France, Paris, Département des Cartes et Plans, ref. Réserve Ge DD 2987 (063).

Novus orbis.[4] However, Fine never actually authored a book in which his world maps were used.

The two sources for Fine's cartography differ in the type of map projections they use, and by their lack of reference to each other: only in his later years did Fine expound his method for framing the heart-shape maps which made him famous.[5] As for the didactic treatises, no map is to be found in them, for there the *Lecteur Royal* only draws the mesh of the mapped parallels and meridians of the sphere, or roughly outlines a country's shape (France). Such a divorce between a man's art and his teaching might be explained by professional secrecy and the need to keep an edge on one's competitors; but the character of novelty of Fine's maps has yet to be questioned in the context of the mid-1530s.[6] In this essay I shall try to show that, beyond the workshop directions of his *Protomathesis*, useful information can be traced about Fine's theoretical thinking in the field of cartography. I shall begin by recalling the technical and scientific background against which Fine acquired and improved his know-how.

I. *The Ptolemaic Heritage*

Relatively early in his career (between 1515 and 1520), Fine was involved in the printing of scientific books. By their demonstrative nature, such books demanded illustrated plates, a craft at which the young man became expert, as is witnessed by Juan Martinez Pedernales Siliceo's *Arithmetica* (1514), or Augustinus Ricius's *De motu octavae spherae* (1513).[7] Years later, he would undertake the reprint of Georg Peurbach's *Theoricae novae planetarum* (1525) and Reisch's celebrated *Margarita philosophica* (1535).[8] Thanks to Radtolt's preface to the *editio princeps* of Euclid's *Elements*, we are well informed about the difficulties printers faced when arranging an engraving amid typography in the early days of printing: inking the sheets in several steps brought about shifting and off-centring of the matrices ('registration').[9] These difficulties were still worsened if several colours were to be

4 Mela (ed. J. Vadian), *Libri de situ orbis tres* (Paris, 1530); J. Hüttich and S. Grynaeus, *Novus orbis regionum ac insularum veteribus incognitárum* (Basel, 1532).

5 The only explanation about an (approximate) cordiform map is provided by Fine in the last (1550) edition of *De mundi sphaera*.

6 On craft secrecy in the Renaissance see e.g. P. O. Long, *Openness, Secrecy, Authorship: Technical Arts and the Culture of Knowledge from Antiquity to the Renaissance* (Baltimore, 2001).

7 J. M. Siliceo *Ars arithmetica* (Paris, 1514). This book is an abacus treatise, the frontispice of which was engraved by Fine. Fine also produced a second edition of this work in 1519, on which see Cifoletti in this volume. A. Ricius, *De motu octavae spherae et canones super instrumentum luminarium* (Trino, 1513).

8 G. Peurbach, *Theoricae novae planetarum id est septem errantium syderum...nuper summa diligentia Orontii Finei emendatae* (Paris, 1525); G. Reisch, *Margarita Philosophica* (Basel, 1535).

9 See Euclid, *Opus Elementorum Euclidis in geometriam artem* (Venice, 1482), Preface.

used. As a result of his experience (to which his longevity added much), Fine's abilities seem to have been in increasing demand among editors, a fact which is reflected by the pervasive use of the 1534 cordiform map in travel and cosmography books.[10]

The 1510s, a decade of apprenticeship for Fine, witnessed the emergence of codices of folded plates, with no less than three editions of Ptolemy's *Geography* between 1505 and 1514. This book indeed played a seminal role in the upsurge of interest in what was then called 'the Sphere', as it expounded the benefits of geometry for measuring and drawing the shape of the earth. The sphere was the main frame of the teaching of astronomy then provided at university, especially at the Sorbonne. Astronomy, as a liberal art inside the quadrivium, was operative in medicine through Galen's doctrine of humours and correspondences.[11] In most universities, the teaching of the Sphere was based on an authoritative sixteenth-century course-book, namely Sacrobosco's *Tractatus de sphaera*. This widely used book starts by teaching that the earth, like the seven concentric orbs surrounding it, is spherical. It then defines those great circles which divide the heavenly sphere, and, in a corresponding manner, the terrestrial globe: the equator, meridians, and colures, the ecliptic tilted by 23° 30' on to the equator; then the secondary circles: tropics and polar circles, dividing the earth into five zones (or 'belts'): two glacial zones surrounding the poles, one torrid zone surrounding the equator, and two temperate zones in between. A 'theory of habitations' introduced the student to the concept of climates: parallel circles spaced according to the mean duration of day on equinoxes under these latitudes. These very definitions were the starting point for Fine's *Protomathesis* and *De mundi sphaera*, as they were endemic in post-mediaeval encyclopaedias such as Reisch's *Margarita*.

Yet by 1500, Sacrobosco's manual, with his quotations of ancient poets and mnemotechnic verses, gradually lost its audience among scholars (although it did not disappear from the curriculum of universities until much later), for, on the one hand, the theory of zones was contradicted by the transgression of the torrid zone by Portuguese sailors and, on the other, mapping sea-routes and coastlines on a spherical model of metal was a cumbersome and costly procedure, which enabled comparison of two routes but did not allow for the measurement of a particular distance. By way of contrast, the *planisphère*, as a plane map of the full sphere, or of a part of a sphere, allowed for a measure of distances using a pair of compasses and a straight edge, provided that a true mapping, that is a mapping preserving respective distances, could be devised.

[10] A. F. Johnson, 'Oronce Finé as an Illustrator of Books', *Gutenberg-Jahrbuch*, vol. 3 (1928), pp. 107–109.

[11] See G. Aujac, *Claude Ptolémée, astronome, astrologue, géographe: connaissance et représentation du monde habité* (Paris, 1993), p. 76. In Early Modern Europe, authors of astronomical treatises were frequently also medical doctors: Fernel, Copernicus, Rheticus to name but a few.

Ptolemy's *Geography* gives precise directions for creating such maps. His method rests on two essential ideas: first, that it is possible to represent coastlines or any line from the sphere on to a plane to an arbitrary precision, provided a definite rule is provided for mapping circles from the sphere (such as equator, tropics, zones and climates, meridians) on to the plane; second, that the circular cone can be unfolded on to a plane. Unfortunately, Ptolemy's *Geography*, like many books from Antiquity, is deprived of all images as it only resurfaced in western countries towards the end of the Middle Ages.[12] In the case of Ptolemy's treatise, however, a reconstruction of the original plates can be attempted since not only has the Alexandrian geographer precisely expounded his methods for drawing maps (Book 1), but he has managed also to list the totality of towns, cities and places with their geographic coordinates (this cumbersome list actually fills Books 2 to 7 of the treatise). Reconstructing and redrawing the complete apparatus of plates mentioned by Ptolemy in his *Geography* (consisting in one world map and twenty-six regional maps) was a challenging task for editors and engravers of the early days of printing, all the more so as the reports and sea-measurements provided by explorers raised new problems, both practical and theoretical.

However, shortly after 1510 the deviations observed between the *Geography* and the sea-charts drawn by sailors led mapmakers to pair the twenty-seven 'historical maps' (Ptolemy's reconstructed maps) with 'modern' maps, while in the 1511 Venice edition some improvements were made directly to the historical maps (which numbered twenty-six as the opening planisphere was extended to encompass 320° instead of Ptolemy's 240° fan), and Waldseemüller's 1513 Strasbourg edition supplemented the twenty-seven historical 'Ptolemean' plates with a set of modern, corrected maps.[13] Oronce Fine was fully involved in the efforts to restore and update Ptolemy's charts, through the design of new surveying instruments (his first device is described in 1525 in the *Aequatorium*),[14] and through the publication of tables of revised geographical coordinates for sea-ports or cities.[15] The minute plotting of distances on maps seems to have been a particular

12 The same can be said for Vitruvius's *De architectura.* The first printed latin edition of Ptolemy's *Geography* is that published in Vicenza (1475), based on Jacopo d'Angelo's Latin manuscript (1409). The Bibliothèque Nationale de France holds an example of the Ulm edition (1482), illustrated with cartographic plates in colour. For further details of the various editions of the *Geography*, see Aujac, *Claude Ptolémée*, pp. 173–178.

13 B. Sylvano of Eboli, *Claudii Ptholemaei Alexandrini liber geographiae* (Venice, 1511). Woodcut, printed in two colors, 42 × 56.5cm. Copy consulted: Osher Collection, University Library of Southern Maine.

14 Fine, 1526.

15 These corrected tables of sundry western European cities are to be found partly in the 1530 edition of *De mundi sphaera*, and partly in the *Prothomathesis* (1532). They were not re-issued in the last (1550) edition of *De mundi sphaera.* See F. de Dainville, 'How

concern to cosmographers through the 1520s, for in that decade Jean Fernel measured the length of a meridian between Paris and Amiens, the results of his survey being published in his *Cosmotheoria* (1528). The care that Fine took in collecting data for his maps can be ascertained by comparing his *Nova totius Galliae descriptio* to the later map of France drawn by Jolyvet, and to a modern map.[16] In this respect, François de Dainville concluded in 1970 that, while Fine had undoubtedly recorded Fernel's data in his map, his plate also reflected a greater emphasis on towns' placement rather than on other geographical data (coastlines, rivers, etc.). Dainville also suspected that a method for longitude-finding based on lunation was used.[17] All in all, while the mistakes of previous maps (especially for the towns of southern and south-western France) are repeated in Fine's map of France, the relative distances between towns in eastern France (Dauphiné, Savoy, Burgundy and Lorraine) are quite accurate.[18]

As has previously been highlighted, a further benefit of Ptolemy's *Geography* is that it provides procedures for building plane maps of the terrestrial globe. Plane maps allow for easier reading of distances than a metal globe (to certain scaling conventions). Moreover, they are more convenient; but for all this, mapping on a plane the meanderings of coastlines and rivers from the sphere is a considerable mathematical challenge, since a sphere cannot be unfolded onto a plane without some tearing. Fine emphasises this in his *Protomathesis*:

> If finally a drawing of the sphere in full should help, you would have to perform this using two hemispheres, and likely circle projections: for it is impossible to encompass the full content within a single plane figure without strain, or without the earth's magnitude undergoing loss of proportions.[19]

Having himself noted this troublesome situation, Ptolemy elaborates, in the *Geography*, an angle- and distance-preserving map from optical considerations, with the extra condition that meridians and parallel circles are transformed into a network of straight lines and circular arcs only. Ptolemy essentially provides two methods, which were common knowledge to mapmakers in the era of Oronce

---

did Oronce Fine draw his large map of France?', *Imago Mundi*, vol. 24 (1970), pp. 49–55, at p, 49.

16 J. Jolyvet, *Nouvelle description des Gaules avec les confins d'Alemaigne et Italye* (Paris, 1578). Copy consulted: Archives Nationales, Paris, ref. NN 9/1.

17 See Dainville, 'How did Oronce Fine draw his large map of France?'.

18 The increase in precision is also noticeable for Picardie. This is probably due to Jean Fernel's survey between Paris and Amiens, even if Fine had made several journies to Noyon (as mentionned by Charles de Bovelles in his preface to the *Géométrie practique* [1542]).

19 'Tandem si juvet integrum Orbem delineare, id duobus hemisphaericis, & similibus circulorum projectionibus absolvas oportet : nam unica figura plana totam habitabilem compredehendere, absque difformittate, ipsiusve Telluris disproportionata magnitudine, est impossibile.' Fine, 1532*b*, p. 154.

Fine: as such, it is useful to explain these methods here. They are, as we shall see, the basis for Fine's mapmaking science as expounded in the *De mundi sphaera*.

Both methods are derived from the fact that (in contrast to the sphere) a straight circular cone can be exactly unfolded onto a plane: if such a cone is cut along a generatrix, the generatrices meeting in the vertex are mapped in the plane as lines converging to a point; further, parallel circles centred in the cone's axis are mapped as arcs of concentric circles (Fig. 9.2). Ptolemy's maps are obtained by applying the northern hemisphere to a straight circular cone, its axis being the sphere's polar axis, and its base being the equator of the sphere (Fig. 9.3). The cone thus cuts the sphere through the equator and through some parallel circle. Ptolemy's first system applies the following rules: the zero-degree meridian is to be mapped onto a vertical (straight) line, while side meridians are mapped as slant lines, meeting at the pole of the cone. Parallels are mapped as arcs of concentric circles, their respective distances being preserved to a certain scaling factor.

This mapping preserves distances along two special parallels only (the equator and the contact parallel between cone and sphere), and further proves inefficient for mapping the southern hemisphere, since distances along parallels are exaggerated. Ptolemy's first solution was to map the southern hemisphere using a second cone, symmetrical with respect to the equator: but this leads to the skewing of meridians along the equator (Fig. 9.4). Ptolemy's second system was designed to preserve distances along the zero-degree meridian (mapped as a straight, vertical line), and along parallels also. The parallels are still mapped onto concentric circle arcs, but with specially arranged radii, which leads to curve meridians, as shown by the following map (Fig. 9.5). In the age of Fine, Ptolemy's two systems were already in use, though only for latitudes between 35° south and 63° north. Waldseemüller's attempt to encompass 360° in longitude made it obvious that meridians displayed a change in curvature and thus could not be drawn as plain circle arcs. Fine was obliged to teach mapmaking according to these premises.

II. *Fine's Teaching*

Oronce Fine's printed text-books should provide more insight into his crafts and knowledge than a map, since in these works the author was free to explain his geometrical constructs at full length. Yet in Fine's case these books tell us little, for both the *De mundi sphaera* and the *Protomathesis* consist, for their cartographic chapters, in simple basic directions, without mathematical arguments. To some scholars, this reflects the mediocrity of Fine's mathematical skills.[20] Such a con-

20 Thus for Emmanuel Poulle, 'Fine's scientific work may be briefly characterised as encyclopaedic, elementary, and unoriginal. It appears that the goal of his publications, which range in subject from astronomy to instrumental music, was to popularise the university science that he himself had been taught.' E. Poulle, 'Fine, Oronce', in C. C. Gillispie (ed.), *Dictionary of Scientific Biography*, 18 vols. (New York, 1970–1990), vol.

clusion, however, should be nuanced, as teaching by directions is none other than Ptolemy's own style of teaching. By imitating his forerunner, it might be argued that Fine wished to conform to the cosmographer's ethos; or perhaps, too, that he was an adept of the method of teaching by practical examples.

Whether in the *De mundi sphaera* or in the *Protomathesis*, Oronce Fine proposes five cartographic methods to students (Table 1), only one of which is clearly rooted in Ptolemy's *Geography*. Furthermore, these maps all stem from different geometrical ideas. Nonetheless, each of these map-making schemes can claim to be rational, as the author duly notes: 'We are to show how [...] to map rationally the net of parallels and Meridians onto the Plane.'[21] Now, as the meaning of 'rational' has to be questioned, a review of the cartographic systems recommended by Fine should be undertaken.

Area to plot	**Map projection**
Regional map (10°×10° area)	Trapezoidal map
Eighth of globe	Octant map
Half globe	Equatorial stereography
More than a half globe	Oblique stereography
Full globe	Two hemispheres using stereographic maps

Table 1 – A Map Projection for Each Scale: or Fine's Cartographic Grammar.

II.1. *Trapezoidal Map*

For this map, using Ptolemy's first projection, a 10°×10° region is mapped on the plane as a quadrilateral, the upper and lower side being circle arcs and the lateral sides being line segments (as can be seen in Fig. 9.4). Moreover, because of the properties of the circular cone used in this scheme, the circle arcs are concentric and the sides, like any meridian, meet at the mapped pole. For drawing regional maps, however, Ptolemy advocates replacing circle arcs with straight lines, provided that arc length is preserved for the upper and lower parallel bounding the map, which yields a trapezoid (Fig. 9.6).[22] Fine starts his chapters on cartography by explaining this accurate and easy-to-draw method, which had been systematically used for Renaissance editions of the *Geography* (such as the 1472 Germanus edition, for instance).

---

15 (Supp.), pp. 153–157.

21 'Superest ostendere qualiter [...] Meridianorum et parallelorum contextura in plano rationaliter extendatur.' Fine, 1532*b*, p. 155.

22 '[...] nec procul a veritate fiet sicut in inicio operis diximus si pro circulis rectas lineas describemus. Praeterea particularibus in tabulis annotabimus meridianos ipsos non inclinatos et flexos, sed invincem æque distantes.' ['We shall not drift astray from truth, if, as we have pointed out in the beginning of this treatise, we draw straight lines instead of circles and besides for regional maps, meridians which, instead of converging, are parallel to each other (...)'.] Ptolemy, *Cosmographia* (Bologna, 1462), book 7, chapt. 1, l. 6. Latin translation by Jacopo d'Angelo.

II.2. *Octant Map*

Fine devotes a chapter to the mapping of the eighth of a sphere (an octant), encompassed between the equator and meridians meeting at a right angle. This area is bounded by three circle arcs of equal length, which are perpendicular to each other. A natural plane mapping might consist in representing these three lines by the three sides of an equilateral triangle, but the parallels could not be mapped as straight lines without losing distances along these lines. Instead, Fine maps the three lines as three circle arcs meeting at 120°. The parallels are mapped as concentric circle arcs, which preserves distances along these lines. Fine then enforces isometry along meridians using regular divisions along these parallels, and joining these divisions. The distances along meridians are however only approximate, since the length of arc AB representing the 45° longitude-meridian (see Figs. 9.7 and 9.9) is to the length of central meridian AD, in the ratio $\pi/3 \approx 1{,}05$ instead of 1, which yields an error less than 5 per cent anywhere in the map for the distances.[23] This mapping, which applies only to an octant, preserves simultaneously distances and convergence of meridians. Fine seems to be the inventor of this method, as he was the first to publish it, writing in the *Protomathesis*:[24]

> But if inside a given circle you describe (keeping the opening of the compasses unchanged) a curvilinear equilateral triangle, and choose one side as the Equator, and the opposite vertex as one of the two poles, the quarter of meridians meeting in this pole; if, further, you describe the particular parallels, as arcs meeting them at a right angle: this will yield a network of indistinct parallels and meridians, which will conform to the spherical globe, and onto which the eighth part of the inhabitable world can be mapped.[25]

No map of this sort drawn by Fine is known: the first attempt at using this method

[23] The meridians are thus mapped as special, non circular curves: they can only be drawn through a point-by-point (i.e. approximate) process. Fine, in his customary authoritative speech, does not dwell on this difficulty. With modern notations, assuming that the extreme meridians of the octant have – 45° and 45° longitude, the spiral curve representing the Ø meridian has the polar equation $r = 2 \times \cos\left(45° \times / + 30°\right)$, which coincides with a circle when Ø = 45°.

[24] This view is supported by Snyder who, however, points out that the octant fan was already drawn (but without grid) in the notebooks of Leonardo da Vinci. See J. P. Snyder, *Flattening the Earth: Two Thousand Years of Map Projections* (Chicago and London, 1993), p. 40

[25] 'Quod si intra datum circulum, curvilineum æquilaterumque (invariato circino) descripseris triangulum, atque unum ejus latus quadranti Æquatoris, oppositum vero punctum alterutro polorum deputaveris, & in ipsum polum convenientes Meridianorum quadrantes, propriosque circunlineaveris parallelos, sese mutuo in 90 gradus intersecantes: resultabit eorundem Meridianorum atque parallelorum haud dissimilis contextura, quæ super globo contingit sphaerico, & in qua describi poterit octava pars ipsius habitabilis orbis.' Fine, 1532*b*, ' De cosmographia', book 5, chapter 7, p. 154.

was published in 1556, in a collection of sea charts by the Normandy seaman Le Testu (1509–1573), once a student of Fine. However, as we shall see, this partial map is in fact a by-product of a more general scheme.

II.3. *The Equatorial Stereogram*

For mapping a hemisphere, Fine suggests the following template :

> It finally remains to show how the net of parallels and meridians of a hemisphere is to be unfolded onto a plane. Now let us describe a circular meridian ABCD, which is divided into four quarters by its two perpendicular diameters AC and BD, each quarter in turn being divided into 90 divisions the usual way; further, let BD be a half Equator, and AC a Meridian passing through the World's axis, and let A and C be the poles of the World. Then apply a ruler across pole A and every division (or every five divisions) of half-circle BCD, and mark up the circles' intersections by the ruler with BD the Equator. The same way, apply the ruler across point B and every division (or every five divisions) of half-circle ADC, thereby dividing line AC. Which things completed, you will describe geographic parallels joining poles A and C, each one through a division of Meridian AC associated with a division of circle ABCD; the centres of these never depart from straight line AC, which is to be extended in both directions for the purpose. You will then describe Meridians, each one rising from a division of the equator and meeting to poles A and C: extend straight line BD in both directions, for the centre of each meridian is to be sought there. Always describe the meridians and parallels by pairs with the same radius. Finally plot both tropics, and both polar circles, together with the proper figures. Having prepared this network, choose at your guise the zero-degree meridian, and by the way cover the map with the corridors of winds: for this very network of circles is ideally suited to hydrography. We leave the rest to your own skill.[26]

26 'Superest tandem ostendere qualiter hemisphaerica meridianorum atque parallelorum contextura in plano rationabiliter extendatur. Figuretur itaque circulus meridianus ABCD, binis dimetientibus AC & BD, in centro E sese orthogonaliter dirimentibus in 4 quadrantes, & quadrans quilibet in 90 gradus solito more distributus; sitque BD recta dimidius Æquator, AC vero Meridianus in rectum axis mundi coextensus, & ipsa A & C puncta, Mundi poli. Applica deinde regulam ex polo A per singulos, vel quinos tantummodo gradus dimidii circuli BCD; & nota circulas intersectiones ejusdem regulae in Æquatore BD. Haud dissimiliter applicata ex puncto B regula, per singulos aut quinos gradus ipsius ADC semicirculi, dividito rectam AC. Quibus absolutis, circumlineabis circa polos A & C geographicos parallelos, per singulas divisiones ipsius AC Meridiani respondentes ABCD circuli partitiones coincidentes; quorum centra non discendunt a recta AC, quæ propterea in directum utrobique venit extendenda. Lineabis consequenter Meridianos, per singula Æquatoris BD distinctiones, in utrumque polum A & C convenientes: producta in directum ex utraque parte recta BD, in qua singulorum meridianorum centra veniunt investigenda. Delineabis autem semper duos aut meridianos aut parallelos, eadem apertura circini. Inscribes tandem tropicos, una cum polaribus circulis, & propriis longitudinum atque latitudinum numeris. His ita paratis, imponito quam volueris ipsius orbis medietatem, & simul exarato ventorum

This construct (see Fig. 9.8) is based on stereography, a mapping method ascribed to the Greek astronomer Hipparchus. Stereography is a one-to-one mapping of a plane tangent to a sphere, onto the sphere itself. It consists in joining by a straight line each point with that peculiar point of the sphere which is opposed to the point of tangency (see Figs. 9.10–9.11); a pairing of each point of the sphere with a point of a plane is then achieved.[27] This mapping possesses the crucial property that any circle on the sphere is mapped as a circle (or possibly a straight line, i.e. a circle of *infinite* radius) in the plane. The directions provided by Fine show how to implement this projection using a minimum of lines. Of interest is the fact that, instead of using the plane tangent to the sphere, he draws divisions along the parallel plane passing through the centre of the sphere, a practice which takes advantage of the property of similar triangles.

In Fine's era, stereography was of common use for the engraving of astrolabes. Ptolemy had described the equatorial projection in a book called *Planisphaera* which has only survived in Arabic language, and which is the likely source for the design of astrolabes in the Arabic world.[28] However, for chronological reasons, Fine certainly did not know this book when he wrote the *Protomathesis*. The polar stereographic projection seems to have first been used by one Walter Lud (1448–1525) for printed maps in Waldseemüller's edition of Ptolemy's *Geographia* (1507). It was copied, together with Lud's comments about the design of the map, in the 1512 edition of Reich's *Margarita philosophica*, with which Fine was acquainted (see Fig. 9.11).[29] It can thus be conjectured that Fine only became familiar with it around that date.

II.4. *Oblique Stereography*

In the 1542 edition of his *Protomathesis*, Fine endeavours to teach how a new map (*nova orbis designatio*) encompassing more than a hemisphere (*plus quam dimidiam comprehendens sphaeram*) can be achieved:

> Lastly another sketch of the terrestrial globe should be added which, failing to encompass the whole sphere, nonetheless enables mapping more than a

lineamenta: nam hæc geographica circulorum contextura, hydrographiæ videtur admodum commoda. Reliqua tuo submittimus ingenio discutienda.' Fine, 1542, book 5, pp. 155–156.

27 Stereograms are obtained by mapping points on the sphere on to a plane tangent to the sphere. A ray emitted from S crosses the sphere at some point M and maps it onto the tangent plane in a point M'. Point T is mapped to itself, whereas point S cannot be mapped (but is of no practical use).

28 The Arabic manuscript of *De planisphaera* was not translated in Latin until the fiteenth century, and was only printed in 1536 (in Basel, translated by Hermannus Secundus).

29 See Pelletier, *Cartographie de la France et du monde*, p. 14. Stereography had already been used by Behaïm (1492). See J.-M. Besse, *Les grandeurs de la Terre: aspects du savoir géographique à la Renaissance* (Lyon, 2003), p. 101.

half of it onto a plane, and preserves for the eye the circle of its outline. Let a circle of arbitrary magnitude be described, which two perpendicular diameters will part into four quadrants, as does for example the tropic of Capricorn, which is to be drawn in the plane; out of these diameters, one stands for the meridian, the other for the horizon. Let both the equator and tropic of Cancer be drawn inside the circle with respect to the proportion of scales, as is taught by the making of the planisphere. Let us then draw the Ecliptic, the north pole of which will be placed along the mid-line. Equator, tropic of Cancer and line of horizon are to be drawn with such care as to be erasable later. From then on, let the ecliptic be divided into twelve signs, with 30 degrees assigned to each sign, or equivalently six divisions of 5 degrees each; great circles rise from the pole of the ecliptic, cross through a division each and terminate to the tropic of Capricorn. Let parallel circles be described about the ecliptic, with a mutual distance of one single degree, or 5 degrees: this way, thanks to the making of the planisphere or thanks to our directions, you will be able to gather them easily. Having prepared this net, let the pole of the ecliptic stand for the north pole of the world; the circles which take rise from it stand for the meridians; finally, the ecliptic itself will stand for the circle of the equator, and the equidistant circles about the ecliptic will beautifully feature the parallels of places. First mark the longitudes of places from the mid line on the longer end of the equator's circle on the right; and finish up drawing the rest, as art commands. This short exposition we have thought through at length while this book was under press, and it has not seemed useful to say more about it, as it can be easily understood, partly through the astrolabe's or planisphere's making, and partly through our own directions.[30]

30 'Aliam tandem juvat superaddere ipsius terrestris orbis designationem: quae etsi non totam, plus tamen quam dimidiam sphæram in plano coextensam, non ineleganter comprehendit, & rotundam oculis gratissimam figuram observat. Describatur igitur circulus quidam liberae magnitudinis: qui binis dimentientibus ad rectos sese dirimentibus angulos, in quatuor quadrantes dividatur, instar videlicet tropici Capricorni, quem in planisphaerio delineare solemus: quorum dimentientium, alter meridianum, alter vero rectum imitatur Horizontem. Intra quem circulum, Æquator una cum Cancri tropico (ut in ipsius planisphaerii docetur compositione) proportionaliter figuretur. Postmodum ipsa delineetur Ecliptica: cujus polus Septentrionalis, in linea notetur meridiana. Debent autem aequator, & Cancri tropicus, atque linea Horizontis recti, adeo subtiliter pingi: ut deleri facile possint. Haec consequenter ecliptica in duodecim signa, & signum quodlibet in 30. gradus, vel in 6. partes aequales, quárum quaelibet 5. gradus repraesentabit: per circulos magnos, ex ipsius Eclipticae polo egredientes, & in Capricorni sese diffundentes tropicum, dividatur. Eidem rursum Eclipticae, aequidistantes seu paralleli circumscribantur circuli, singulis, aut quinis tantum gradibus distributi: quemadmodum ex ipsius planisphærii, atque directorii nostri constructione, colligere vel facile potes. His in hunc modum præparatis, polus Eclipticæ, Mundi polum Articum, & prodeuntes ab eo circuli Meridianos, ipsa vero Ecliptica circulum Æquatorem, & eidem Eclipticæ aequidistantes locorum parallelos pulchre repraesentabunt. Initiabis demum locorum longitudines a linea recta Meridiana versus dextram, in longum Æquatoris circuli: & absolves reliqua, velut ars ipsa requirit. Hujus autem brevitatem, quo dum haec imprimerentur, illam excogitavimus: neque multum id nobis visum est necessarium,

This mysterious plotting (of which no picture is provided) must refer to an oblique stereographic mapping. In essence, the technique is the same as the one preceding, save that it is applied to a tilted globe (see Fig. 9.12). Thus, the equator is no longer mapped as a straight line, but as a circle. The interpretation of Fine's directions for oblique stereography stems from the fact that circles are mapped as circles, and, more specifically, that the tropic of Cancer is a straight line. This situation is reached only when the tilted sphere assumes the position of the earth, that is with an angle of 22° 30', a natural choice indeed. This angle of inclination of the earth in space had been previously discussed in the *Protomathesis* and deemed to be 22° 30'.

Now, a closer reading of Fine's directions reveals that the reader is left with several difficulties: some crucial steps like positioning the pole, drawing the ecliptic and parting the meridians are hardly explained. There, the reader is advised to turn to 'the planisphere's making' or alternatively to 'the doctrine of the astrolabe'. These short cuts, for which a pitiful excuse is provided (the book is under press) were not meant to really teach, but were more likely intended as milestones set out for some future claim for priority: obviously, the cartographer did not wish a challenger workshop to use this mapping technique.[31]

II.5. *A Grammar of Map-making*

The striking fact regarding the four techniques provided by Fine's *Protomathesis* is that they are based on entirely different mathematical ideas: the trapezoidal map stems from the unfolding of a cone tangent to the sphere; the octant map takes advantage of some symmetries to frame an orthogonal grid of lines; stereography is a circle-preserving sphere-to-plane mapping. Furthermore, Fine's *De cosmographia* far from provides full coverage of the mapmaking techniques of the day: the orthographic projection, for example, which maps the hemisphere onto a rectangular grid using latitude and longitude as coordinates, and which had already been applied in Antiquity by Marinus of Tyre, has been discarded.

The question then arises: why did Fine select these four particular mapping techniques for teaching? The answer may lie in a consideration of the features common to all four techniques: first and foremost, the meridians always verge to

cum illa partim ex Astrolabi sive planisphaerii, partim vero ex directorii nostri compositione deprehendi vel facile possint.' Fine, 1542, p. 156.

31 Though an old token of astrolabe construction, oblique stereography was revived in the late fifteenth century by one Johannes Werner, in *De quatuor orbis terrarum in plano figurationibus* (1514). However, Werner, like Fine, only explained how to frame the network of parallels and meridians: oblique stereography would not be used for maps until the works of Jacques de Vaux (1583) and John Blagrave (1596). According to Snyder, Johannes Werner (1468–1528) studied under the Austrian humanist Stabius, a professor in Vienna and close friend of Pirckheimer and Conrad Celtis. See Snyder, *Flattening the Earth*, p. 27.

the poles, a practical fact which sums up the very nature of a pole, and was a compulsory feature of geographical maps to Renaissance minds (indeed, it echoes the winding paths taken by Renaissance painters to contemplate using a point at infinity in perspective representation). For meridians, when mapped as parallel lines (as is the case in orthographic or in the later Mercator maps) imply the idea of an infinitely remote pole, by then a brain-twisting novelty. Second, the four techniques keep parallel and meridian lines meeting at a right angle. Third, these maps can be drawn using exclusively straight lines and circles, which refer to the geometric tradition of 'ruler-and-compass' methods widely practiced throughout the Renaissance. Other features, such as trustworthiness according to length scale, or conformity of angles, seem to have possessed lesser significance in Fine's mind, since not all four methods comply with them accurately. Indeed, despite a parsimony in mathematical content, Fine's grammar of maps appears to be based on definite principles and in this respect brings a clear improvement on Ptolemy's achievements.

III. *Fine's Engraved Maps*

Fine, however, applied none of the aforementioned methods to the design of his engraved plates. As has already been mentioned, early on in his life Fine was involved in the re-printing of sundry mathematical books, almost certainly as an engraver, and pursued this activity throughout his career.[32] Engraved maps were meant to be leased to printers for the purpose of editing a number of different books; as such, for Fine as for other mathematicians, they would naturally provide an alternative source of income, and one which would sometime benefit the engraver for decades.[33] Thus, between 1530 et 1536, Fine designed and engraved an important series of world maps.

The first one, bearing the title *Nova et integra orbis descriptio*, is dated 1531. It is a double heart-shape map, each 'heart' mapping a hemisphere (a 180°-sector in longitude). The two heart-shape maps are tangent at some point on the equator, and the zero-degree meridian crosses the Caspian sea (*Hyrcanum*) in the middle, a choice which centres the map on the ancient world and splits apart the east and west coastlines of North America. The second one, and the last now

32 In addition to the plates previously engraved, Fine reprinted Reisch's *Margarita* in 1535, Georg Peurbach's *Theoricae novae planetarum* in 1543, while Charles de Bovelles warmly thanks him for his help in the preface of the 1542 reprint of his *Géométrie practique*.

33 Suffice it to cite here Geoffroy Tory, whose polyvalence mirrors Fine's decades later. An architect like Salomon de Caus would go so far as to gain a privilege for the plates illustrating his books, so as to maintain exclusivity. See A. Marr, '"A Duche graver sent for": Cornelis Boel, Salomon de Caus, and the Production of *La perspective auec la raison des ombres et miroirs*', in T. Wilks (ed.), *Prince Henry Reviv'd: Image and Exemplarity in Early Modern England* (London, 2008), pp. 212–238.

extant, is dated 1534 (but was in fact published by Grynaeus only in 1536) and is entitled *Recens et integra orbis descriptio* (Fig. 1). Still a heart-shaped map, it now successfully encompasses a full 360° equator within a single map. In his 'address to the reader' (*Orontius Delphinatis Studioso lectori*), Fine provides an account of the publication:

> It has nearly been fifteen years, impartial reader, since I first rendered the aspect of the whole Earth in this picture of a human heart: and this thanks to our lenient Maecenas, the most Christian and almighty Francis, King of France. As I realised that this description of the Earth charmed the King, himself a learned man and outstanding Geographer, and that everyone (even foreigners) would praise it: I came to wish sharing this picture of the Earth with all those who indulge themselves in mathematics. Which I succeeded into at my own expense, after sundry tribulations and interruptions of my studies (which until now would hinder me). It is therefore with a sense of wisdom and generosity that I offer you, careful reader, this geographic heart-shaped map, now corrected and augmented through several most recent remarks of hydrographers. You are only left with appreciating the pain and labour I have taken, with a light, fair and benevolent spirit. Finally, take advantage of this desirable map, which we have brought to you through the favour and generosity of our almighty and most Christian King.[34]

As there is no other evidence of a heart-shaped map drawn by Fine prior to 1520, the starting date (1520 or 1531) is an open question.

A heart-shaped map turns out to be the ultimate consequence of Ptolemy's second system. Recall that this second system is based on the following premises: first, that distances are preserved along the zero-degree (or central) meridian (mapped as a straight, vertical line), and along parallels too; second, the parallels are mapped onto concentric circle arcs centred around the north pole. As a result, this second Ptolemaic system preserves distances all across the map, as Fine observes:

[34] 'Decimus Quintus Circiter agitur annus, candide lector, quo universam Orbis terrarum designationem in hanc humani cordis effigiem primum redegimus: idque in gratiam Christianissimi ac potentissimi Francisci Francórum regis, Maecenatis nostri clementissimi. Quam dum videremus ipsi Regi Polyhistori ac non vulgari geographo, valde placere, ab omnibus quoque (etiam exteris) laudari plurimum: desiderabam eandem Orbis descriptionem, universis mathematicorum studiosis aliquando communicare. Quod, post varia fortunae, ac studii nostri (quae hactenus nobis impedimento fuere) discrimina, tandem nostro effecimus periculo. Itaque, plurimis recentiorum hydrographorum observationibus auctam, & emendatam ipsius geographici cordis imaginem, tibi studiose lector, cunctisque bonae voluntatis hominibus, cordato ac liberali, praesentamus animo. Reliquum est igitur, ut hunc laborem nostrum & industriam, humano vultu non graveris accipere, & aequi bonique consulas: Ipsi demum Christianissimo ac magnifico Regi nostro, prosperam exoptes faelicitatem, cujus favore atque munificentia, haec (interea dum molimur graviora) tibi communicavimus.' O. Fine, *Recens et integra orbis descriptio* (Paris, 1534).

> Thanks to this map, it is possible to find almost exactly the distance between two places of given latitude and longitude (provided they do not exceed 90 degrees). For having numbered out the latitudes and longitudes of both these points, or having plotted these places on the map: put one leg of a compasses onto one of the points, and the other tip to the other point. Then keeping the opening of the compasses unchanged, align the two legs along that line which parts the map into two halves, and which is divided along with degrees: and count up the degrees between the legs of the compasses. Multiply this number by 62 if you need [Roman] miles, by 31 if you need French leagues, by 20 for common leagues: you will obtain the distance between the two places.[35]

In the process of extending Ptolemy's 180° aperture map to 270°, or even 360°, the changing curvature of meridians becomes more and more obvious (and the furthermost meridian produces a 'heart' shape). Waldseemüller's 1507 attempt at extending Ptolemy's map to 180° in longitude is based on three-arc meridians, that is to say meridians made of three arcs of circles with different centres and radii, patched together, so as to comply with the prescribed divisions along the parallels. This map (see Fig. 9.5) already outlines part of a heart.

Three works, in fact, exhibit an unambiguous heart-shaped map prior to Fine's 1531 attempt. The oldest is a map of the world due to Bernardus Sylvanus of Eboli (1511).[36] It seems to be purely graphical (no explanation is provided as to how it was designed), displays some approximate features, and does not seem to have reached a considerable audience (Fig. 9.13).[37] The second is a treatise on mapmaking (*De quatuor orbis terrarum in plano figurationibus,* 1514) by Johannes Werner, which we have already mentioned in relation to oblique stereography. A theoretical work, it did not gain wide recognition.

The third, Peter Apian's stand-alone engraved map *Tabula orbis cogniti* (1530), appeared almost at the same time as Fine's *Nova et integra orbis descriptio*. Though Apian's source for cordiform maps is still debated, Mangani recently pointed to the knowledge of coastlines reflected by both maps to argue that the German and French geographers worked independently of one another.[38] Con-

35 'Ex hac plana terrarum orbis descriptione, duorum quorumcumque locorum datarum longitudinum atque latitudinum directum itineris intervallum (modo illud nonaginta non superet gradus) prope verum supputare licebit. Numeratis itaque eorundem locorum longitudinibus atque latitudinibus, eisdemve locis in Charta coassumptis: imponito unum circini pedem super altero locorum, alium vero extendito in reliquum. Dein traducito circinum invariatum in eam rectam, quae figuram bifariam dividit, & in suos gradus distributa est: & animadvertito, quot gradus capiat ipse circinus. Hos enim si per LXII miliaria, aut gallicas leucas XXXI, seu XX. Communes, quindecimve majores multiplicaveris: viatoriam eorundem locorum distantiam obtinebis.' *Idem*.

36 See Snyder, *Flattening the Earth*. p. 32.

37 Sylvanus's 1511 heart-shaped map, though purely graphical, is a clear improvement on Waldseemüller's 1507 attempt at developing Ptolemy's second system: it encompasses a 320° wide sector.

38 G. Mangani, *Il 'mondo' di Abramo Ortelio: Misticismo, geografia e collezionismo nel*

trary to Werner's and Sylvanus's maps, Fine's engraved maps were an instant success and set new standards in map-making for the rest of the century: Mercator culled the scheme of his *Orbis imago* (1538) from Fine's double cordiform *Nova et integra orbis descriptio*; the cordiform maps of Hondius (*De cosmographicis rudimentis*, 1561) and Ortelius (*Nova totius orbis terrarum ... descriptio*, 1564) are based on the doctrine put forth in the *Recens et integra orbis descriptio*.[39] Likewise, Fine's engraved maps were reprinted for two decades in numerous books.

As we have seen, Ptolemy's second system is absent from Fine's teaching, at least as it appears in the *Protomathesis*. Only in the 1555 edition does a first (and last) allusion to the construction of cordiform maps appear. Unexpectedly, the curious 'octant map' explains why (Fig. 9.14):

> Besides, we remarked that this mapping of the eighth of the terrestrial globe can easily be extended (as we have already mentioned in the second and the third section of the present chapter) to map a quarter of the globe, or a full hemisphere, either northern or southern. So, having drawn triangle ABC, standing (as we have already said) for a eighth of the globe, we are to complete equator BDC by two thirds of a circumference around pole A: for the sake of simplicity, let us denote GDEFH this Equator. Now, leaving the opening of the compasses unchanged, let us describe arc EAF around centre D: so that the four arcs GE, ED, DF, FH will be equal to each other, and will stand for the four quadrants of the equator. Next we have to divide arc BE and arc CF into as many divisions as there are in BD and DC: and keeping the same opening of the compasses as before, we are to draw the other meridians radiating from pole A, by pointing the other end of the compasses at each division along arc EBDCF in succession. We are then to continue each parallel circle from triangle ABC, together with tropic and polar circle, to their boundaries AG and AH. This way, surface AEDF will properly map a quarter of the Earth, and the whole surface AGDH a full hemisphere, whether northern or southern: which is adequate enough for what looks to conform to the spherical nature. This is obvious in our example, where it can be seen that each interval contains five degrees, regardless of direction.[40]

Rinascimento dei Paesi Bassi (Ferrara, 1998), esp. chapter 7: '*Cor cœli*'.

39 See Snyder, *Flattening the Earth*. p. 37.

40 'Hanc porró octavae partis ipsius globi terrestris descriptionem (ut ad secundum et tertium hujusce capitis institutum deveniamus) in quartam, aut dimidiam partem ejusdem globi terrestris, boream quidem vel austrinam, vel facile coextendi posse tandem animadvertimus. Descripto itaque abc triangulo, octavam partem ipsius globi terrestris (ut dictum) repraesentante, complenda sunt circa polum a duo tertia circunferentiae ipsius aequatoris bdc: voceturque idem Æquator, facilioris intelligentiae gratia, gedfh. Et invariato circino, describendus est circa punctum d arcus eaf: hic enim dimidiam partem fixi meridiani (a quo locorum numerantur longitundines) repraesentabit. Invariato rursum circino, describendus est arcus ag circa punctum e, similiter arcus ah circa punctum f: hoc enim modo, quatuor arcus ge, ed, df, fh, aequales erunt ad invicem, & quatuor Æquatoris quadrantes repraesentabunt. Dividendi sunt consequenter arcus be, atque cf in tot gradus in quot

As Johannes Werner was the first to publish explanations about cordiform nets, yet without producing any map, the originality of Fine's *Recens et integra orbis descriptio* must be called into question. In fact Fine, although he does not explicitly credit Werner for heart-shaped maps, was certainly acquainted with Werner's book, for in another passage from the *Protomathesis*, the German astronomer is praised for his skills:

> Several different observations have been made concerning the sun's maximum angle of declination. Ptolemy, following the way he has been writing about (as you can see in the first book of his great construction) found it to be 23° 51'. Afterwards Alcmeon claimed a slightly lower value: that is 23° 33'. But Peurbach asserts in the 17th chapter of the 1st book of his Epitome, that he has himself found 23° 28'. Then, recently, some highly learned Italians together with a German going by the name of *Joannes Werner*, a man as profound in philosophy as in mathematics, have said that, above 23° declination, they found precisely 29'. As for me, I agree with Regiomontanus in believing it amounts to 23° 30'. Therefore, you will discern the best of these observations: of which we now show you the art.[41]

This passage, one of the very few in which Fine condescends to quote his sources, proves that he, possibly through his connections with Italy (his birthplace, Dauphiné, was then certainly a bilingual area) and Basel printers, was quite aware of recent developments in mathematics in Italy and the Holy Roman Empire.

---

partiti sunt ipsi arcus bd, atque dc: & eadem qua prius apertura circini, caeteri meridiani ex a polo prodeuntes delineandi, traducto circini pede in singulas divisiones ipsius arcus ebdcf, suo ordine. Continuendi sunt demum in arcus ag & ah, singuli paralleli ipsius abc trianguli, una cum tropico, atque polari circulo. In hunc ergo modum descripta figura aedf quartam partem, totalis porro figura agdh dimidiam, boream quidem vel austrinam ipsius globi terrestris pulchre repræsentabit: utpote, quae ad sphaericam rationem magis videatur accedere. Quemadmodum ex ipsa figura, in exemplum depicta, sit manifestum: in qua singula intervalla, quinque gradus (ut in praecedenti) quaquaversum includunt. Poterit & eadem globi terrestris medietas, alia ratione delineari, utrunque mundi polum comprehendens: in hunc qui sequitur modum […].' Fine, 1555, Book 5, Chapter 7, pp. 55–56.

41 'De ipsius autem maximae declinationis quantitate, variae repertae sunt observationes. Hanc enim Ptolemaeus, supra scripta via (ut videre licet primo suæ magnæ constructionis) reperit graduum 23, et minutorum 51. Post quae Alcmeon, paulo minorem eandem asseruit declinationem: utpote graduum 23, & 33 minutorum. Purbachius vero testatur 17 primi Epitomatis sese hanc invenisse graduum itidem 23, sed 28 tantum minutorum. Novissime autem quidam Itali doctissimi, una cum *Joane Vernero germano*, viro in utraque lingua, Philosophia et mathematica admodum erudito, praeter 23 gradus, aiunt sese reperisse minuta 29: quorum observatio parum differt a Purbachiana. Ego autem cum Ioanne Regiomontano, existimo eam fore 23 graduum et 30 minutorum. Tu ergo veriorem omnium harum observationum diligenter examinato: ea qua nunc monstravimus arte.' Fine, 1532, Book 5, Chapter 4, p. 3.

A distinction has to be made about Fine's works: while his teaching of geography remains classical in form, his engraved plates demonstrate a far more innovative use of geometry, the practice of which may have been revealed to private students only. In the *Protomathesis*, the method for locating town places on a plane map is rooted in Sacrobosco's teaching of the Sphere, which claims that the skies (and the earth) are spherical in shape, and which divides the terrestrial globe in special circles, zones and climates. In his books, Fine's directions to the reader for the design of planispheres are basically descriptive, and deprived of theoretical insight: no proof supports the constructs, which are taught as a step-by-step process. The author's style here parallels that of Ptolemy himself in his *Geography*.

Fine's apparent disregard for mathematical consistency in map design is further supported by the lack of unity of his methods, at least from a mathematical point of view. Indeed, he relies on a different system for each scale: Ptolemy's trapezoidal mapping for regional maps stems from different mathematics than the octant map, which itself is foreign to stereography. However, these mappings share common features, which provide telling evidence of Fine's consistency in method: first, in all of the systems, distances are preserved along parallels and meridians; second, they all use only straight lines and circles; third, mapped meridians, like meridians on the sphere, meet at the pole. This property would have been lost if the orthogonal projection, already used by Waldseemüller and others before Fine, had been included in the corpus. We therefore conclude that, to Fine, a plane map had to enforce both these properties: to keep true distance scales along meridian and parallel circles, and keep meridians meeting at poles.

For drawing his world maps (or planispheres), Fine makes use of heart-shaped projections. While the principles underlying such maps had already been provided by Ptolemy, the Alexandrian master had not pushed his idea so far as to map the full globe. Moreover, in mapping the inhabited part of the world only, Ptolemy could provide only a fair approximation for the mapped parallels and meridians using circles and straight lines. For decades, his Renaissance followers – Sylvanus, Waldseemüller and others – would try to elaborate on his directions to encompass a broader area in their maps; while Werner first published an accurate description of heart-shaped maps using non-circular curves as meridians, Fine drew the first world maps based on this system. Though heart-shape projections fulfil Fine's requirement of isometry along coordinate circles and convergence of meridians, he would use these maps for printed colour plates only, and would partly explain the technique in the last edition of his *Protomathesis*. Whether Oronce Fine wished to refrain from sophistication in his books, or wished to keep his craft a secret, these maps reflect the rigour of the artist, as their design implies the use of trigonometric tables. To conclude, the characteristic features of Fine's practice of mapmaking provide an outline of his methods in one specialised area: not only do they mirror a craft held by geometry to fulfil the requirements of precise properties, but also a quest for standard representations irrespective of scale, and at the same time a conservative, traditional view of teaching.

Pedro Nunes against Oronce Fine: Content and Context of a Refutation

HENRIQUE LEITÃO

I. *Virtus virescit vulnus*

In 1546 the Portuguese mathematician Pedro Nunes (1502–1578) published the *De erratis Orontii Finaei*, a devastating critique of Oronce Fine's attempts at solving some classical problems of mathematics (duplication of the cube, squaring of the circle, etc.) and other scientific matters (Fig. 10.1). Besides exposing Fine's errors, the book contains other explanatory digressions, making it an interesting piece of sixteenth-century mathematics. The attack by Nunes was the first in a series of refutations by famous mathematicians that seriously undermined Fine's scientific reputation. The Frenchman Jean Borrel followed Nunes with books against Fine published in 1554 and 1559, Niccolò Tartaglia attacked in 1560, and Adrian van Roomen fired his salvo in 1597.

Oronce Fine was no stranger to polemics.[1] On the contrary, harsh criticism and the attacks of adversaries seem to have been frequent during his life. It is known that he was in prison for some time in 1523–1525 and that his adversaries were legion. His motto 'virtus virescit vulnus' (courage [or virtue] grows stronger from a wound) testifies to the constant attacks he suffered and to his determination not to succumb to critiques.[2] His early biographers, Antoine Mizauld, André Thevet and Bernardino Baldi all referred to the polemics in which he was involved, and more recent students of his life have consistently highlighted the polemical traits of his personality.[3] Therefore, the attack that Pedro Nunes launched against

1 For biographical information on Fine I follow A. Rochas, *Biographie du Dauphiné* (Paris, 1856–1860); L. Gallois, *De Orontio Finaeo gallico geographo* (Paris, 1890); R. P. Ross, 'Studies on Oronce Finé (1494–1555)', unpublished PhD. diss. (Columbia University, 1971).

2 The motto appears in the ornate figure in the title page of Fine's *Protomathesis* (Paris, 1532). William Cunningham, in his well known *Cosmographical Glasse* (1559), borrows from Fine's motto and includes the words 'virescit vulnere veritas' in the title-page. See W. Cunningham, *The Cosmographical Glasse, conteinyng the Pleasant Principles of Cosmographie, Geographie, Hydrographie, or Nauigation* (London, 1559), title-page.

3 A. Mizauld, 'Vita Orontii', in Fine, 1556*b*, sigs. v^{r}–vi^{r}; A. Mizauld (ed.), *Funebre symbolorum virorum aliquot illustrium, de optissimo et doctissimo viro Orontio*

him in 1546 was not something completely new, but it was surely the first fully-fledged assault on Fine's scientific credibility.

Although Nunes asserted that his intentions were strictly scientific, the timing and the general circumstances surrounding the publication of his book suggest that other factors may also have been at play. It appears that, to his advantage, Nunes targeted not only Fine's mathematics but also his fame. Indeed, that such a prominent mathematician as Oronce Fine had exposed a very weak flank was almost a blessing to a mathematician in need of establishing his credentials, as Nunes was around 1546.

II. *Oronce Fine's Work in Pure Mathematics and Its Critics*

Fine's literary and scientific production is truly impressive. He wrote, edited or was involved in the making of more than thirty different books, some of which were edited more than once.[4] Although several of his books attracted criticism, three in particular seem to have been the most problematic. These were the books that we will refer to by their abbreviated titles: the *Protomathesis* (1532), the *Quadratura circuli* (1544) and the *De rebus mathematicis* (1556).[5]

The publication of the *Protomathesis* was an important piece in the establishment of Fine's reputation and it is surely related to Fine's recent appointment to the chair of mathematics at the Collège Royal, in 1531.[6] It revealed a mathematician competent enough to address a broad spectrum of topics, mostly of applied or practical mathematics, cosmography, gnomonics, and related matters. All of these were subjects of the highest importance for sixteenth-century mathematicians and mathematical practitioners, but the book signals also Fine's higher ambitions: in it he presents his first attempt to solve one of the most famous problems of Greek mathematics, the squaring of the circle, 'solved' in the second book of the 'De geometria'. The solution is wrong, needless to say, but this does not seem to have affected Fine very much. In fact, a decade later, Fine was even surer of his talents as a mathematician. In 1544 he presented an ambitious collection of

---

Finaeo (Paris, 1555); A. Thevet, *Les vrais pourtraits et vies des hommes illustres Grecz, Latins, et Payens…* (Paris, 1584); B. Baldi, *Cronica de' Matematici overo Epitome dell'Istoria delle Vite Loro* (Urbino, 1707) (see also B. Baldi, *Le Vite de' Matematici. Edizione annotata e commentata della parte medievale e rinascimentale*, ed. E. Nenci [Milan, 1998], pp. 442–456).

4 For the bibliography of Fine's works see Gallois, *De Orontio Finaeo gallico geographo*, pp. 71–81; Ross, 'Studies on Oronce Finé', pp. 398–449; D. Hillard and E. Poulle, 'Oronce Fine et l'Horloge Planétaire de la Bibliothèque Sainte-Geneviève', *Bibliothèque d'Humanisme et Renaissance*, vol. 33 (1971), pp. 311–350; R. P. Ross, 'Oronce Fine's Printed Works: Additions to Hillard and Poulle's Bibliography', *Bibliothèque d'Humanisme et Renaissance*, vol. 36 (1974), pp. 83–85.

5 Ross lists a 1543 edition of the *Quadratura circuli*, although no copy seems to be extant. See Ross, 'Studies on Oronce Finé', p. 442.

6 On which see Pantin in this volume.

works that, among other matters, addressed the problem of squaring the circle and inscribing any regular polygon in a circumference.[7] In this new book, the *Quadratura circuli*, solving the classical problems of mathematics was not a detour but indeed the core of the whole work.

Fine's errors and mistakes in the various works that make up the *Quadratura circuli* are obvious; in most cases they are shockingly basic, bordering on the grotesque. Surprisingly though, he seems to have been very satisfied with his accomplishments and returned once again to the problem of squaring the circle in his later years. The result of these investigations was published posthumously, in 1556, in *De rebus mathematicis hactenus desideratis libri IV*, the extraordinary subtitle of which announces that he had been able to demonstrate the squaring of the circle by more than one hundred different ways: *Quibus inter caetera, circuli quadratura centum modis et supra, per eundem Orontium recenter excogitatis, demonstrator*.[8] Not surprisingly, this far-fetched claim was enough to upset several mathematicians.

Fine's various attempts to solve some of the most famous problems in mathematics, such as squaring of the circle or cube duplication, were the object of severe criticism by contemporary scholars. The first sustained critique of Oronce Fine's mathematics appeared in Pedro Nunes's 1546 work, *De erratis Orontii Finaei* which we will analyze in more detail in the next section. A few years after the publication of Nunes's book, the French mathematician Jean Borrel (1492–1572), presented his *Opera Geometrica* (1554), in which Fine is attacked.[9] This

7 The *Quadratura circuli* (Fine, 1544*c*) contains the following: (i) Quadratura Circuli, tandem inuenta & clarissime demonstrata; (ii) De circuli mensura, & ratione circunferentiae ad diametrum, Demonstrationes duae; (iii) De multangularum omnium & regularium figurarum descriptione, Liber hactenus desideratus; (iv) De inuenienda longitudinis locorum differentia, aliter quam per Lunares eclipses, etiam dato quouis tempore, Liber admodum singularis; (v) Planisphaerium geographicum, quo tum longitudinis atque latitudinis differentiae, tum directae locorum deprehenduntur elongationes.

8 The book (Fine, 1556*b*) contains the following material: (i) Liber de inventione duarum rectarum inter datas extremas continue proportionalium; (ii) Liber de ratione circumferentiae ad circuli diametrum; (iii) Liber de inventione lateris cuiuslibet polygoni regularis in dato circulo descripti; (iv) Liber de omnimoda solidorum transmutatione, cum ipsa sphaerae cubicatione, et versione cubi in spaeram aequalem.

9 Borrel gives no indication of having known Nunes's book. Relatively little is known about Jean Borrel (Buteus, Buteo). He was born in 1492, became a friar, and died in 1572. He studied mathematics and apparently attended Fine's classes in Paris. See G. Allard, *Bibliothèque du Dauphiné, contenant l'histoire des habitants de cette province qui se sont distingués par leur génie, leurs talents et leurs connoissances*, nouvelle édition revue et augmentée (Grenoble, 1797), pp. 86–87 ; B. Boncompagni, 'Intorno ad un trattato di aritmetica del P.D. Smeraldo Borghetti Lucchese, etc.', *Bulletino di bibliografia e di storia delle scienze matematiche e fisiche*, no. 2 (1869), pp. 257–269; J. Verdonk, 'Borrel, Jean' in C. C. Gillispie (ed.), *Dictionary of Scientific Biography*, 18 vols. (New York, 1970–1990), vol. 2, p. 618.

was followed by another book in which Borrel uses a much more aggressive tone: *De quadratura circuli libri duo* (1559). Borrel, a competent if only modestly talented mathematician, was especially upset by two facts: that Fine had insisted on errors made years before and, most of all, that Fine had utterly corrupted Archimedes's brilliant approach to the problem of squaring the circle.[10]

A few years later, in Part IV of his *General Trattato* (1560), the Italian mathematician Niccolò Tartaglia introduced several critical remarks directed against Fine's mathematics. Tartaglia followed other adversaries of Fine, noting that the French mathematician always wanted to give the impression that he knew much more than he really did and that many of his demonstrations were ridiculous.[11] Finally, by the end of the sixteenth century, one other refutation of Fine was published by Adrien van Roomen in the *Exercitationes cyclicae contra Iosephum Scaligerum* (1597).[12]

Nunes, Borrel, Tartaglia and van Roomen were undoubtedly the most important critics of Fine, but other scholars expressed their deep repulsion at his lack of mathematical knowledge. For example, on 8 August 1556, the Italian mathematician Francesco Maurolico, in a letter to his protector, the Vice-Roy of Sicily, Juan de Vega, vented his fury against Fine in the most vehement terms. Like several others before and after him, Maurolico was angry on account of Fine's incapacity to grasp the subtlety of Archimedes's procedures, thus presenting a distorted caricature of Archimedes's methods.[13]

Furthermore, although Fine's attempts at solving the classical problems of mathematics were the issues that motivated most of the refutations, other topics also ignited controversies. Nunes, for example, besides his attacks on Fine's attempts at solving the problems of squaring of the circle and duplication of the cube, also criticized Fine's techniques for the determination of longitude and some errors he committed while studying sundials. Francesco Barozzi wrote a 'Digressio contra Orontium', directed against Fine's *De speculo ustorio* (1551);[14]

10 Borrel, *De quadratura circuli*, pp. 92–94.

11 '[...] volendo il detto Orontio dar a credere al mondo, che lui sapesse molto piu di quello che sapeua'. N. Tartaglia, *La quarta parte del general trattato de' numeri et misure* (Venice, 1560).

12 A. van Roomen, *Exercitationes cyclicae contra Iosephum Scaligerum* (Wurzburg, 1597). On van Roomen see the detailed biographical note in U. Baldini and P. D. Napolitani (eds.), *Christoph Clavius: Corrispondenza*, 4 vols. (Pisa, 1992), vol. 1, p. 87; H. Bosmans, 'Note sur la trigonométrie d'Adrien Romain', *Bibliotheca Mathematica*, no. 3 (1902), pp. 342–354.

13 Maurolico's letter is quoted in R. Moscheo, *I Gesuiti e le matematiche nel secolo XVI. Maurolico, Clavio e l'esperienza siciliana* (Messina, 1998), p. 295. The manuscript of this letter, which was printed several times, is extant in Bibliothèque Nationale de France, Paris, MS Par. Lat. 7473, fols. 1^r–16^v.

14 'Digressio contra Orontium' in F. Barozzi, *Admirandum illud geometricum problema tredecim modis demonstratum* (Venice, 1586), pp. 192–196.

Giovanni Battista Benedetti, in his *Diversarum speculationum mathematicarum et physicarum liber* (1585), included a text against Fine,[15] and, more generally, Federico Commandino denounced that Fine did not know Greek, a critique that Borrel had also expressed.[16]

These critiques had very different impacts. Despite the fact that some mathematicians attacking Fine were of the caliber of Tartaglia and Benedetti, the refutations by Nunes and Borrel seem to have been the most influential, in part, surely, because they were the first. Furthermore, since Fine died in 1555, Nunes's critique was the only one of which he could have been aware. Also, Nunes's criticisms were the most interesting from a strictly technical point of view. According to Marshall Clagett, Nunes was 'the most acute and learned of the critics' of Fine, who 'not only corrected the errors of the French mathematician but revealed himself as the most penetrating student of Archimedes' technique of approximations yet to write in Latin.'[17] Thus, whether due to circumstantial or to more profound reasons, the refutation by Nunes has a prominent place among all those directed against Fine.

Definitely stained by these harsh critiques, Oronce Fine became the object of derision and contempt. Bernardino Baldi, while acknowledging Fine's fast mind, considered that his intellect was not very sharp and that he was excessively confidant in his own capacities. This, said Baldi, made him commit many mistakes which were used against him by his adversaries.[18] The opinion of the Jesuit mathemati-

15 The *Diversarum speculationum mathematicarum et physicarum liber* was published in Turin in 1585 and re-published later with the title *Speculationum mathematicarum et physicarum fertilissimus pariterque utilissimus tractatus* (Venice, 1586), and *Speculationum liber: in quo mira subtilitate haec tractata continentur* (Venice, 1599). In the 1586 edition the text 'Notabiles errores Orontii et Tartaleae' is at pp. 360–361.

16 This criticism was serious because in his edition of the first six books of Euclid's *Elements*, Fine had presented the definitions of the propositions in Greek. Commandino, in the letter to the Duke of Urbino in his *Euclidis Elementorum libri XV* (Pesaro, 1572), complained: '[...] Orontius quidem Phinaeus haud obscuri nominis autor priores tantum sex libros nulla graeci codicis ratione habita edidit.' Likewise, Jean Borrel denounced Fine's pseudo-scholarship: 'Post Zambertum Orontius sex libros Elementorum priores ab aliis detruncatos aedidit, propositiones Graeco sermone, cuius erat ignarus, Zamberti uersionibus interferens'. Borrel, *Quadratura circuli*, p. 208.

17 M. Clagett, *Archimedes in the Middle Ages*, 5 vols (Philadelphia, 1976–84), vol. 3, p. 1246. In particular, Clagett considered Borrel's criticism 'more strident (but less subtle)' than Nunes's.

18 'Era d'ingegno prontisimo, ma non acuto, onde scrisse molto e comesse di molti errori; persuadevasi troppo del proprio sapere, e la fortuna, che lo favoriva, accresceva in lui il diffeto naturale; il perché molti valent'huomini furono, che scrivendogli contro, scopersero li suoui errore'. [He had a very quick, but not sharp, mind, in that he wrote much and committed many errors; he was too convinced of his own knowledge, and fortune, which favoured him, increased this natural defect; this is why many great men, who wrote against him, exposed his errors]. Baldi, *Cronica de' Matematici*, pp. 126–127.

cian Claude Milliet de Chales reflects a degree of ambiguity. Fine is included in his list 'De illustribus mathematicis', but de Chales summarizes his opinion about him with the dry comment: 'Non uidetur multum promouisse Geometriam'.[19]

Historians of mathematics have agreed with these critiques. Delambre stated that Fine 'imagined himself to have solved the squaring of the circle, the trisection of an angle [...] which was rather scandalous on the part of of a professor of the Collège Royal of France',[20] and Montucla, while noting Fine's interesting work in elementary mathematics, stated that the 'solutions' of the famous problems 'are nothing but a tissue of lies [...] one could say that he never saw nor understood a geometrical demonstration'.[21] David Eugene Smith, in a brief remark in *Rara Arithmetica*, summarized his limited appreciation for Fine in the following words: 'He wrote extensively on astronomy and geometry, but he was not a genuine scholar', and concluded by saying that Fine was 'perhaps the most pretentious French mathematician of his time'. Emmanuel Poulle, in the notice in the *Dictionary of Scientific Biography*, stated that 'Fine's arrogance about his own accomplishments undoubtedly made his errors of logic all the more intolerable to his opponents', and concluded with a severe judgment: 'Fine's scientific work may be briefly characterized as encyclopedic, elementary, and unoriginal'. The same opinion is shared by Marshall Clagett who spares no critiques to what he calls Fine's 'geometrical ineptness.'[22]

III. *Pedro Nunes's* De erratis Orontii Finaei *(1546)*

In 1546, from the small printing shop of João de Barreira and João Álvares (the printers of the University of Coimbra) there appeared a book with the following title:

> *De Erratis Orontii Finaei Regii Mathematicarum Lutetiae Professoris qui putauit inter duas datas lineas, binas medias proportionales sub continua proportione inuenisse, circulum quadrasse, cubum duplicasse,*

19 C. Milliet de Chales, *Cursus seu Mundus Mathematicus* (Paris, 1690).

20 '[...] s'imaginait avoir trouvé la quadrature du cercle, la trisection de l'angle [...] ce qui est un peu scandaleux de la part d'un professeur du Collège Royal de France.' J.-B. Delambre, *Histoire de l'Astronomie du Moyen Age* (Paris, 1819), p. 400.

21 '[...] ne sont qu'un tissu de paralogismes [...] on diroit qu'il n'a jamais vu ni connu une démonstration géometrique.' J. E. Montucla, *Histoire des Mathematiques*, 4 vols. in 2 (Paris, [1799–1802]), vol. 1, p. 574.

22 D. E. Smith, *Rara Arithmetica. Catalogue of the Arithmetica Written before the Year MDCI with a Description of those in the Library of G. A. Plimpton of New York* (Boston and London, 1908), p. 160; E. Poulle, 'Fine, Oronce', in C. C. Gillispie (ed.), *Dictionary of Scientific Biography*, 18 vols. (New York, 1970–1990), vol. 15 (Supp.), pp. 153–157; Clagett, *Archimedes in the Middle Ages*, vol. 3, p. 1176. It should be noted that, despite this general low appraisal of Fine's work, in recent years attempts have been made to re-evaluate his contribution to intellectual culture in the Renaissance, on which see Marr's introduction to this volume.

> *multangulum quodcunque rectilineum in circulo describendi, artem tradidisse, & longitudinis locorum differentias aliter quam per eclipses lunares, etiam dato quouis tempore manifestas fecisse, Petri Nonii Salaciensis Liber unus.*[23]

The title is completely clear about the book's objectives. It reads (simplifying somewhat):

> On the Errors of Oronce Fine, Professor of Mathematics at the Parisian Collège Royal who thought to have discovered how to find two mean proportionals in continuous proportion, the squaring of the circle, the duplication of the cube, how to inscribe on a circle any rectilinear polygon, and how to find the difference of longitude of two places, at any time, by a method other than lunar distances. By Pedro Nunes from Alcácer do Sal...

Any learned reader would immediately realize that this title, despite its serene appearance, addressed serious mathematical matters; a mathematically trained reader would probably realize that it was intimately connected with the contents of Fine's *Quadratura circuli*.

The refutation of Fine's errors was the declared objective of Nunes's book. Immediately after the title page, however, he announced that besides the refutation of Fine's errors the book treated other related matters and listed them: (i) The process invented by Plato to find two mean proportionals and duplicate the cube; (ii) The very clear demonstration of Archimedes on the ratio of the circumference to the diameter; (iii) The method to obtain longitude from the motion of the Moon; (iv) An explanation of the definitions in Book 5 of Euclid's *Elements*; (v) On the theory and construction of horizontal and vertical sundials; (vi) Demonstration and use of the principal tables of directions of Regiomontanus.[24] That is, although presented as a book of refutation, Nunes took the opportunity to present some original material of his own. All of this material was somehow related to

[23] All quotations from Nunes's *De erratis Orontii Finaei* (Lisbon, 1546) are taken from H. Leitão (ed.), *Pedro Nunes. Obras, Vol. 3: De erratis Orontii Finaei* (Lisbon, 2005). For a detailed and more technical analysis of the book's contents see the notes in pp. 249–398 of this work.

[24] 'Praeter argumentorum Orontii confutationes in hoc libro continentur: (i) Platonis inuentum de duobus mediis proportionalibus inueniendis, et cubo duplicando. (ii) Archimedis demonstratio perquam lucida de ratione circunferentiae ad diametrum cum ueris numeris. Nam qui in libro ipsius Archimedis nuper impresso continentur, corrupti sunt. (iii) Qua ratione differentia longitudinis locorum ex motu Lunae sit elicienda. (iv) Definitionum quinti libri elementorum Euclidis explicatio. (v) Horizontalium et Verticalium horologiorum ratio, atque constructio. (vi) Praecipuarum tabularum directionum Joannis de Regiomonte demonstratio, et usus'. Leitão, *Pedro Nunes. Obras, Vol. 3*, p. 4. These subjects are discussed, respectively, in chapters II, XI, XV, XVII, XIX and XVIII. All other chapters are strictly devoted to criticizing Fine's claims.

Fine's work but it included new explanations prepared by the Portuguese mathematician.

In the letter 'Ad lectorem' that opens the *De erratis*, Nunes explained the objectives and the origin of his book. He started by stressing that he was not motivated by anything besides the loftiest goals of science. Even more than that, he explained that his desire was to ensure that less informed persons could not be led astray by Fine's errors. That is, a noble combination of desire to explain true knowledge and the charitable mission of preventing others from falling into the traps of error.[25]

Having explained his *animus* and after declaring that Fine's errors are 'not many', but 'so extraordinary they must be exposed,'[26] Nunes detailed the circumstances that led to the publication of the book:

> Thirteen years ago I had decided to write a letter to Oronce warning him to ground his inventions more prudently and more wisely before publishing them publicly. I changed my mind, however, having considered that this was especially an obligation of those learned men that live in the same city where he publicly lectures on mathematics. But, realizing that he did not retract himself – neither prompted by warning from others nor by personal resolution – from the errors he had committed before and that he had even added more absurdities, I thought I could not let these errors propagate any longer.[27]

The motives behind, and the chronology of, the composition of the *De erratis* thus seem clear. Writing in 1546, the expression 'thirteen years ago' refers to the moment immediately after the publication of Fine's *Protomathesis* (1532). Thus, according to Nunes it seems that the publication of the *Protomathesis* made him think of writing to Fine warning him to be more cautious and prudent. In fact, from a passage in a book Nunes published in 1537 it is evident that by this date he already held Fine in very low esteem.[28] At this juncture, however, Nunes decided not to write after all. According to his own words, he felt this to be a duty of other Parisian scholars.

[25] See Leitão, *Pedro Nunes. Obras, Vol. 3*, p. 5.

[26] *Idem.*

[27] 'Quem ego iam ante annos tredecim, per literas admonere statueram, ut consultius et maturius inuenta sua probaret, ante quam foras emitteret. Sed mutaui consilium, quoniam id magis eorum officium esse putaui, qui in eadem urbe, in qua idem Orontius Mathematicas publice docet, iisdem artibus, et disciplinis instructi sunt. Caeterum cum nondum uideam illum, uel aliorum admonitione, uel sponte sua, ab institutis erratis esse reuocatum: sed potius nouorum accessione, pristina peccata cumulasse: non id dissimulandum ulterius existimaui.' Leitão *Pedro Nunes. Obras, Vol. 3*, pp. 5–6.

[28] *Ibid.*, p. 45. Nunes continued to denounce the errors of Fine in his later works: sporadic criticism against the French mathematician can be found in *Petri Nonii Salaciensis Opera* (Basel, 1566) and in the *Libro de Algebra* (Antwerp, 1567).

Whether Nunes was thinking about anyone in particular, or just manifesting a vague desire, is something we will never know, but it is important to realize that his indication of familiarity with Parisian mathematical circles was most likely authentic. The University of Paris had long been a destination for Portuguese students. In the first decades of the sixteenth century there were strong intellectual relations between Portugal and France centred, above all, on the Collège de Saint-Barbe, in Paris. Many examples could be adduced to testify to these relations but it is perhaps sufficient to recall two cases, that of the French doctor and mathematician Jean François Fernel (*ca.* 1497–1558) and that of the Portuguese mathematician Francisco de Melo (1490–1536).[29] Fernel studied at the Collège de Saint-Barbe from 1519 onwards. There he came into contact with the large group of Portuguese masters and students and thus became aware of events and news about Portugal. Fernel's books show that he was very well informed about the cosmographical novelties Portuguese sailors were discovering and in his books he expresses his admirations for Portuguese maritime expeditions.[30] Francisco de Melo studied in Lisbon but went to Paris to complete his studies in 1513 at the latest. At the Collège de Montaigu he studied under Gaspar Lax and Pierre Brissot and also taught mathematics there.[31] In 1521 he sent to the Portuguese King, D. Manuel, some of his mathematical works (commentaries on Euclid's *Optics* and *Catoptrics*, on Archimedes, etc.) showing that he was a competent and even original mathematician.[32]

29 On Jean Fernel see C. S. Sherrington, *The Endeavour of Jean Fernel. With a List of Editions of his Writings* (Cambridge, 1946); A. Herpin, *Jean Fernel. Médecin et Philosophe* (Paris, 1949); *The Physiologia of Jean Fernel (1567)*, trans. J. M. Forrester (Philadelphia, 2003). On Francisco de Melo see A. Ribeiro dos Santos, 'Da vida e escritos de D. Francisco de Mello', *Memórias de Literatura Portuguesa publicadas pela Academia Real das Ciências de Lisboa*, no. 7 (1806), pp. 237–249; L. de Matos, *Les Portugais a l'Universitè de Paris entre 1500 et 1550* (Coimbra, 1950), *ad indicem*; R. Hooykaas, 'Science in Manueline Style. The Historical Context of D. João de Castro's works', in A. Cortesão and L. de Albuquerque (eds.), *Obras Completas de D. João de Castro*, 4 vols. (Coimbra, 1968–1981), vol. 4, pp. 231–426 (esp. pp. 412–414).

30 See the opening letters by Fernel in his *Monalosphaerium* (Paris, 1526 [1527]), and *Cosmotheoria* (Paris, 1528). On the importance of Sainte-Barbe as a meeting place for Portuguese and French scholars see J. Quicherat, *Histoire de Sainte-Barbe. Collége, Communauté, Institution,* 3 vols. (Paris, 1860–1864), esp. vol. 1, pp. 348–349, 352–355.

31 Melo's talent was of some note since Gaspar Lax dedicated to him his *Arithmetica speculativa duodecim libros demonstrata* (Paris, 1515), and the famous humanist Luís Vives, in a letter to Erasmus of 4 July 1520, mentions having met him in Paris.

32 These works survive in a manuscript in Biblioteca Nacional, Lisbon, Cod. 2262. It can be deduced that the manuscript was written in Paris between 1514 and 1517. Interestingly, although Francisco de Melo abandoned his mathematical activities upon returning to Portugal in 1521, he was the rector at the University of Lisbon when Pedro Nunes arrived in Lisbon after his period at Salamanca.

Fernel and Melo are just two examples of scholars with mathematical interests whose careers reflect the relations between Portugal and France (Paris in particular) during the period with which we are concerned. Nunes, naturally, was well informed about the scientific scene in Paris and knew of Fine's standing there, hence his expectation that someone might challenge Fine's erroneous mathematical claims. But then, in 1544, Fine published the *Quadratura circuli*. Having remained silent for several years, Nunes was aghast when he received this new book, where Fine insisted on old errors and added new ones. As he explained with evident irritation:

> I have just received the new book of the mathematician Oronce Fine, *De quadratura circuli*, in which he claims to have solved those very difficult problems that throughout the ages and for a very long time have never been solved by very wise men despite the great ingenuity, hard labour and continuous meditation they have put into it.[33]

He then briefly described those very difficult five problems, listing the prominent mathematicians of ancient times that had tried – and failed – to solve them. This led him to conclude:

> In my opinion, Oronce Fine has gone mad. If it were not so he would have recognized the first errors committed twelve years ago and he would be shocked with the new and great ones that he now commits, as I will clearly show in this book.[34]

Therefore, he decided to act and within a few months his *De erratis Orontii Finaei* was ready to go to the printer's shop.

Nunes's *De erratis* is an interesting mathematical book. Although firmly limited to discussing aspects of Fine's publications, as already noted Nunes took the opportunity to digress and present his own ideas. The most important of these digressions can be found in Chapter 11, where Nunes correctly reproduces and explains Archimedes's subtle method of calculating the ratio of perimeter to diameter of a circumference as explained in *De mensura circuli*. This very fine digression led Marshall Clagett to note that Nunes 'revealed himself as the first Western author (except perhaps Regiomontanus) to understand or, at least, to explain the nuances of Archimedes's use of approximations which had eluded Fibonacci, Pacioli and Finé, all of whom had fashioned caricatures of the Archimedean proof.' The

33 'Perlatus est ad me modo orontii Finaei Mathematici nouus quidam liber de circuli quadratura inscriptus. In quo quinque illa problemata difficillima se dissoluisse iactat, quae per omnes aetates et aeuo longissimo a doctissimis uiris, magna industria assiduoque labore atque meditatione conquisita, nondum tamen sunt inuenta.' Leitão, *Pedro Nunes. Obras, Vol. 3*, p. 7.

34 'Ego uero eum puto insanisse, aliter enim primos errors suos ante duodecim annos commissos agnouisset, et proinde hos nouos ingentesque formidasset, quos in hoc libello apertissime explicabo'. *Ibid.*, p. 9.

importance of this achievement led Clagett to state that 'Nunes deserves an honored place among the students of Archimedean mathematics in the first half of the sixteenth century'.[35]

However, for the most part Nunes is engaged in attacking Fine's arguments. Since it is impossible to analyse in detail all the arguments, a classification of his critiques will show what his major objections were. Nunes accuses Fine of committing the following types of errors:

(a) *Basic errors due to ignorance of the most elementary aspects of mathematics and astronomy*. There are many places where Nunes denounces Fine's ignorance of even the most basic notions of mathematics and astronomy. This type of criticism is especially harmful because it creates a general impression that Fine's works are incompetent.

(b) *Errors arising from inept transcription of other authors*. A second type of error, according to Nunes, derives from Fine's uncritical use of other authors' texts. Furthermore, Nunes hints at the more serious accusation that Fine's procedure is sometimes dishonest.

(c) *Ignorance of the 'art of demonstration'*. A third type of criticism is not directly related to the mathematical content of Fine's arguments but, more generally, to what Nunes calls his ignorance of the 'art of demonstration'. It is again a very serious attack since it applies to all of Fine's scientific work and not just to the work under examination.

(d) *Repetition and insistence on errors*. Finally, Nunes shows irritation with the fact that Fine repeated the same errors in books published twelve years apart; an irritation that other scholars critical of Fine, namely Borrel, would also express.

Three more aspects are worth mentioning. Firstly, Nunes does not claim to have corrected all of the errors found in Fine's books; he says that he considered only those that were most relevant, and that he omitted altogether other aspects that were deserving of reproach. Secondly, although Nunes's criticism is sometimes harsh, it cannot be accused of being *ad hominem*.[36] And, finally, the aspect that most infuriated Nunes – and other mathematicians – was Fine's absurd confusions and errors on the Archimedean techniques used to determine the ratio of perimeter to diameter of a circumference and his preposterous notion that he could 'correct' or 'improve' the mathematics of Archimedes.

35 Clagett, *Archimedes in the Middle Ages*, vol. 3, p. 1222.

36 A point already noted by Ross who characterizes the criticism by Nunes (and Borrel's and Romanus's) as 'quite fair and accurate but also quite severe'. Ross, 'Studies on Oronce Finé', p. 259.

Nunes's criticisms of Fine seem to have reached a wide audience. Although *De erratis* appears to have been printed in a limited number of copies, it was rapidly known throughout Europe. In 1571 a second edition was published in Coimbra and in 1592 a third edition appeared in Basel (Fig. 10.2). As was to be expected, Nunes's *De erratis* was known in France. Élie Vinet, a famous editor of scientific texts, the mathematician Jacques Peletier and the celebrated Petrus Ramus were acquainted with book.[37] Outside France the book was also known; one may note, for example, that John Dee had a copy in his famous library.[38] The interest the book created can be measured by the fact that some mathematicians made manuscript copies of the whole work, apparently because they were unable to find a printed copy. This was the case with the German mathematician and astronomer Johannes Praetorius (1537–1616)[39] and with the Flemish cartographer Gerard Mercator (1512–1594).[40]

As with Nunes's other publications, *De erratis* seems to have been highly esteemed by mathematicians of the Society of Jesus.[41] The copy of *De erratis* from

37 Vinet cites Nunes's *De erratis* in *Definitiones Elementi Quinti et Sexti Euclidis* (Bordeaux, 1575), and in *La maniere de fere les solaires, que communement on appele Quadrans* (Bordeaux, 1583). Peletier refers the book in his *In Euclidis elementa geometrica demonstrationum* (Lyon, 1557) and also in his *De l'usage de la geometrie* (Paris, 1573). Ramus's copy of Nunes's *De erratis* is presently in Paris, at the Bibliothèque de la Sorbonne (R XVI B 3=45). It contains some manuscript marginal annotations. Interestingly, it is bound together with a copy of the 1544 edition of Fine's *Quadratura circuli*.

38 See J. Roberts and A. G. Watson, *John Dee's Library Catalogue* (London, 1990); *De erratis* is item 100. Dee's library also contained Nunes's *Opera* (1566), *De crepusculis* (1542), and *Libro de algebra* (1567), which are, respectively, items 189, 674 and 769 in this catalogue.

39 A manuscript copy of *De erratis* by Joannes Praetorius is presenty at the Stadtbibliothek Schweinfurt (AvS Ha 2 [H 76]). It was made between 21 and 30 August 1573: 'Librum describere ex impresso exemplari coepi anno 1573 die 21. Augusti Witebergae.' (fol. 1^r); and, at the end: 'Absolvi die 30. Augusti. 1573. Witebergae' (fol. 79^r). See U. Müller (ed.), *450 Jahre Copernicus "De revolutionibus". Astronomische und mathematische Bücher aus Schweinfurter Bibliotheken* (Schweinfurt, 1993), pp. 44, 342–343.

40 Mercator's library was dispersed but a manuscript copy of its catalogue survived and has recently been published. See A. Cherton and A. Watelet (eds.), 'Catalogus', in M. Watelet (ed.), *Mercator Cosmographe* (Paris, 1994), pp. 403–413. It shows that Mercator owned several of Nunes's books and a manuscript copy of *De erratis*.

41 Ugo Baldini, a leading historian of Jesuit science, states that 'among the mathematicians of the mid sixteenth century, Nunes was one of the more influential in Clavius' School'. U. Baldini, *Legem impone subactis. Studi su Filosofia e Scienza dei Gesuiti in Italia, 1540–1632* (Rome, 1992), p. 243. In fact, even before Clavius, *De erratis* already appears in Jesuit programs of mathematics, for example in Biblioteca Apostolica Vaticana, Collezione Barberini, Fondo Latino, MS 304, fol. 174^v. See A. Romano, *La Contre-Réforme Mathématique. Constitution et Diffusion d'une Culture Mathématique Jésuite à la Renaissance (1540–1640)* (Rome, 1999), p. 78.

the Roman College – most likely the copy used by Clavius – is still extant.[42] In fact Clavius cites the book in several occasions. The Jesuit mathematician Giuseppe Biancani, in his influential *Aristotelis loca mathematica*, notes that Nunes must be used as an 'antidote' to Fine's errors,[43] and another famous Jesuit mathematician, Claude Milliet de Chales, in his celebrated *Cursus seu mundus mathematicus*, refers also to Nunes's refutation when mentioning Fine.[44]

IV. *The Context of a Refutation*

When considering Pedro Nunes's writing style one may say that, like many of his contemporaries, he had 'the aggressiveness of Renaissance scholarship, which did not hesitate to point out weaknesses wherever they could be found.'[45] Indeed, in his books, Nunes never hesitates to correct other mathematicians' errors. However, despite this character trait, and despite the pledge that his intentions in writing *De erratis Orontii Finaei* were nothing but scientific, there are good reasons to believe that other factors were at play in his decision to publicly attack Oronce Fine. At least, there are very good reasons to believe that Nunes had much to gain from such an attack beyond mere intellectual gain. To understand the reasons for this it is necessary to consider his socio-professional position in 1544–46.

Very little is known about Pedro Nunes's origins, besides the fact that he was born in 1502 in a small town of Alcácer do Sal some fifty kilometers from Lisbon.[46] The first documented aspects of his life show him studying medicine in the University of Salamanca, in Spain. Having returned from Salamanca in 1529, without having completed his studies, he was appointed (somewhat surprisingly) cosmographer and tutor of scientific matters of the king's younger brothers. Clearly at this time he was already renowned for his competence in science, but nothing is known of his intellectual pursuits at this time. Nunes's first publication was only in

42 Presently in the Biblioteca Nazionale, Rome (Res. 14. 12. Q. 34). It has the mansucript inscription: 'Ex Biblioteca majori Coll. Rom. Societ. Iesu'.

43 'Orontium Finaeus, Parisiis Mathematica docuit, varia composuit quae passim reperiuntur, legenda tamen cum antidoto Petri Nonii de erratis Orontii 1530.' G. Biancani, *Aristotelis Loca Mathematica ex universis ipsius Operibus collecta, & explicata* (Bologna, 1615), p. 60.

44 'De Geometria libros duos egit, item de circuli quadratura, de descriptione figurarum regularium, quae quidem bona sunt: Scripsit et alia in aliis materiis. Nunc Nonius in multis reprehendit.' C. Milliet de Chales, *Cursus seu mundus mathematicus*, p. 14.

45 O. Neugebauer, 'On the Planetary Theory of Copernicus', *Vistas in Astronomy*, no. 10 (1968), pp. 89–103, at p. 89.

46 Unfortunately there is no biography of Pedro Nunes in English. Joaquim de Carvalho's notice is still the most concise biographical sketch: 'Nunes, Pedro', in *Grande Enclopédia Portuguesa e Brasileira* (Lisbon and Rio de Janeiro, 1935–), vol. 19, pp. 53–65. For the first half of Nunes's life, see my 'Para uma biografia de Pedro Nunes: O surgimento de um matemático, 1502–1542', *Cadernos de Estudos Sefarditas*, no. 3 (2003), pp. 45–82.

1537: a book containing the translation of some introductory scientific texts – Sacrobosco's *Sphere*, Book 1 of Ptolemy's *Geography*, the initial chapters of Peurbach's *Theorica novae planetarum* – and also two original works on nautical matters. The book was written in Portuguese and aimed at a local audience. Consequently, it was not particularly noticed outside Portugal. In 1542, however, Nunes published a small book on the difficult problem of the duration of twilight, *De crepusculis*, which was a remarkable piece of mathematical research. The book, written in Latin, and using sophisticated mathematical techniques clearly targeted learned circles throughout Europe.

It was surely in view of Nunes's achievement with *De crepusculis* that in 1544 he was appointed by the king to the recently created chair of mathematics at the University of Coimbra. Nunes's arrival at the University, and to the most prestigious position a mathematician could wish for in Portugal, was unconventional. He did not compete for the position against other candidates, as the University required in normal circumstances. Instead, he was directly appointed by the king, a royal interference that the University did not suffer gladly. Thus it is clear that in 1544, when he was appointed to the chair of mathematics, Nunes still needed to solidify his position and establish his scholarly credentials. His only book of advanced mathematics, *De crepusculis*, was undoubtedly a piece of first rate astronomical mathematics but in 1544 it was still too early to measure its real impact. Nunes needed to display his mathematical competence by publishing more books.

Publishing a book on an advanced topic in mathematics was nearly impossible in Portugal at this time. An analysis of the production of the Portuguese book market in the sixteenth century shows that the only works with mathematical content that national printing houses took the risk of publishing were works on commercial arithmetic or popular introductions to astronomy and cosmography.[47] Those seeking more advanced or specialized texts had to import them, as no local printer would engage in the enterprise of publishing an advanced mathematical work unless very favourable circumstances were in place. In fact, there is evidence that in the early 1540s Nunes was having difficulty in publishing his treatises. At the end of *De crepusculis* he announced that 'very soon' he expected to publish his other works, presumably finished or very near completion by that date, and he listed them: On the astrolabe; On spherical triangles; On the geometric planisphere; On the theory of proportions; On the preparation of the globe to navigation; and others.[48] Yet these promised publications never appeared and it was not

47 On scientific books in Portugal in the sixteenth century see my 'O livro científico antigo (séc. XV e XVI). Notas sobre a situação em Portugal', in *O Livro Científico Antigo dos séculos XV e XVI. Ciências físico-matemáticas na Biblioteca Nacional* (Lisbon, 2004), pp. 15–53.

48 'Reliqua opuscula nostra breui (ut speramus) in lucem edemus. De astrolabio opus demonstratiuum. De triangulis sphaericis. De planisphaerio geometrico. De proportione in quintum Euclidis. De globo delineando ad nauigandi artem, et nonnulla

until 1544 when, unexpectedly, Fine's *De quadratura circuli* reached Pedro Nunes that he was spurred into publishing a very different type of book to those promised.

In a sense, the *De quadratura circuli* was a blessing to Nunes. Here was Oronce Fine, a professor of wide fame, the holder of the most prestigious chair of mathematics in France, recklessly displaying his incompetence in geometrical matters. It is obvious that a direct assault would be immensely profitable to Nunes. Not only could he expose and correct Fine's errors, thus preventing others from making similar mistakes (which, Nunes claimed, was his sole objective), he could also, by showing himself to be more competent than the celebrated Parisian mathematician, establish his own scientific credentials. Furthermore, he could claim priority as the first scholar to catalogue and denounce the errors of Oronce Fine – a considerable advantage in the competitive mathematical culture of the sixteenth century. Finally (and as something of a bonus), this publication offered him the opportunity to present some of his own work.

There were, however, some 'rules of engagement' to be respected. Publishing books of refutation was not, and is not, without its risks. 'Exposition, criticism, appreciation', a leading mathematician of the twentieth century reminds us, 'is work for second-rate minds' and scholars in the sixteenth century thought the same.[49] In 1524, in his well known 'Letter against Werner', before starting his demolition of Johann Werner's *De motu octavae spherae*, Nicolaus Copernicus took great care in explaining that he was aware that 'faultfinding is of little use and scant profit, for it is the very mark of a shameless mind to prefer the role of the censorious critic to that of the creative poet'. He explained the dilemma faced by every critic: 'Hence I even fear that I may arouse anger if I reprove another while I myself produce nothing better'.[50] Thus, the art of refutation required that the attacker showed his capacity as a creative scientist, distancing himself from the role of mere 'censorious critic'. Moreover, every criticism or refutation should be made with dignity, avoiding the personal attacks and the insult. All this Nunes achieved in his book.

De erratis Orontii Finaei stands as an isolated case in Pedro Nunes's scientific production. It is the only polemical work he ever wrote, and it is the only book in which he addressed at length the classical problems of geometry. Twenty years would pass before he published another book and when he did he returned to his more cherished pursuits: theoretical navigation, mathematical astronomy, alge-

---

alia quae hodie molimu'. H. Leitão (ed.) *Pedro Nunes. Obras, Vol. 2* (Lisbon, 2003), p. 120.

49 G. H. Hardy, *A Mathematician's Apology* (Cambridge, 1990 [first edition 1967]), p. 61.

50 N. Copernicus, *Complete Works. Minor Works*, trans. and commentary by E. Rosen with the assistance of E. Hilfstein (Baltimore and London, 1992), pp. 127–165, at p. 145.

bra. For Nunes, the positive outcome of the refutation against Fine seems beyond any doubt. Immediately after the publication of *De erratis* in 1546 great honours were bestowed upon him and he rose to the highest positions in Portuguese society. In 1547 he was appointed Chief Cosmographer of the Kingdom – a position that seems to have been created especially for him. In the following year, 1548, he rose to the rank of Knight of the Order of Christ, a social distinction of great importance. But perhaps the most telling confirmation of his success comes from the University of Coimbra. In 1548, when the king's brother, D. Luís, visited the University, *rhetor* João Fernandes gave the welcome speech. On this important occasion, before the complete body of the University's professors, Nunes was singled out as one of the glories of the University. Above all (and far more significantly) just two years after the publication of *De erratis* he was hailed by Fernandes as a new Archimedes:

> But for what wrong have I overlooked you, most erudite Pedro Nunes? Did I think that medicine was beneath the notice of that intellect you have been blessed with? Indeed, no discipline is beneath the notice of a man, no matter how exalted his intellect. But love of divine learning took you from the earth to the sky where, because there is no place for diseases, you naturally repudiated not science as a whole, but the use of medicine. They are happy men, whose concern it has been to pass on the secrets of heaven to their descendents as an inheritance. The past has had many Archimedes, but our generation has contented itself with one Pedro, for men of such worth (lit. 'men of adamant/diamond') are not born often, so that their rarity may be prized. What shall I say about your knowledge, which is complete in every way in the divine field of universal learning? I shall embrace it all in one phrase. It was through you that our Prince Ludovic, for whom the earth is a mere pinprick, has been able to study those broadest circles of the universe.[51]

51 'At quo te crimine tacebam, Petre Nuni eruditissime? Putabamne inferiorem rem medicam ista tui ingenii felicitate? Certe nulla disciplina est hominis quamlibet sublimi ingenio inferior. Rapuit te tamen diuinae Matheseos amor a terris in Caelum, ubi cum non sit morbis locus, merito non scientiam, sed medicinae usum repudiasti. Felices animi, quibus curae fuit caelum hereditate posteris transmittere. Multos habuit antiquitas Archimedes, nostra tamen aetas uno Petro contenta est, non enim nascuntur frequenter adamentes, ut raritas in pretio sit. Quid dicam de tua in uniuersae matheseos diuinitate omnibus numeris absoluta eruditione? Vno uerbo omnia complectar. Per te factum est, ut princeps noster Ludouicus, cui terra punctus est, latissimos illos mundi orbes contempletur.' J. Fernandes, *A Oração Sobre a Fama da Universidade (1548)*, trans. and annotated by J. A. Osório (Coimbra, 1967), p. 146. The editor is grateful to Drs Alice and Jason König for their assistance in the translation of this passage.

Oronce Fine's Legacy in the French Algebraic Tradition: Peletier, Ramus and Gosselin

GIOVANNA CIFOLETTI

> Though not a great mathematician, Finé was an inspiring teacher; his humanist friends called him the 'restorer of mathematics'. Among those who undoubtedly heard his lectures were three young men whose works were to have important consequences for the upgrading of commercial arithmetic in the business life.
>
> – Natalie Zemon Davis

In a well-known article published in 1960, Natalie Zemon Davis attributed to Oronce Fine not only the high status deriving from the official role given to him by François Ier (namely Royal Lecturer in Mathematics), but also the historically important role of having instructed a generation of mathematicians and authors of scientific books.[1] My perspective in this essay is teleological, demonstrating the ways in which Fine's legacy in three main realms – arithmetic, the vernacular scientific book project and Euclid's *Elements* – affected the main French contribution to mathematics of the sixteenth century: symbolic algebra. In this context, the relevant students of Fine are Jacques Peletier du Mans and Petrus Ramus (Pierre de la Ramée). One other individual – Guillaume Gosselin – will provide an opportunity to view Fine's long-term influence on French mathematical style.

I. *Arithmetic*

Fine's first publication was not a work by him, but the second edition of the *Arithmetica* by Juan Martinez Pedernales Siliceo.[2] Fine was twenty-five at the time and employed as a young professor at the Collège de Navarre in Paris. It is probable that he had learned mathematics from Martinez himself, an Iberian mathematician who had come to the university of Paris in 1498, had taught mathematics while being a student (of dialectic and theology) and had then published his *Arithmetica* in Paris in 1514, before being called to Salamanca as professor of natural philosophy.

1 N. Z. Davis, 'Sixteenth-Century French Arithmetics on the Business Life', *Journal of the History of Ideas*, vol. 21, no. 1 (1960), pp. 188–222.

2 Martinez, ed. Fine, 1519. Fine's edition of Martinez's work went through two more Parisian editions, in 1526 and 1540.

Martinez's *Arithmetica* included both kinds of arithmetic – theoretical and practical – as its title made explicit: *Arithmetica ... in theoricen et praxim*. Arithmetic appeared at this time in three main genres: first, the Nicomachean-Boethian theoretical arithmetic, where the structure is determined by the classification of numbers (even, odd, prime, perfect, etc.); second, the *algorismus*, inaugurated by Sacrobosco to introduce Arabic numbers to the West, consisting of the four operations on integers, the four operations on fractions and the four operations on sexagesimal numbers. Sacrobosco discussed also the operations of mediation (the finding of the arithmetical mean), duplication (i.e doubling), the summation of arithmetical series, and the extraction of square and cube roots. Finally, in the abacus schools there existed, since Fibonacci's *Liber Abaci*, a third genre of arithmetic which excluded sexagesimal numbers (considered useful for astronomers but not for merchants), and included instead the rule of three, the basis for the main merchant's rules (company, exchange, alloys, interest) and algebra.

Martinez's book retains the Nicomachean-Boethian structure in the theoretical part and the practical structure of the *algorismus* in the second part. But we can find also, at the end of the second part, a simple version of the rule of three, belonging to the abacus tradition.[3] This was a novelty in university mathematics, but the relevant innovation has disappeared from Fine's elegant *in folio* edition. It is also noticeable that Fine's own *Arithmetica practica* appeared, always in Latin, in several editions or reprints (1535, 1542, 1544 and 1555) [Fig. 11.1]. Yet, there was clearly a public for an arithmetic that fitted into the mould of college teaching of mathematics and at the same time included some commercial results.

In 1531, a year after the foundation of the Collège Royal, François Ier appointed Fine as Royal Lecturer.[4] It is legitimate to think that in the famous *Protomathesis*, written by Fine in the same year, we can find an indication of what he intended to teach at the Collège, and that his proposition interpreted effectively the desires of the king. We know that mathematics appeared at the Collège after the establishment of two initial chairs, Greek and Hebrew, and was followed by French law, Latin and Medicine. The official purpose that made the Collège Royal different from the other colleges of the University of Paris was to teach certain key subjects which were not taught in the latter. Certainly mathematics was not well taught at the University of Paris, indeed the poor standard was underlined by Fine's most famous student, Ramus, later in the century.[5] Generally, mathematical teaching at the university consisted, at most, in about a year of reading Euclid's *Elements* and Sacrobosco's *Sphaera*. Therefore, even the idea of a formal univer

3 This is the arithmetical proportion, of everyday use in calculating percentages.

4 See A. Tuilier (ed.), *Histoire du Collège de France. I: La création 1530–1560* (Paris, 2006) and Pantin in this volume.

5 See e.g. I. Pantin, 'Ramus et l'enseignement des mathématiques', in K. Meerhoff and M. Magnien (eds.), *Ramus et l'Université* (Paris, 2004), pp. 71–86.

sity course in mathematics was an innovation. But the contents covered were also new, especially in two ways: they were inspired by recent humanist theories of education and they were chosen from among the disciplines which were considered priorities for the king and for the court.

Insofar as the contents are concerned, arithmetic and algebra of the abacus schools were among the typical 'popular' arts, i.e. the arts not connected to the university and not in Latin, which could have been chosen as topics for the Collège Royal. But these disciplines were not introduced at the Collège during Fine's time as Lecturer. Rather, they were introduced through the efforts of three of his disciples: Jacques Peletier du Mans, Petrus Ramus and Pierre Forcadel. Fine limited the introduction of new disciplines to practical geometry, some new aspects of cosmography and sundials (gnomonics).[6] These arts were to be approached at the Collège Royal in the same manner as the newly introduced classical languages, Hebrew and Greek – that is to say in the new philological way, implying an immersion in ancient culture but also an acknowledgement of a discontinuity with respect to the present, and the assumption that mathematics was likely to be reinterpreted as a result. According to the teachings of the humanist Regiomontanus, ancient mathematics should be studied by contextualizing mathematical discoveries, in order to understand the origins and purpose of present mathematics. Therefore, the two priorities in mathematics were, on the one hand, to reconstruct classical treatises and disciplines and, on the other, to develop their utility for the public (as opposed to their bookish tradition).

In the dedicatory letter to François Ier in the *Protomathesis*, and even more in the first section of the *Arithmetica*, Fine takes up the *topos* of the utility of mathematics to show it *a contrario*: what would exist without mathematics?[7] Without numbers we would have no music, no geometry, no philosophy (divine and human), and no laws. Mathematics is thus not only useful, it is necessary.[8] Consistent with these statements, Fine chose to present four 'practical' disciplines in the *Protomathesis* (which should be thought of as a sort of mathematical encyclopaedia). The first book comprises practical arithmetic, the second book theoretical (*ad intelligentiam Euclidis*) and practical geometry, the third book is devoted to cosmography, and the fourth book to sundials.[9]

In the first section, devoted to practical arithmetic, we find in part the contents of Martinez's *Arithmetica*. This could have been an excellent opportunity for Fine to follow Martinez's example and to present the rule of three, but he did not

6 On which subjects see P. Brioist, Besse and Mosley, and Eagleton in this volume.

7 For further discussion of this *topos*, see Axworthy in this volume.

8 Indeed, Fine asserts, like Plato, that mathematics is essential for the education of the young. For further discussion of the practical content of the *Protomathesis* see P. Brioist in this volume.

9 It is worth noting that implicit in the study of geometry was fortification and the *ars militaris*, topics quite appropriate for the *noblesse*.

do so. He followed Martinez's text in many ways, but not on this point, because where Martinez had introduced the rule of three Fine instead provided a substantial presentation of the theory of proportions. This seems to indicate two things: first, that Fine did not want to move away from the scheme of practical arithmetic; second, he considered theory of proportion a crucial aspect of mathematics of nature: i.e. the fourth or the sixth proportional is the way to solve problems, to find the unknown *in naturalibus.*[10] It is notable that some major seventeenth-century figures, such as Galileo and Kepler, thought in the same way, even after the appearance of Viète's symbolic algebra and Fermat's and Descartes's mathematics. Both Galileo and Kepler were quite explicit in giving the theological and metaphysical reasons for favouring proportion: it should be given priority in interpretations of the world because the world has been created according to proportions. The heavens and the sub-lunary world share this theoretical structure, are written in this language and according to these laws. University mathematics, which had always been connected to astronomy, had only to accommodate the compatibility between the heavenly spheres and the sub-lunar world, and in this context Fine was prepared for change.[11]

On the other side of the spectrum, the theory of proportions was included in the mathematics of the abacus schools, which concerned the human world, the *commercium*: business (*négoce*), gifts, inheritances, contracts, and so on. In human affairs proportion has a place, as Aristotle had explained in the *Ethics*, but the rules of algebra proved also to be useful.[12] What appeared at the beginning as shortcuts for finding results, the rule of three as a part of the *ars rei et census*, was to be celebrated at the end of the century as 'the divine algebra', an art to solve all problems. So, the Euclidean theory of proportion went though the transformations necessary to conform to the parallel developments in arithmetic, in other words, the numerical interpretation of the theory of proportions became dominant in geometry. Fine accepted this innovation, but did not include the rule of three and algebra in his treatise, as his disciples would do. However, these disciples were prepared to do so, I claim, as a result of Fine's teaching: their different choice was certainly shaped by new social needs and perspectives, but also from

10 We have further information about Fine's readings in the recent theory of proportions: he owned a manuscript of Jean de Murs's *Quadripartitum numerorum*, a work on all the parts of arithmetic written by a Master of Arts of the University of Paris in 1343, which included a book on speculative arithmetic, a book on practical arithmetic having the structure of an *algorismus*, a book on proportions, and a book on algebra. The editor states that between the two works 'there is a similarity of conception'. See G. L'Huillier (ed.), *Le Quadripartitum numerorum de Jean de Murs. Introduction et édition critique* (Geneva, 1990).

11 The structural relationship between the heavens and the sub-lunary world is also of relevance to Fine's cosmography, on which see Mosley and Besse in this volume.

12 I am thinking of distributive justice, as in *Nicomachean Ethics*, 1129a–1133b.

a familiarity with a previous program of widening the spectrum of the disciplines in the university curriculum, most importantly practical arithmetic.

It should be noted also that what I have presented as absolute opposition could, in fact, subsist perfectly well within the same research program. A theory of proportions was transmitted by the abacus schools and the more literate traditions that emerged from these schools, such as the Franciscan mathematical tradition, had been in search of a theoretical framework for both human and natural things.[13] Furthermore, the conceptual connection between the rule of three and the theory of proportions had been clear for centuries and made explicit by, among others, Luca Pacioli. References to the theory of proportions were present in the abacus teaching of 'commercial' rules, but references to commercial rules rarely existed in practical arithmetic (Martinez is one of the few examples of this). As regards Fine, we may say that he extended the contents of practical arithmetic, but did not mix the genres by neglecting to include algebra.

It is important to consider to what extent humanist education as a *paideia* should be connected to Fine's choices in the *Protomathesis* and his teaching at the Collège Royal. It is notable that humanist education included training in the use of classical languages as living languages, reading of classical books and, in the afternoon, some physical education. 'Par temps pluvieux ou intempéré', as Rabelais writes, it recommended also the observation of the practice of the arts and crafts. This principle should not be confused with the encouragement of a real training in arts and crafts. The purpose of this experience was rather to inculcate in students a frame of mind allowing them to shape theoretically – according to that frame – all sorts of activity. Students had to emerge from school as masters of arts in the sense that they were masters of an *idea* of the encyclopaedia of arts and sciences.[14]

Mathematical teaching at the Collège Royal, proposed by Oronce Fine at the moment of his nomination had to distil this general view to form a concrete programme, which needed to demonstrate the unity of the ideal encyclopaedia. To show this unity meant, for mathematics, to ascribe to Euclid what had historically belonged to the abacus tradition. In fact, the integration of the abacus tradition into the university had started elsewhere, in Venice, with the *Summa de arithmetica* by Luca Pacioli (1494). Though much richer in abacus-arithmetical content, this text had constructed the autonomy of the algebraic discipline by finding for it a demonstrative foundation in Euclid. It is quite probable that Fine was inspired by Pacioli in stressing the importance of proportions, with an implicit connection between the theory of proportions and the rule of three or rule of algebra.

13 See M. Bartolozzi and R. Franci, 'La teoria delle proporzioni nella matematica abachista da Leonardo Pisano a Luca Pacioli', *Bollettino di Storia delle Scienze Matematiche*, vol. 10, no. 1 (1990), pp. 3–28.

14 See my 'L'Utile de l'entendement et l'Utile de l'action: Discussion sur l'utilité des mathématiques au XVè siècle', *Revue de Synthèse,* nos. 2, 3 and 4 (2001), pp. 503–520.

In sum, Fine did give a new importance to practical aspects of mathematics, including practical arithmetic, and went as far as being interested in Martinez's book. When it was his turn to publish an arithmetical book, he published only practical aspects and included some mathematical results from the abacus tradition, but he did not follow Martinez's methods.[15] Instead of importing the rule of three in its original form, he gave this topic a new, yet classical, shape (and this general approach was adopted by his followers). A consequence of this, however, was that in his book the algebraic results were hidden and presented as Euclidean results.[16]

II. *Jacques Peletier du Mans and the French Vernacular Scientific Book Project*

For Fine, algebra (the greatest success of the abacus tradition) could not be included as such in college teaching, but only in its impact on the theory of proportions. However, this was not the case for other aspects of the mathematical sciences. On the contrary, Fine was probably the first in France to launch the programme, completed by the next generation of mathematicians, to publish in French a new genre of mathematical book, in a new style, which I call here the 'French vernacular scientific book project'. This new genre took the form of manuals rather than textbooks connected to teaching practice, and was aligned with the tradition of arts and crafts in the vernacular. Textbooks for college teaching were traditionally published in Latin, and often the translation into Latin indicates, in France, the passage of a book to college teaching. Oronce Fine published three works in French, and not only in the disciplines which did not have a classical tradition or a history of university teaching. The purpose was, as for many later sixteenth-century French authors, to launch a new literature which followed new criteria, according to the taste of the day. The first book, published in 1528, was *La Théorique des Ciels, mouvements, et termes practiques des septs planètes*.[17] In 1543, Fine published *Les Canons et documents très amples, touchants l'usage et practique des communs Almanachz que l'on nomme ephemerides. Briefve et isagogique introduction sur la judiciaire astrologie*, reprinted in 1551, in 1556 and after his death in 1557. Also in 1556 there appeared his *La composition et usage du quarré géométrique, par lequel on peut mesurer fidèlement toutes les*

15 Fine published his arithmetical section of the *Protomathesis* in an autonomous form in 1542, 1544 and 1555.

16 For an understanding of Fine's significance for sixteenth-century French mathematicians, see the classic study by J. C. Margolin, 'L'enseignement des mathématiques en France (1540–1570): Charles de Bovelles, Finé, Peletier, Ramus', in P. Sharrat (ed.), *French Renaissance Studies, 1540–70: Humanism and the Encyclopedia* (Edinburgh, 1976), pp. 109–155.

17 Note that Fine published a Latin translation of this book, *Nova universi orbus descriptio*, in 1532, just after his appointment at the Collège Royal.

longueurs, hauteurs et profunditez. The first of these works, *La Théorique des Ciels* (an up-to-date astronomical treatise, based on Peurbach's version of Sacrobosco's *Sphaera*), offers a good example of the *fortuna* of Fine's vernacular publications.[18] First, a new form of discipline is appreciated and promoted among vernacular readers through a vernacular treatise, which made possible its introduction at the level of universities and colleges, and the book was thus appropriately translated into Latin.

All these vernacular publications appear as works written in a cultivated style, i.e. in humanistic French and in a compact and stylised form. But what was the readership for these texts? Immediately we think of the court, members of the elite academies who were not fluent in Latin, the aristocracy, merchants and their children, who were not sent to university. However, there were also students who could appreciate mathematical books in the vernacular.[19] Astrology, for instance, had a vernacular audience and we know that readership of these works included medical students.[20] Furthermore, some of the new 'academies' (i.e. the colleges outside the university) to which young aristocrats were sent, devised a curriculum entirely in the vernacular.[21]

In order to assess the putative impact of Fine's first steps in vernacular scientific publishing, I will now turn to one of his disciples, Jacques Peletier du Mans, one of the main promoters of the vernacular scientific book project. Peletier began, in 1541, by encouraging the use of French in literature through his translation of Horace: the theory of imitation presented therein contributed to the *manifesto* of the Pléiade. As he explains in the text, the means by which the writing of science in French may be promoted are various, but the most basic and fundamental is the development of a new orthography. In the letter to the reader, he explains that, 'Just as words signify thought, so orthography signifies words.'[22] This approach is clearly presented in the introduction to the *Dialogue de l'ortografe et prononciation françoese* (1550). The dialogue itself represents a debate between Peletier's friends, Jean Martin, Théodore de Bèze, Denis Sauvage, and

18 For the scientific curriculum expected at the time, which new books appeared in the sixteenth-century and for which public, see I. Pantin, *La Poésie du ciel en France dans la seconde moitié du seizième siècle* (Geneva, 1995), especially pp. 15–53.

19 See, for instance, E. Eisenstein, *The Printing Press as an Agent of Change: Communication and Cultural Tranformation in Early Modern Europe* (Cambridge, 1979).

20 For the vernacular audience for astrological publications see e.g. S. J. Schechner, *Comets, Popular Culture and the Birth of Modern Cosmology* (Princeton, NJ, 1997), *passim*.

21 See the classic study by R. Chartier, M. M. Compère, and D. Julia, *L'Éducation en France du XVIe au XVIIIe siècle* (Paris, 1976).

22 'Tout ainsi que la parolle est significative de la pensée, semblablement l'orthographe de la parolle'. *L'Art Poetique d'Horace*, trans. J. Peletier du Mans (Paris, 1541), fol. 2r.

Jean Paul Dauron, at the house of the publisher Vascosan.[23] Only two points about the *Dialogue* need be mentioned here. First of all, Peletier's insistence on the notion that language comes from the people, and is transmitted by contact with the people. Across time and space, only writing can provide a substitute for the normal transmission of language, which is oral. However, writing is more than a mere substitute for speech. In the *Dialogue* he explains that writing should be submitted to meaning rather than to speech, given that writing aids the grasping of sense.[24] In this way, Peletier states clearly that writing has autonomy with respect to speaking but, in particular, writing does not convey pronunciation, and this is why the study and improvement of orthography is so important as a tool for bridging the gap between written and spoken language.

The second point is at a different level, and concerns more directly the project which Peletier was to pursue throughout his life: the writing of scientific books in French. As he explains:

> Our mathematics has never being doing as well as now, nor in a better position to be understood in its perfection. And given that its truth is manifest, infallible and constant, think of the immortality it could give to a language, if it were written about according to a good and true method. Let us look even at the Arabs, who in spite of being very far from us and almost in another world, yet some Europeans have wanted to learn their language, especially because of Astrology, and other secret things they have dealt with in their vernacular, quite unfortunately indeed. For we know what sophistry they have mixed with Medicine and even with Mathematics. Still, they made their language necessary for the study of all this.[25]

23 On occasion, Conrad Badius and Jean Corbin also participated in the conversation. In fact, Peletier's programme for language reform was not carried out by this circle because by the time the *Dialogue* was written the discussion group had already dissolved, largely as a result of religious differences. The *Dialogue* was actually written in Lyon, after the religious wars and the fragmentation of his circle had made Peletier du Mans leave Paris. See N. Z. Davis, 'Peletier and Beza Part Company', *Studies in the Renaissance*, vol. 11 (1964), pp. 188–222.

24 'Voela commant elle ne doet point ètre tant sugette a la prolation qu'a l'antandemant, vu que le plus que nous retirons de l'Ecritture cét l'intelligance du sans.' [This is why it should not so much depend on enunciation but rather on understanding, given the extra we get from writing is the understanding of meaning.] J. Peletier du Mans, *Dialogue de l'Ortografe et Prononciation Françoese* (Poitiers, 1550), p. 75.

25 'Noz Matematiques ne furet jamais mieus au net, qu'elles sont depresant, ni an plus belle disposition d'étre a an leur perfection. E par ce que leur vérité et manifeste, infallible et constante, pansez quele immortalite elles pourroét apporter a une langue, i étans rédigees an bonne et vreye metode. Regardons méme les Arabes, léquez ancores qu'iz soét reculez de nous, e quasi comme an un autre monde: touteffoes iz s'an sont trouvez en notre Europe qui ont voulu apprandre le langage, an principalle consideration pour l'Astrologie, et autres choses secrettes qu'iz ont traittees en leur vulguere, combien qu'assez malheureusemant. Car on set quele sophisterie iz ont

Peletier indicates quite clearly that his plan is to impose French as the language of science or, to paraphrase him, to make that language necessary for the study of science ('rendre notre langue requise en contemplation des sciences'). Peletier took this program seriously. He had already produced, five years before the publication of the *Dialogue*, an edition (with commentary) of an extremely successful work by Gemma Frisius – *Arithmeticae practicae methodus facilis* (1545) – which employed the reform of language proposed in his later works.[26]

This very short book does not go particularly far in integrating abacus arithmetic to university *algorismus*. However, he begins the process of this integration: somewhat like Martinez's book, it belongs to the genre of the practical arithmetic, and is mostly meant to teach elementary arithmetic as a basis for astronomy. But it goes farther than Martinez in the treatment of the rules of commerce, and it even contains a chapter on simple and compound interest under the title of 'De usura'. In this sense it is a proper introduction to the mathematics of business life in the context of university arithmetic.[27] Furthermore, Frisius introduced some algebraic elements without notation. By this I mean that after the 'vulgar rules' he talks about the rule of algebra. In fact, he worked on examples in which the application of the rule amounts to the extraction of the square root. This simple technique could make the readers of Frisius's arithmetic open to the use of algebraic rules in equations. Apparently, this particular mixture of content made it a successful work directed to mathematical practitioners such as court astronomers and calculators of ephemerides for cartography, navigation (calculation of longitude), astrology, and almanacs, and medical doctors, which at the time were growing social groups, becoming connected with a larger group of 'artisans' and interested in the mathematical tools developed in the abacus schools. However, the use of Latin

---

mêlée parmi la Medecine, e les Matematiques mémes. E touteffoes iz ont randu leur langue requise an contamplation de cela. Avisons donq a quoe il peut tenir que nous n'an facions non pas autant, mes sans compareson plus de la notre?' *Ibid.*, pp. 117–118.

26 G. Frisius, *Arithmeticae practicae methodus facilis. Cum Jacobi Peletarii Cenomani annotationibus* (Paris, 1545). This first edition (published by Richard) was followed by a 1563 edition (published by Cavellat). The book was subsequently reprinted several times. In fact, this work was one of the great successes of its time. Another of Fine's disciples, Pierre Forcadel, provided a French translation: *L'Arithmétique de Gemma Phrison* (Paris, 1585). Peletier's *Annotationes* concern mostly astronomical fractions and the translator of the Italian edition, Orazio Toscanella, was particularly interested in reproducing this aspect of Peletier's contribution. See G. Frisius, *Aritmetica pratica facillissima...Con l'aggiunta dell'abbreviamento de i rotti astronomici di G. Pelletario*, trans. O. Toscanella (Venice, 1567). It is worth noting that the main subject of Frisius's text – cosmography – is a discipline with which Fine was particularly involved.

27 The importance of this innovative feature is stressed in Davis, 'Sixteenth-century French Arithmetics'.

indicates that the readership was mostly confined to university studies.[28]

Peletier added to the original text a substantial commentary, largely concerned with sexagesimal fractions, in the section on astronomy – a subject which was totally absent from the abacus schools but existed as a discipline in the colleges – and, on the opposite side of the spectrum, he provided proofs for the extraction of square and cubic roots. At the end of the *Annotationes*, Peletier developed into demonstrations the intuitive arguments Gemma had provided for the algebra section, but again without notation. After this prestigious beginning, Peletier moved on to his own mathematical works, which are always marked by his concern for a perfection of the form of presentation, i.e. by an explicitly reasoned rhetorical, logical and orthographical strategy.

After his annotated edition of Gemma Frisius's work, Peletier wrote his own book on arithmetic, *L'Arithmetique departie en quatre liures, à Theodore de Beze* (1547).[29] It seems that this edition of Gemma Frisius was meant to be used in teaching at the colleges. This means that after introducing a new topic at the French colleges, Peletier introduced it to the court by means of a newly conceived book of arithmetic in French. Peletier introduced many innovations in this and other mathematical texts, starting with a separate manual in which he restructured the domain of practical and commercial arithmetic and algebra. As we have seen with Fine's choice of Martinez, this was new in France. Through changes in the rhetoric of the manuals, Peletier made them more acceptable to the wide audience of the court and to the *noblesse de robe* connected with it. In fact, Peletier's arithmetic contains many more commercial rules than Martinez's and even Frisius's book, and the introductions to the four books orient the cultivated reader towards a new conception of mathematics in which this newly conceived practical arithmetic (not only astronomical, but containing abacus arithmetic and algebra) is presented as a crucial part of the quadrivium.[30] For Peletier, abacus arithmetic and algebra were legitimate fields of knowledge, representing the good part of commerce. In this he was ahead of his time, since a positive view of commerce and interest was not common before the time of Richelieu.

Here it is important to stress that he consciously theorized the features of a scientific book. This, I argue, is an indication of Peletier's increasing interest in the process of printing and the possibilities of dissemination offered by it. A confirmation of this thesis comes from the fact that Peletier lived at the house of the

[28] This text was the most popular Latin arithmetic of the time, with sixty-four editions up to 1595, so that even this simple introduction to algebra played an important role. See F. Van Ortroy, *Bio-bibliographie de Gemma Frisius* (Brussels, 1920).

[29] While the writing belongs clearly to the Parisian period, it was published after another change in Peletier's life, while he was living in Poitiers at the house of the Marnef, in 1549.

[30] This may be compared to the rhetorical strategies Fine employed in his prefaces, on which see Axworthy in this volume.

Marnef family of printers in Poitiers. The Marnef household was a lively intellectual centre: here Peletier encountered Elie Vinet, the humanist astronomer, whom he had met in Bordeaux a couple of years earlier. But, more importantly, the Marnef were particularly aware of the ongoing reform of orthography, and ready to be leaders in the field, even competing with the main Parisian publishers of scientific books, such as Cavellat and Wechel. Thus, they were the main disseminators of Peletier's orthographical reform. In fact, the *Arithmétique* is the first book that might have satisfied Peletier from the point of view of fidelity to his orthography.[31]

III. *Euclid's* Elements

The purpose of editions of Euclid in the sixteenth-century is a complex matter, where theological principles are at least as relevant as mathematical agendas. However, it is important to recognise two key aspects of this first phase of the impact of the *Elements*: the Platonic descent of this most crucial mathematical text, and the common roots of arithmetic, algebra and geometry found in both university and abacus tradition. Stressing the first of these had been the goal of Stapulensis's version; promoting the latter had been the aim of Pacioli's version.[32] By including the text of the great authority of Proclus, Grynaeus had managed to shift the focus of debate to the former's *Commentary*. Until Francesco Barozzi's translation into Latin, Grynaeus's remained the *editio princeps* of Proclus's *Commentary* and the main source through which to discuss Euclid with or against Proclus's arguments. By providing his own version of Euclid's *Elements*, published in 1536, Fine set the matter within the context of the colleges.[33] In fact, he was the first to

31 Peletier's list of publication is in itself an illustration of the inextricability of his interests in mathematics from those in French language and orthographic reform. We have seen that he published the *Dialogue* in 1550, with the Marnef. Around the end of 1553, Peletier left Poitiers for Lyon, where he lived at the house of Jean I de Tournes. While his position was that of teacher for the publisher's son, Peletier managed to publish a series of works with him, all with carefully reformed orthography: *L'Algèbre* (1554), *L'Art poëtique* (1555), *L'Amour des amours* (1555), *In Euclidis elementa geometrica demonstrationum libri sex* (1557), and *Disquisitiones geometricae* (1567), as well as a series of second and third editions of previous works. His relationship with Jean de Tournes is especially significant, not only because this publishing house would continue to publish Peletier's works into the seventeenth century, but also because of Peletier's influence on the activity of the house itself. For about five years, until 1558, all the publications of de Tournes were influenced by Peletier's orthography, and in part corrected by him. See N. Catach, *L'Orthographe française à l'époque de la Renaissance. Auteurs – Imprimeurs – Ateliers d'imprimerie* (Geneva, 1968), pp. 104–107.

32 L. Pacioli *Euclidis megarensis philosophi ... Opera a Campano interprete fidissimo translata* (Venice, 1509).

33 *In sex priores libros geometricorum elementorum Euclidis demonstrationes, quibus ipsius textus graecus suis locis insertus est, una cum interpretatione latina Bartholomaei Zamberti, ad fidem geometricam per eundem Orontium recognita* (Paris, 1536; reprinted 1551).

have the idea of publishing a collection of the first six books of the *Elements*. In this he was followed by a few French authors, but this format also appeared in other countries: one might consider, for example, the German professor of mathematics at Tübingen, Johann Scheubel's edition of the first six books: *Sex libri priores de geometricis principii* (1550).[34]

While this is not the place for a thorough comparison of Fine's work to previous editions, let us stress that he based his work on two editions: Lefèvre d'Etaples's 1516 text, and the *editio princeps* of the Greek text published in 1533 by the theologian Simon Grynaeus.[35] Lefèvre d'Etaples's edition contained Campanus's as well as Zamberti's commentaries and he took seriously Zamberti's view, according to which Theon was responsible for most of the proofs, while Campanus had added some others, with mistakes. In order to produce this first edition of the complete Greek text of the *Elements*, Grynaeus used the Latin translation made from the Greek manuscript by Zamberti in 1505 and two Greek manuscripts supplied by Lazare de Baif and Jean Ruel.[36] Fine accepted Zamberti's and Lefèvre d'Etaples's judgement on the authenticity of the proofs. He therefore took the statements of Euclid's propositions from the Greek text, gave a Latin translation of them based on Zamberti's version, added Zamberti's demonstrations, and completed them with his own proofs and commentaries. Among these, we find remarks on theorems, problems, and some axioms, which show his dependence on Proclus's commentary. This implies that Fine has not only actively used Grynaeus's edition of Euclid, but also Proclus's *Commentary* on Euclid, presented in the same volume. This reinforces the suggestion that he perceived continuity between his enterprise and that of Lefèvre d'Etaples and Grynaeus before him.

Fine's version, in turn, had an impact on Peletier, who composed his own work on the *Elements*: *In Euclidis Elementa geometrica Demonstrationum libri sex* (1557). This work, published by Jean de Tournes, seems to have been influenced by his experience as a private mathematical tutor, in which capacity he instructed Jean de Tournes's son.[37] Peletier edited the first six books of the *Ele-*

34 Scheubel also produced the first German translation of any part of Euclid's work into German: *Das sibend, acht und neünt Büch, des hochberühmbten Mathematici Euclidis* (Augsburg, 1555).

35 Euclid, *Elementa geometriae* (Basel, 1533).

36 To this volume Grynaeus appended the first publication of the four books of Proclus's *Commentary*, taken from a manuscript provided by John Claymond, President of Magdalen College, Oxford. In a long introduction Grynaeus dedicated his translation to Cuthbert Tunstall (1474–1559), Bishop of Durham, and author of a book based on Luca Pacioli's *Summa,* in fact the first strictly mathematical book printed in England: *De arte supputandi libri quattuor* (London, 1522).

37 See above, n. 31. Peletier also taught the son of the Maréchal de France Charles de Cossé-Brissac, Lieutenant Général du Roi en Piémont, in whose household he served also as a physician. For his teaching Peletier used Theon's version of Euclid, a version closer to algebraic interpretations. During his employment as tutor in the Cossé-Bris-

ments using Zamberti's translation.[38] He stressed the 'algebraic' role of Book 2 and provides a technique at least as old as Pythagoras's mathematics, which starts with the division of segments and develops a computation on segments where the product is a plane figure, thereby allowing mathematicians to calculate the respective value of figures. He thus provides the basis for an algebraic interpretation of some of Euclid's books, starting with the geometrical proof of special products. This algebraic role can be seen as the sixteenth-century version of the 'geometric algebra interpretation' provided by some twentieth-century historians.[39]

But how exactly did Peletier navigate between philology and mathematical doctrine when it came to the *Elements*? Stapulensis's edition was already established when Peletier, fifty years later, took up the project. The *Elements* was still textually unstable with respect to the authenticity of proofs and the content of Books 14 and 15. Peletier assumed the task of providing new proofs, at least for a good part of the text he dealt with, i.e., the first six books. Re-examining Euclidean proofs was not Peletier's key motivation, though. What he found highly important about the Euclidean text was the concatenation of mathematical statements, the order of which was a challenge for the mathematical reader. Peletier suggested a few rearrangements in the flow of theorems, problems, lemmas, and corollaries. He developed, moreover, a philosophy of mathematics and a specific vocabulary, which were to have a strong impact.

In Book 1 Peletier revealed his dependence on Proclus's *Commentary* by introducing Euclid's text with a short discussion of principles, that is, of definitions, postulates and common notions.[40] Summarising Proclus, he stressed that both definitions and postulates are such that, once they are proposed, one admits them without effort, while common notions are immediately evident, so that the person who does not grasp common notions is also lacking common sense. Paraphrasing Proclus, he also added philosophical commentaries that had been absent in previous versions of the *Elements*, but the actual words he used, the definitions and propositions, were those of Stapulensis's text. Similarly, he gave a list of common notions, which he calls *animi notiones* (notions of the mind). His list, which

sac household he dedicated his *L'Algèbre* to the Maréchal.

38 J. Scheubel's *Euclidis Megarensis, philosophi et mathematici excellentissimi, sex libri priores* (Basel, 1550) must also have been important for Peletier, who had read Scheubel's *Algebrae compendiosa facilisque descriptio* (Paris, 1552).

39 On this issue, see in particular M. N. Fried and S. Unguru, *Apollonius of Perga's Conica: Text, Context, Subtext* (Leiden, 2001); A. Bernard, 'Ancient Rhetoric and Greek Mathematics: A Response to a Modern Historiographical Dilemma', *Science in Context*, no. 16 (2003), pp. 391–412.

40 The only available edition of Proclus was Grynaeus's of 1533. It was replaced by Barozzi's version: *Proclus Diadochus, In Primum Euclidis Elementorum librum commentariorum ad universam mathematicam disciplinam principium eruditionis tradentium libri IIII* (Padua, 1560).

differs from that accepted today, followed, however, contemporary opinion: the same list is found in the Stapulensis edition and, under the heading *communes animi conceptiones* in Campanus's edition, and as *communes sententiae* in both Zamberti's and Fine's versions. Peletier modified slightly the order of the list, to which he added his own general remarks, yet his most important change was that he called these 'common notions' by the name of *animi notiones*, thereby identifying them with Cicero's *notiones communes*. In fact, given his dependence on Cicero, he could not have missed this connotation of the term.[41] In other words, though he chose to make use of Campanus's list of axioms, Peletier offered a 'new' interpretation of them in neo-Platonic terms, as is indicated by his references to Proclus's text and the use he made of the Ciceronian term '*notiones*'. It is no coincidence that, later in the century, Proclus's *Commentary* and the status of Euclid's axioms were to become important topics of debate. The rediscovery of Proclus's *Commentary* in fact led to two main philosophical debates in sixteenth-century mathematics, the first pertaining to *mathesis universalis* and the second to *de certitudine mathematicarum*.[42] In fact, only three years after Peletier's work on Euclid's *Elements* had appeared, Francesco Barozzi published his edition of Proclus's *Commentary* as well as a separate text with two 'questions', one concerning *de certitudine* and the other *de medietate*.[43]

After the discussion on principles, and before treating the first problem of Book 1 of the *Elements* (the construction of an equilateral triangle), Peletier inserts a most interesting section on hypotheses, demonstrations, problems and theorems. For instance, hypotheses are, in general, not necessarily 'real' (*hanc re existere non est necessarium*), but they should not be absurd nor give rise to absurd implications. The main point is that they are assumed to be true. Peletier then moves on to demonstration:

> Demonstration is called by Dialecticians 'Syllogism which makes knowing', that is to say that which concludes from demonstrated premises. And this takes its origins from Geometry: actually any proof leading to truth is geometrical. This is why it is rightly said that nobody who is not familiar with

41 The role of Cicero in Peletier's work is discussed in my 'Mathematics and Rhetoric: Jacques Peletier, Guillaume Gosselin and the Making of the French Algebraic Tradition', unpublished PhD. diss. (Princeton, 1993). On Peletier's role in the literary tradition, and especially in connection with astronomy, see Pantin, *La Poésie du ciel* and 'La représentation des mathématiques chez Jacques Peletier du Mans: Cosmos hiéroglyphique ou ordre rhétorique?', *Rhetorica*, no. 20 (2002), pp. 375–389. On Cicero's philosophy, see James M. May (ed.), *Brill's Companion to Cicero: Oratory and Rhetoric* (Leiden, 2002).

42 See G. Crapulli *Mathesis universalis: Genesi di un'idea nel XVI secolo* (Rome, 1969) and C. Sasaki, *Descartes' Mathematical Thought* (Dordrecht, 2004).

43 F. Barozzi *Opusculum, in quo una Oratio, et duae Quaestiones: altera de certitudine, et altera de medietate Mathematicarum continentur* (Padua, 1560).

Euclid can distinguish true from false. If somebody investigates more attentively why in the demonstration of propositions the form of a syllogism does not appear clearly, but only some parts of syllogisms are shown concisely, then that is because it is not dignified, when facing a real situation, to conform strictly to what is taught in the schools, according to the prescribed schemes. For the Orator, when he ascends to the forum, does not explicitly deliver what he has learned from his teacher of rhetoric. Rather, he acts in such a way that, while he does his best to remember the rhetorical precepts, he appears to think about anything except rhetoric. So, in doing geometry, given that we do not look for anything but to reach our goal in a perfect way, we dissimulate entirely the figure of the syllogism. However, if it is required, it can be extracted from the heart of geometrical demonstrations. We cut them out, because repetitions would not only produce boredom, but also obscurity. They can easily be filled in once the conclusion is obtained. The conclusions of demonstrations are problems and theorems.[44]

This section depends directly on Proclus's discussion of these topics in the second part of the prologue of his *Commentary*. The 'dialecticians' Peletier has in mind are medieval authors, but Proclus himself had also been a dialectitian, interested in the integration of Stoic logic and the reconciliation of Plato's with Aristotle's philosophy. In fact, in the passage cited above, Peletier reveals his strong commitment to Proclus's philosophy and to an interpretation of the *Elements* in which what counts is their logic. But 'logic' must here be understood as the use of cogent arguments in discourse, not scholastic logic, in which a unique structure validates utterances by means of its isomorphism with a unique ontology. With his words about the ability to hide techniques in the fine texture of the discourse, Peletier hints at the debate concerning the identity between mathematical and Aristotelian logic, that is, about the compatibility between Euclidean demonstrations and syllogisms. The difficulty in articulating this identity, or compatibility, had already been clear to Proclus, and sixteenth-century authors were also quite aware of it.[45]

44 'Demonstrrationem verò appellant Dialectici, Syllogismum qui faciat scire: nempè qui ex probattisimis concludt. Atque haec à Geometria otium habet. Immo omnis quae ad verùm perducit probatio, Geometrica est. Ut verissime dictum sit, neminem scire verum à falso distinguere, cui Euclidis non fuerit familiaris. Quòd siquis attentius scisciabitur, quur in Demonstrandis Propositionibus non eluceat forma syllogismi, sed tantum concisa quaedam membra syllogis – morum appareant: is sic habeat. praeter dignitatem esse, quae in Scholis docentur, ea quum in rem praesentem ventum fit, ex praescriptis formulis observare. Neque enim Oratur, quum ad forum accedit, quae à Rhetore excepit dictata, in digitis collocat: immo id agit, vt quum praeceptorum maxime meminit, nihil minus quàm Rhetoricen cogitare videatur. Sic in opere Geometrico, quum nil aliud spectamus quàm ut scepum exquisitè assequamur: syllogismi figuram omninò dissimulamus. Quae tamen siexigatur, è probationibus Geometricis ad vivum exprimitur. Sed nos ea rejecamus, quae repetita non modò tedium, sed etiam obscuritatem parerent. Quod inter Demonstrationes facilè perspicient qui iudicio praediti erunt. Demonstrationum autem conclusiones, sunt Problemata & Theoremata.' J. Peletetier, *In Euclidis Elementa* (Lyon, 1557), p. 12.

Thus, Peletier appears as the representative of a particular sixteenth-century view of mathematics: he did not believe that Euclidean demonstrations were syllogisms, at least not in the sense of traditional syllogisms of the first figure. He thought that the Euclidean demonstrative system needed to be revised by reordering several theorems, so as to show the mutual dependence of propositions more clearly and to follow more directly the natural flow of human understanding. But he also believed that mathematics was the primary science and that 'nobody can distinguish true from false without being familiar with Euclid.' The value of mathematics was intrinsic, since 'any proof leading to truth is geometrical.' The basis for this belief is explained in the book's preface, addressed to Charles de Lorraine. Peletier there begins by saying that the main reason to recommend mathematics, above and beyond its utility and value, is the fact that while other arts consist of probable opinions, mathematics consists of true statements, and that nothing counts in it but order and *ratio*. He concludes with the claim that there is no point in ascribing the origin of geometry to a particular people, be they Egyptians, Chaldeans or Phoenicians, because geometry is in fact a sort of theory of the world and connected to creation itself, and:

> [...] in the same way in which the constitution of the World has been inscribed in the divine Mind from eternity, so the disciplines are a sort of celestial seed implanted in us, which yield fruit according to the care we put in making our portion grow.[46]

Peletier expressed this belief many times in terms of 'flames situated in the human mind', that is, in terms of the light of nature within the human mind which, as Melanchthon had explained, consisted in the faculty of counting and in basic notions.[47] What was usually called mathematics is a more articulate product of the same faculty: given that Euclid's *Elements* are the core of the classical mathematical corpus, they are here taken as a synonym of mathematics as a whole.

IV. *Ramus and Gosselin*

I have tried to indicate some of the factors that prompted Peletier, after publishing his algebra, to write a book on Euclid's *Elements.* There are certainly several reasons for this, and they most probably have something to do with John Dee's famous and greatly successful lectures delivered in Paris in 1550. Two main ele-

45 The arguments have some affinity with those of O. Harari, *Knowledge and Demonstration: Aristotle's Posterior Analytics* (Dordrecht, 2005).

46 'Ut in mente divina ab aeterno infixam fuisse Mundi constitutionem: sic disciplinas, caelestia quadam semina esse. quae in nobis insita, et pro rata cuiusque portione exculta, fructum edunt.' J. Peletier, *In Euclidis Elementa*, Preface.

47 See P. Melanchthon, *Liber de anima* in C. G. Bretschneider (ed.), *Philippi Melanchthonis Opera*, 28 vols. (Braunschweig, 1846), vol. 13, p. 138. In fact, for Melanchthon the *lumen naturae* proves the existence of the mind qua *architectatrix sapiens*.

ments emerged in my discussion: the Proclean idea of a mathematical 'natural light', made explicit through 'common notions', and the idea of a science common to both arithmetic and geometry, which was just beginning to be identified with algebra. These two ideas motivated Peletier throughout his career and propelled him to study those features of mathematics that make it different from any other science and make of it a privileged path to superior knowledge.

Peletier shared these views with Petrus Ramus, his contemporary and another of Oronce Fine's pupils, who published a work on Euclid's *Elements* in Paris in 1545.[48] This text may seem disappointing to the modern reader, for it is a list of all Euclidean postulates, axioms and propositions. Not only are the demonstrations missing, but so are the figures. However, I think that both absences indicate only that he believed strongly in the need for students to reconstruct Euclidean mathematics in a dialogical process accompanied by figures drawn in the sand, as he himself explains.[49] It would be useful to know in what ways this book was used. It was certainly conceived as a teaching tool, but was it actually used in any college or even at the Collège Royal? Regardless, it is noteworthy that the dedicatory letter to Charles de Lorraine – who was also, as we have seen, Peletier's patron – contains a sentence that points in the same direction as the preface to Peletier's *In Euclidis Elementa*, by declaring that (according to Proclus) for Pythagoras and Plato these 'notions' were innate and should for this reason be called by the name of 'mathesis', or, as it were, 'reminiscence' or 'remembrance'.[50] Furthermore, if we look at Ramus's *Dialectique*, we can find some elements that place a number of Peletier's mathematical ideas in a broader context. Indeed, Peletier and Ramus both worked on the logic of Euclid's *Elements* and both considered the popular project of trying to translate Euclid's demonstrations into Aristotelian 'scientific' arguments (first figure syllogisms) to be useless.

Finally, it is worth considering another mathematician who took up some aspects of Fine's legacy: Guillaume Gosselin. Gosselin published an Algebra – *De arte magna* (1577) – in which he commented on Diophantus's *Arithmetic*. He also published a translation and commentary of the first two books of Tartaglia's *General Trattato*: *L'Arithmétique de Nicolas Tartaglia Brescian* (1578). Both of these works fed into his third publication, *De ratione discendi docendique mathematics* (1583), but certain elements of this work can be also traced back to Fine. First of all, in what concerns the theoretical frame of the work, we can find some

48 P. Ramus, *Euclidis Elementa mathematica* (Paris, 1545).

49 Isabelle Pantin takes this lack of figures less seriously. See Pantin, 'Ramus et l'enseignement des mathématiques', p. 73.

50 '[Pythagoras and Plato] believed that in the soul there are many excellent notions from the first intelligence, situated by nature in non generated and eternal examples.' ['(Pythagoras and Plato) in anima tam excellentes notitias ab intelligentiae primae ingenitis et aeternis exemplis insitas, et ingeneratas crediderunt [...]].' Ramus, *Euclidis Elementa mathematica*, dedicatory letter.

aspects of Fine's neoplatonic introduction to the *Protomathesis*, even though Gosselin was aware of Peletier's, Ramus's and even Tartaglia's philosophical theories on mathematics. Furthermore, Gosselin includes a long section of *propositiones hactenus desideratae* which directly evokes, in the title as well as in the content, Fine's posthumous work *De rebus mathematicis hactenus desideratis*, published just after his death, in 1556.

It certainly cannot be said that Oronce Fine should be considered among sixteenth-century French algebraists. However, his position as a mathematics professor at the Collège Royal meant that he was the teacher of a generation of mathematicians who constructed a French algebraic tradition immediately prior to Viète and Descartes. As has been noted, two of the most important members of this group of algebraists, Peletier and Ramus, studied at the Collège de Navarre in Paris, where Oronce Fine taught mathematics, and attended his lectures at the Collège Royal. Peletier and Ramus shared many interests and theoretical theses, both working on grammar, orthography, the geometry of Euclid, algebra, and rhetorical tradition. We can legitimately interpret their common interests as a legacy of Fine's teaching. In spite of the few points in common, there is no famous *querelle* dividing them, in contrast to those between Peletier and other students of Fine, such as Jean Borrel, for example. Indeed, the affinities between Peletier and Ramus where apparent to some of their contemporaries: from the point of view of a German thinker such as Jakob Schegk, Peletier and Ramus belonged to the same 'party': those who reduced mathematical demonstrations to arguments of common language, that is dialectical arguments.[51]

Yet what was the impact of these two figures and thus, ultimately, of Oronce Fine? While it seems now that on most mathematical topics Peletier preceded Ramus, Ramus had an enormous effect on education, within France as well as abroad. During their lifetimes, Ramus had a far greater influence on the colleges and on generations of teachers, while Peletier, instead, aimed his reform at the intellectual milieu of the academies and, even more, at the public he could reach through publication.[52] Improving the practice of printing was in fact his main goal, and some experimentation was allowed. Accordingly, he published many different sorts of works: translations of classical authors, poetry, mathematical and medical theoretical treatises, scientific poetry, manuals of practical geometry and astrology. Even though in part these works were concerned with teaching, they were really conceived in order to improve the genres to which they belonged, and developed their topics with much greater attention to the reader than Ramus's

51 See G. Cifoletti 'From Valla to Viète: the Rhetorical Reform of Logic and its Use in Early Modern Algebra', *Early Science and Medicine*, vol. 11, no. 4 (2006).

52 Peletier was very aware of the possibilities of printing for mathematics. See e.g. his comments in the Proème of the second book of his *Arithmétique*, sig. Gii^{v}.

works. The disciplines therein were treated in detail, not merely with a list of topics and a discussion of the sources, full of digressions, as in many of Ramus's works. Peletier's works had a greater impact on mathematics itself, while Ramus influenced logic. We can say, though, that both followed the way opened by Oronce Fine in understanding mathematics not only as an important intellectual heritage, but also as an encyclopaedia of practical disciplines, and mathematical texts as susceptible to a new and more effective structure.

Dropped Out of Sight: Oronce Fine and the Water-clock in the Sixteenth and Seventeenth Centuries

ANTHONY TURNER

In most recent histories of horology, the water-clock, or clepsydra, disappears at the Renaissance.[1] In some earlier works it holds a slightly more prominent place. If Willis Milham ends his seven page summary of the development of the instrument with the words 'but the clepsydra at best was an unhandy timekeeper', F. J. Britten, following Hamburger, speaks of a 'revival of clepsydrae' in the seventeenth century and he gives a few details.[2] In the still earlier works of Planchon and Dubois there is a fuller discussion of the water-clock in the form that it was best

1 For example, D. S. Landes, *L'Heure qu'il est. Les horloges, la mesure du temps et la formation du monde moderne* (Paris, 1987) makes no mention of water-clocks after the sixteenth century; Dohrn-van Rossum suggests implicitly that water-clocks dried up in the late thirteenth/early fourteenth century: 'The reason why the tradition of hydraulic drives came to an end can only be connected with rising demands as to reliability – as would have been demanded by the astronomers – or mechanical efficiency'. G. Dohrn-van Rossum, *History of the Hour: Clocks and Modern Temporal Orders* (Chicago and London, 1996), p. 91. After a survey of water-clocks in Antiquity and the Middle Ages in both East and West, Bruton tells us about a revival of interest for Heronic mechanisms during the Renaissance when hydraulic gardens were created full of 'fountains, water-gushing from animals, birds singing, organs and moving figures such as Neptune appearing out of the water, but no timekeeper'. E. Bruton, *The History of Clocks and Watches* (New York, 1979), p. 28. After confusing the true water-clocks of the seventeenth-nineteenth centuries with some early twentieth-century imitations, Von Bassermann-Jordan says no more about them. See E. von Bassermann-Jordan, *Montres, Horloges et pendules*, 2e édition française par Hans von Bertele (Braunschweig, s.d. [*ca.* 1964]), pp. 339–40. Hill mentions seventeenth- and eighteenth-century European water-clocks only in passing to end his account of them in the Middle Ages. See D. R. Hill, 'Water Clocks' in A. Smith (ed.), *The Country Life International Dictionary of Clocks* (Feltham, 1979), p. 135. Salles does mention eighteenth- to twentieth-century water-clocks, but gives no details. R. Salles, *Si le temps m'était comptait ... Mesure et instruments* (Rennes, 1991), p. 61.

2 W. I. Milham, *Time and Timekeepers, Including the History, Construction, Care and Accuracy of Clocks and Watches* (New York, 1923), p. 57. F. J. Britten, *Old Clocks and Watches and their Makers. Being an Historical and Descriptive Account of the Different Styles of Clocks and Watches of the Past in England and Abroad...*, 6th edition much enlarged (London, 1932), p. 14.

known during the Ancien Régime.[3] Both authors use some late seventeenth-/early eighteenth-century texts and both, like Britten, were familiar with some material examples – experience which their twentieth century successors may not have had. Whatever the case, all these accounts of water-clocks are inadequate and one has the impression that for their authors the water-clock has hardly any importance compared with *proper* mechanical clocks, that is to say, those powered by springs or weights. In consequence the history of the water-clock in the Early Modern and Modern periods is largely ignored in general histories of horology. The aim of the present essay, therefore, is to show, by means of an outline of its development in the sixteenth to eighteenth centuries, the interest of this neglected phase in the history of time-keeping. But before beginning, the myth that a water-clock is not a truly mechanical clock should first be dismissed. Whether the motive force for the mechanism of a clock is the weight of water, of stone, of lead or brass, or be supplied by the energy of an uncoiling spring or even by electricity changes nothing in the nature of the machine that measures time.[4] The true, fundamental, distinction to be made in the history of horology is between fixed clocks and portable ones.

The materials exist to write a history of water-clocks during the Early Modern period, and there have even been recent discoveries. The machines inherited at the Renaissance from Antiquity and the Middle Ages all depended on the inflow or outflow of water. Renaissance scholars would add some new models – siphon clepsydras, compartmented-cylinder clepsydras and a few developed during researches on the weight of air, on its temperature and on the movement of water.

Siphon clepsydras are true products of the Renaissance. They emerge from that plenitude of playthings – automata, fountains, trick drinking vessels, singing birds, mechanised nefs and walking figures – devised and realised by scholars and craftsmen following the rediscovery of the writings of the Greek mechanicians of Hellenistic Antiquity and in particular of Hero.[5] If some knowledge of Hero had traversed the High Middle Ages it largely disappeared during the thirteenth cen-

3 M. Planchon, *L'horloge, son histoire retrospective, pittoresque et artistique* (Paris, s.d. [1899]), chapter 1. P. Dubois, *Histoire de l'horlogerie depuis son origine jusqu'à nos jours...* (Paris, 1849), chapter 5.

4 Montucla was of the same opinion: 'Je ne connois point de trace', he writes 'd'horloge mécanique avant Ctesibios'. J. F. Montucla, *Histoire des mathématiques,* nouvelle édition, 4 vols. (Paris, An VII [1798/99]), vol. 1, p. 532.

5 A recent survey of the Hellenistic origins is B. Gille, *Les Mécaniciens grecques: la naissance de la technologie* (Paris, 1980). The classic study, now in need of rewriting, is A. Chapuis and E. Gélis, *Le Monde des automates, étude historique et technique*, 2 vols. (Paris, 1928). For recent orientations see A. Marr, '*Gentille curiosité*: Wonder-working and the Culture of Automata in the Late Renaissance', in R. J. W. Evans and A. Marr (eds.), *Curiosity and Wonder from the Renaissance to the Enlightenment* (Aldershot, 2006), pp. 149–170.

tury. An isolated reference in the mid-fifteenth century apart, it was not until the last decades of the fifteenth century, even the beginning of the sixteenth century that fuller studies began.[6] The first manuscript copies extant of the Greek text date from this period as do the first extracts to appear in print.[7] Thereafter the work was used throughout the sixteenth century, Latin and Italian translations appearing in its closing decades.[8]

I. *Oronce Fine's Water-clock*

The extant texts of Hero do not include a description of a water-clock, but the numerous ways of manipulating water by siphons that he describes provided both craftsmen and scholars with a starting point for the conception and realisation of new and various models.[9] One such was that developed by Oronce Fine. At the end of the treatise on sun-dials which makes up the fourth part of his *Protomathesis* (1532), he added a short text entitled: 'To make a water-clock showing equal hours recently devised with marvellous art by the author'.[10] The text, in a summary translation, is as follows:

> Before closing, allow me to add the description and explanation of a water-clock that I recently devised. For even if several investigators of not despicable things have explained various hydraulic machines, up to now no one has

6 In a work entitled *Proteus* of which the authorship remains doubtful. For discussion see M. Clagett, 'The Life and Works of Giovanni Fontana', *Annali dell' Istituto e Museo di Storia della Scienze di Firenze*, vol. 1, no. 1 (1976), pp. 5–67, at pp. 25–6. The work is mainly devoted to problems related to hydraulics and to siphons. For a text entitled 'Horologium aqueum' by Fontana see *ibid.*, pp. 11–13.

7 Giorgio Valla, *De expetendis et fugiendis rebus* (Venice, 1501), Lib. XV, cap i, abbreviates into eight pages part of Hero's Preface and his theory of siphons. For the reception of Hero see M. Boas, 'Hero's *Pneumatica*, a Study of its Influence and Transmission', in O. Mayr (ed.), *Philosophers and Machines* (New York, 1976), pp. 90–100; A. Marr, 'Understanding Automata in the Late Renaissance', *Journal de la Renaissance*, vol. 2 (2004), pp. 205–222.

8 Among those who interested themselves in Hero can be cited Regiomontanus, Leonardo, Cardano, Ramus, Gesner, Buontalenti, Baldi and John Dee. For details see Boas, 'Hero's *Pneumatica*', pp. 92–3, n. 9; Marr, 'Understanding Automata', p. 210, n. 41, and for Leonardo, C. Zammattio, 'The Mechanics of Water and Stone', in L. Reti (ed.), *The Unknown Leonardo* (London, 1974), pp. 196–197.

9 For the text of Hero see the edition of the Greek with German translation by W. Schmidt, in *Heronis Alexandrinus Opera quae supersunt omnia*..., 5 vols. (Lepizig 1899–1914), vol. 1. There is an English translation, by Bennet Woodcroft, *The Pneumatics of Hero of Alexandria* (London, 1851) and a French version by Albert de Rochas, *La Science des philosophes et l'art des Thaumaturges dans l'Antiquité* (Paris, 1882).

10 The separate title page for *De solaribus horologiis*, which was re-published on its own in 1560 by Cavellat, is dated 1531. See Eagleton in this volume for further discussion of this text.

> described what we have found, [a machine] which shows equal hours by the flow of water [...]. Firstly [Fig. 12.1] a rectangular tower should be constructed ABC about 3 cubits high. Inside it a reservoir, D, should be placed filled with very clean water. Above this place the arbor AB carrying a cylinder E and a dial A which shows the twelve equal hours. After that we construct a ship FG in gilt brass which the water easily carries. A curved tube [H I] passes through the mast and the hull so that the end H is in the water. A cord, running from the top of the rigging, is coiled round the cylinder E. At the end of this cord we put a [counter]-weight K. The diameter of the tube I is calculated so that in a whole hour as much water falls into the barrel [L] as is necessary to allow the hand on the dial to traverse the distance of an hour [...]. So we completed with art this beautiful machine which is shown in the illustration as it was presented to François Ier [...]. When the reservoir D is full, the ship and the [counter]-weight set in place, and the hand positioned at the end of the hour, the air in the siphon comes out of the opening I and, because it has a horror of a vacuum, the water follows it. And because the part of the tube outside [the ship] is longer that the part contained in the mast, I is lower. So one understands that the water continues to flow [...]. The water flows, the ship drops down and the hand turns. But, because the ship continues to float in the water falling at a constant speed, the part of the tube GH which is beneath the [surface of] the water remains always at the same depth. In consequence, the water runs out uniformly and the hour shown is also uniform [i.e., is an equal hour].[11]

This is only a short text, but one which suggests questions and invites commentary. Firstly why is it placed at the end of a treatise on sundials? What exactly had Fine 'recently devised'? If his idea was realised did it have any influence?

That a description of a water-clock should be placed at the end of a sun-dial text comes probably from the fact that consistently since Antiquity, water-clocks were seen as the natural complement of solar clocks – the time-*keeper* which could function at night and when the sky was overcast, when the time-*finder* that is a sun-dial could not operate. Progressively in the decades to come this role would be taken over by weight- and spring-driven clocks, although change in the motive force changed nothing in the task.[12] What is more interesting is to investigate in what ways Fine was really original. His basic siphon mechanism running through the ship is not novel as it derives from Hero of Alexandria's 'siphon with uniform debit'.[13] Nonetheless, in using it, Fine was to some degree original for when he developed his clepsydra and wrote his text no complete edition of Hero existed in print, only the extracts offered by Giorgio Valla. Although, as is the case

11 Fine, 1532*b*, cited from Fine, 1560, pp. 190–192.

12 For a more detailed discussion of the dial/clock relation see A. Turner, 'Essential Complementarity – the Sun-Dial and the Clock', in H. Higton (ed.), *Sundials at Greenwich: A Catalogue of the Sundials, Nocturnals, and Horary Quadrants in the National Maritime Museum* (Oxford and Greenwich, 2002), pp. 15–24.

13 See de Rochas, *La Science des philosophes*, p. 94, n. 11.

for Leonardo and for Cardano, the exact source of Fine's knowledge of the *Pneumatica* remains to be discovered, at the least his name can be added to the list of those who knew and used the text before a complete version of it appeared in print.

If the 'marvellous art' of Fine is thus reduced to having drawn information from an ancient text which was little known at the time – the late 1520s – when he wrote, he had at least the merit of understanding the text and knowing how to apply it to the construction of a time-keeper – something that no one else had done. But his originality is limited to that. That Fine's water-clock shows equal hours is not because of his own ingenuity but simply because the siphon arrangement causes the water to flow out at a uniform rate. That he nonetheless insists on the point may suggest that the distinction between equal and unequal hours, and use of the latter, had not yet been fully absorbed into the daily life even of educated society.

Was an example of Fine's clock ever made? To this question he gives us the answer himself. He states without equivocation that 'this beautiful machine [...] is shown in the illustration as it was presented to François Ier', and there is no reason to doubt this affirmation of a fact as easy for his contemporaries to verify as it is difficult for us.[14] If he made other water-clocks perhaps they were not as luxurious as that presented to his monarch, but the dates of his first activity may be important. Fine, who had taught mathematics from 1516 at the Collège de Navarre, must have developed his *Protomathesis* during the 1520s and thus have conceived and had his clepsydra built during the same period. Perhaps it is not unreasonable to postulate that he presented it to the king towards the end of the decade. In 1530 he was appointed to the chair of Mathematics at the Collège Royale, receiving his first salary in March 1531.[15] Was it the presentation of such an elaborate machine – tangible evidence of his capacity in the realm of practical mathematics reinforcing the pedagogical reputation that he had established at the Collège de Navarre – that won him the post? If, even partially, such were the case

14 Were there other water-clocks at the court of François Ier? A clock that the king sent to Süleyman the Magnificent in 1547 is said to have been one. That it used water is certain, that the water was used as the motive-force is less certain, as the text of Jean Chesneau is ambiguous: 'Un grand horloge faict à Lyon où y avoit une fontaine Qui tiroit par l'espace de douze heures de l'eau qu'on y mettoit, qui est un chef d'œuvre et de hault pris'. J. Chesneau, *Le Voyage de Monsieur d'Aramon, ambassadeur pour le Roy en Levant*, cited in O. Kurz, *European Clocks and Watches in the Near East* (London and Leiden, 1975), p. 24, n. 2.

15 See D. Hillard and E. Poulle, 'Oronce Fine et L'Horloge Planétaire de la Bibliothèque Sainte-Geneviève', *Bibliothèque d'Humanisme et Renaissance*, no. 33 (1971), pp. 311–351, at p. 321, n. 12; I. Pantin, 'Teaching Science and Astronomy in France: The *Collège Royal* (1550–1650)', *Science & Education*, no. 15 (2006), pp. 189–207, at p. 191. See also Pantin in this volume for further discussion of Fine's teaching activities.

then it is a good illustration of one of the social roles of water-clocks in the Renaissance (as indeed of scientific instruments in general), that of being an unusual and ingeniously contrived device that offered concrete evidence of the skills of a mathematical client in search of a patron.

II. *Water-clocks after Fine: the Compartmented Cylinder Model*

How many examples of Fine's water-clock might have been constructed we do not, and probably cannot, know. That the design was picked up in the bookish tradition of machines we do know since the second clepsydra proposed by Salomon de Caus, 'Another way of making a water-clock', the form of the float apart, is virtually identical.[16] But water-clocks could also be conceived outside the Heronic tradition of siphons.[17] An example is provided by the device known as the 'drum' or 'compartmented cylinder clepsydra', a model which would have a long life. The essential part of the compartmented cylindrical clepsydra, 'the soul of these clocks',[18] is:

> [...] a drum, or round box of well soldered metal [...] in which a determined quantity of water is prepared, and there are several little compartments which communicate one with the other [by a small orifice] close to the centre, and which let pass just as much water as is needed to let [the drum] fall gently little by little.[19]

When the amount of water and the size of the openings between the compartments are properly adjusted the movement of the drum can be used to mark time on an hour scale.

An ingenious hybrid combining the principles of both the inflow and the outflow clepsydra as the water flows into and then out of the compartments in the cylinder, the origins of this device remain in doubt. It may derive from the mer-

16 S. de Caus, *Les Raisons des forces mouvantes avec diverses machines tant utiles que plaisantes. Aux quelles sont adionte plusieurs dessins de Grotes et Fontaines* (Frankfurt, 1615), section VIII. It should be noted that among his predecessors de Caus mentions Hero and Fine.

17 See for example a model devised by Franciscus Linus, an English Jesuit at Liège, of which Peiresc preserved a brief description now conserved in Bibliothèque Nationale de France, Paris, MS Français 9531, fol. 150^{r-v}.

18 'Traité des horloges Elementaires, ou de la manière de faire des Horloges avec l'Eau, la Terre, l'Air & Feu', translation of D. Martinelli, *Horologi elementari divisi in quattro parti...* (Venice, 1669) in J. Ozanam, *Récreations mathématiques et physiques* (Paris, 1694), I have used the edition published at Amsterdam in 1697 where the phrase quoted occurs on p. 478.

19 '[...] un tambour ou boëte ronde de métal bien soudée [...] dans laquelle il y a une certaine quantité d'eau preparée, & plusieurs petites cellules qui ont communication les une avec les autres proche du centre, & qui ne laissent écouler l'eau qu'autant qu'il est necessaire pour [le] faire descendre doucement & peu à peu'. Ozanam, *Récreations mathématiques*, p. 311, n. 19.

cury clock of Isaac b. Sid of *ca*. 1276, described in the *Libros del Saber* in Castilian and translated into Italian in 1341 by Gueruccio Federighi in his *Trattato del Sfera*. But it could also have an independent origin elsewhere in Europe for what is perhaps a water-clock of this type is depicted in an illustration to a moralised bible executed for Louis VIII of France in *ca*. 1223–1226.[20] Whatever the case, evidence is lacking for the influence of these medieval devices on the [re-]invention of the drum clepsydra during the Renaissance. This could have occurred towards the end of the sixteenth century when Attilio Parisio published a brief account of the subject to accompany a water-clock using a drum which he had made and for which he sought a papal patent. His text was deliberately obscure, but it was explained and commented on in 1626 by an anonymous author writing in Treviso. It was not until 1648 that a clear, complete description was published by Mario Bettini (1582–1657).[21]

It was from this time onwards that the compartmented cylinder clepsydra began to play a part in the mainstream of European horology. Immediately on publication the work of Bettini was taken up by Francesco Eschinardi.[22] In 1656, the Campani brothers in Rome working on the development of a silent clock for use at night used the principle of the compartmented cylinder. Pier Tommaso Campani stated explicitly that the idea came from remembering 'a silent clock of which the invention was not new, that operated by means of water, which flowed in a certain drum variously divided.'[23] In England, the *virtuoso*, amateur of the sciences, Sir Anthony Cope (d. 11 June 1675) had a drum clepsydra based on that described

20 For a general survey of the history of such water-clocks see S. A. Bedini, 'The Compartmented Cylindrical Clepsydra', *Technology and Culture*, no. 3 (1962), pp. 115–142. For Isaac b. Sid see M. Rico y Sinobas (ed.), *Libros del Saber de Astronomia del Rey Don Alfonso X de Castilla,* 5 vols. (Madrid, 1863–66), vol. 4, pp. 67–76; D. R. Hill, *Arabic Water-clocks* (Aleppo, 1981), pp. 126–30; A. A. Mills, 'The Mercury Clock of the *Libros del Saber*', *Annals of Science*, no. 45 (1988), pp. 329–44. For the clepsydra in the moralised bible, C. B. Drover, 'A Medieval Monastic Water-Clock', *Antiquarian Horology*, vol. 1 (1953–56), pp. 54–8; C. B. Drover, 'The 13th Century "King Hezekiah" Water-Clock', *ibid.*, vol. 12 (1980), pp. 160–64; J. H. Combridge, 'The 13th Century "King Hezekiah" Water-Clock', *ibid.*, p. 300.

21 For Parisio see Bedini, 'The Compartmented Cylindrical Clepsydra', pp. 118–24; for the anonymous commentary, which has never been published, see *The Honeyman Collection of Scientific Books and Manuscrcipts Part III: Manuscripts and Autograph Letters of the 12th to the 20th Century*, Sotheby Parke Bernet, London (2 May, 1979), lot 1251. Bettini's description is in his *Aerarium philosophiae mathematicae* (Bologna, 1648), pp. 45–60: 'Epinomis post partem II, tomi II [...]'.

22 F. Eschinardi, *Horologium hydraulicum ex Aerario P. Mario Bettini cum appendice P. Francesco Eschinardi* (no place or date), and in M. Bettini *Apiaria philosophiae mathematicae*... 3 vols (Bologna, 1645–54), supplement.

23 *Lettera di Pier Tommaso Campani ... Nella quale le dimonstra l'Origine e l'artificio dell'Oriolo de lui Fabbricato l'anno 1656...* (Rome, 1660), cited in Bedini, 'The Compartmented Cylindrical Clepsydra', p. 129.

by Bettini with which he made various experiments.[24] Perhaps it was news of this example that provoked the Royal Society in London to order the making of a similar one in January 1664.[25]

In Italy and England therefore the compartmented cylindrical clepsydra was already known when in 1669 Domenico Martinelli published the fullest description of it that we possess.[26] For the machines that he describes, Martinelli claims no originality, his work is purely descriptive. He explains in detail the devices described by the authors mentioned above, and among them the model which would impose itself and be manufactured throughout the eighteenth and nineteenth centuries. This is made up of a wood frame, of which one or both of the vertical members carry a 24 hour scale divided 8/9 to 12/1 and thence back to 8/9 with sub-divisions of 30, 15 or 5 minutes. In the centre of the frame the drum is suspended from two cords attached at the top to hooks on the frame, and with their lower ends coiled round a long arbor passing through the centre of, and carrying, the cylinder. When the cylinder is set into motion, all being properly adjusted (size of the compartments, size of the orifices, thickness of the cord, weight of water), the position of the ends of the arbor marks the hour on the scale on the frame (Fig. 12.2).

Use of the arbor to mark the hour differentiates this form of falling drum water-clock from the model described by Parisio, Bettini and Eschinardi in which time was shown by numerals attached to one of the flat faces of the cylinder. Gradually it came into general use. At the end of the seventeenth century it received considerable publicity in France. Writing from Nîmes on 12 March 1691, François Graverol (1636–1694) sent a description to the *Journal des Sçavans* in which he stated that the clock had been invented by a Jesuit in Bologna (Bettini would fit this description) and perfected by the sons of the Professor of Law, Taliaisson, at Toulouse. What Graverol described was the clock previously described by Martinaelli (and illustrated in his second plate, p. 44), which fact was noted by the editor of the *Journal des Sçavans*, who further added that an example had functioned very well for eight years at the *Accademia fisico-matematici* domiciled in the palace of Giovanni Giustini Ciampini at Rome.[27] In 1734, Jacques Allexandre

24 R. Plot, *The Natural History of Oxfordshire* (Oxford, 1705 [2nd edition]), p. 240.

25 '6 January 1663/4 Mr. Hooke was desired to take care, that the instrument for the measure of time, consisting of only one wheel with hollow camera's in it, moving either with sand or water for a good space of time, be made'. T. Birch, *The History of the Royal Society of London*..., 4 vols. (London, 1756), vol. 1, p. 367. 'To have a Wheel made with hollow Caverns in it, for a Clock or a Jack to be moved, either by Water or Sand'. 'Experiments and Observations recommended to Dr. Hooke', *Royal Society Classified Papers*, vol. 20, no. 26, item 9.

26 D. Martinelli, *Horologi elementari*, in Ozanam, *Récreations mathématiques*, pp. 473–498.

27 This information probably came from Wilhelm Homberg (1652–1715) who had just

claimed that the instrument had been invented by one of his colleagues in the order of the Benedictines of St Maur, Charles Vailly (1646–1746), who 'had one made in the town of Sens in Burgundy by Regnard, pewterer of the said town, in 1690', whence the device became known in Paris.[28] Elsewhere he modifies this statement, saying that Vailly got his idea from the *Technica Curiosa*, where in the forty-first proposition Schott described a clock working on the same principle but using sand, for which Vailly 'successfully substituted water'.[29] Ozanam, writing in 1694, had his own version: 'These clocks are not as new in France as some would want us to believe: & there are more than twenty years since some of our curious [enquirers] have had them in their cabinets'. Following Martinelli, he affirms that 'there are more than thirty years that they were invented in Italy'. 'But' he adds:

> [...] as a secret and a mystery has always been made of the way in which they are constructed internally, I thought I would give pleasure to the many persons who would like to make them, by making this secret public with the translation of an uncommon book, and give occasion to our nation, of which the talent is the perfection of the arts, to push the construction of this automate to its ultimate perfection.[30]

Even so, that this type of clock was still really a novelty in Paris finds some confirmation by the publication of a prospectus announcing their fabrication by the 'sieur Duflos' who, having obtained a royal patent, sells them at his house 'rue de la Vieille-Monnaie, attenant au Singe-Vert'.[31]

Even if it is not possible to be precise about the origins of the compartmented cylinder clepsydra, the essential part of its development seems to have taken place during the seventeenth century. In its first form, with hour indications on the flat surface of the cylinder, relatively few examples seem to have been made.

---

returned from Rome where he attended meetings of this academy precisely in 1691.

28 J. Allexandre, *Traité général des horloges* (Paris, 1734), pp. 73–87. The term used by Allexandre – *étaminier* – means literally someone who applies a thin coat of pewter to another metal, but the sense is probably of a general worker in pewter.

29 Allexandre, *Traité général des horloges*, p. 293.

30 'Ces Horloges ne sont pas si nouvelles en France que l'on a voulu le faire croire : & il y a plus de vingt ans que quelques-uns de nos Curieux en ont dans leurs Cabinets [...] qu'il y a plus de trente ans qu'elles ont été inventées en Italie. Mais, comme on a toujours fait un secret & un mystère de la manière intérieure dont elle est construite, on a cru faire plaisir à bien de gens qui voudront l'exécuter, de rendre ce secret public par la traduction d'un Livre peu connu, & de donner occasion à nôtre Nation, dont le talent est de perfectionner les Arts, de pousser la construction de cet Automate jusqu'à sa dernière perfection' Ozanam, *Récreations mathématiques*, p. 475. What he says here in his commentary on the work of Martinelli contradicts what he had earlier written, where he recounts the appearance of the instrument in Burgundy in 1693: 'Si je sçavois qui est l'Inventeur d'une Montre si simple & si extraordinaire, je lui rendrois ici justice' (p. 311).

31 Cited in Planchon, *L'horloge*, p. 8.

Nevertheless this arrangement was not without importance for, thanks to the work of the Campani brothers, this dial arrangement was adopted in the weight- and spring-driven night clocks which became popular in Italy and England during the second half of the seventeenth century. In the form in which the hours were indicated by the cylinder arbor, the drum water-clock had considerable success. Throughout the eighteenth and nineteenth centuries it was made by pewterers, mainly in the centres of this industry at Chartres and Sens although no doubt elsewhere, being popular as a low price, attractive and reliable time-keeper (see Appendix III). That examples are found relatively rarely today does not mean that they were rare in their time. As Planté explained in 1890, 'foreign collectors and [...] the founders of spoons have, for their different purposes, made such chase after them that it has become difficult to find examples'.[32]

III. *Water-clocks after Fine: the Ideas of Perrault and Amontons*

The siphon clepsydra was, as we saw above, devised by the learned and connected with the world of the court. The drum clepsydra, also devised by the learned, rapidly left their world to become integrated into everyday life. From the learned world also stemmed water-clocks based on models from Antiquity, and others conceived in the context of researches effected around barometers, thermometers and the weight and elasticity of the air. For an idea of activity in this domain we can examine the water-clocks of Claude Perrault (1613–1688) and Guillaume Amontons (1663–1705).[33]

Perrault was led to the study of water-clocks by his taste for mechanics in general and by his particular need to comment on the descriptions of them included in the *De architectura* of Vitruvius for the magnificent translation with commentaries that he published in 1673. In his commentary he notes that all water-clocks are subject to two main defects:

> The first [...], is that the water flows with more or less facility according as the air is more or less dense, warmer or cooler: for that prevents the hours from being exact. The other is that water runs out faster at the beginning when the vessel from which the water falls is full, than at the end because the weight of the water is greater at the beginning than at the end'.[34]

32 J. Planté, *Gnomons et clepsydres* (Laval, 1890), p. 24.

33 I leave out of consideration here the table-clepsydras elaborated by Athanasius Kircher (1602–1680) and Gaspar Schott (1608–1666) and their successors of which two seventeenth-century examples have survived. For a general survey see S. A. Bedini, 'The 17th-Century Table Clepsydra', *Physis*, no. 10 (1968), pp. 25–52.

34 'Le premier [...], est que l'eau s'écouloit avec plus ou avec moins de difficulté selon que l'air estoit plus ou moins epais, ou plus froid ou plus chaud: car cela empeschoit que les heures ne fussent justes. L'autre est que l'eaux s'écouloit plus promptement au commencement lorsque le vaisseau d'ou l'eau tomboit estoit plein, que vers la fin, à cause que la pesanteur de l'eau estoit plus grande au commencement qu'à la fin'. C.

It was in order to remedy these faults, he continues, that Oronce Fine had devised his siphon clepsydra. A few years earlier (probably between 1667 and 1669) Perrault had himself devised a balance water-clock, 'the advantage [...] which is found in the equality of the water-course, which can be adjusted, is not something entirely to be scorned'.[35] An example of Perrault's clepsydra was set up in the garden of the Bibliothèque du Roi, an example with a striking train still exists.[36]

Perrault's remarks in his Vitruvius were the starting point for his younger contemporary, Guillaume Amontons (1663–1705). He, in the context of researches into barometers, thermometers and hygroscopes produced an entirely original clepsydra (see Appendix I). He intended it for use at sea.[37] It was made up of a glass tube three (French) feet long attached to a vertical plank carrying a double twelve hour scale. At its lower end the tube became thicker to finish in a small reservoir worked at right angles to the tube itself. Inside the tube, which is filled with water there is a short capillary tube carried by a metal disc resting on an air-bubble introduced into the column. The capillary tube passes through the air bubble to rest on a second metal disc. To use the instrument, the disc was lowered

---

Perrault (ed. and trans.), *Les Dix Livres d'Architecture de Vitruve corrigez et tradvits nouvellement en François...* (Paris, 1684 [2nd edition, 1674]), p. 287, n. 13.

35 '[...] l'avantage [...] qui se trouve dans l'égalité du cours d'eau qui peut être réglé, n'est pas une chose tout-à-fait à mépriser'. 'Horloge à pendule qui va par le moyen de l'eau' in J.-G. Gallon, *Machines et inventions approuvées par l'Académie Royale des Sciences*, 6 vols. (Paris, 1735), vol. 1, pp. 39–40. Perrault transmitted the design by letter to Huygens, and demonstrated it to the Académie des Sciences.

36 A. Picon, *Claude Perrault, ou la curiosité d'un classique* (Paris, 1988), p. 98; Collections of the Musée du Conservatoire des Arts et Métiers, Paris, inv. 299.

37 See [G]. Amontons, *Remarques et Experiences Phisiques sur la Construction d'une Nouvelle Clepsidre, sur les Barometres, Termometres & Higrometres* (Paris, 1695). Fontenelle's remarks on the subject in his éloge of Amontons in 1705, are not without interest: 'Quoique les Clespsidres, ou Horloges à eau, si usitées chez les Anciens, ayent été entièrement abolies parmi nous par les Horloges à rouës infiniment plus justes & plus commodes, M. Amontons ne laissa pas de prendre beaucoup de peine à la construction de sa Clepsidre, dans l'esperance qu'elle pourroit servir sur mer ; car de la manière dont elle était faite, le mouvement le plus violent que pût avoir un Vaisseau ne le déregloit point, au lieu qu'il dérègle infailliblement les autres Horloges. On a pû voir dans le livre de M. Amontons avec combien d'Art sa Clepsidre étoit construite ; il n'y a guere d'apparence qu'il se soit rencontré avec aucun des anciens Inventeurs'. ['Although the clepsydrae, or water-clocks, so common among the Ancients, have been totally abolished among us by the infinitely more exact and convenient wheel-clocks, M. Amontons did not spare taking of a great deal of trouble in the construction of his clepsydra in the hope that it would be useful at sea; for given the way it was constructed, the most violent motion that a ship could have deranged it not at all, whereas it would infallibly derange other clocks. With what art his clepsydra was constructed can be seen in M. Amontons's book; it is hardly seems that there is any contact with the ancient inventors'.] B. Fontenelle, *Histoire de l'Académie Royale des Sciences. Année MDCCV. Avec les Mémoires ... pour la même année* (Paris, 1730), p. 152.

until it was set against the time known. Water passing through the capillary tube causes the air-bubble to rise. If the amount of air is properly adjusted to the weight of the tube and its disc, as they all rise together, time is indicated by the position of the disc against the hour-scale. Temperature effects can be compensated by reading the hour against the line on the scale corresponding to the temperature read from a thermometer set horizontally at the top of the supporting plank from which a grid of oblique compensation lines is drawn.

Amontons, like Perrault and probably thanks to him, was aware of Fine's water-clock, the siphon of which fitted in with his own researches on air. He was not alone. In England the architect William Winde devised a double tube clepsydra which is described in Appendix 2. A night clock invented and made by M. Beausard, sometime naval lieutenant in 1790, presented to the Academy of Sciences in 1791 seems to have a direct relation with that of Amontons.

Interest in water-clocks, then, was long lasting. The research devoted to them from the sixteenth to the eighteenth centuries, and even later, was part of larger, more urgent and more immediate investigations in hydraulics, pneumatics and mechanics. But such investigations were not purely theoretical. The drum-clepsydra moved rapidly from the world of ideas to the practical manufacturing world, to become a time-keeper made and used for two centuries. It is a time-keeper, however, generally ignored in histories of time-measurement because it was made, outside the normal circuits of time-piece manufacture, by pewterers. But between water-clocks and traditional horology, we have seen, there were some reciprocal influences. Research into clepsydrae was an integral part of the efforts made by superior craftsmen and *savants* in the Renaissance and Early Modern period to improve all the possible methods of time-measurement, researches which were made throughout Europe. In the twentieth century they have become forgotten because relatively few objects resulting from the research have survived, because many of those that were made did not depend upon traditional mechanical horology, and because, in the end, other methods of time-measurement predominated. Even so the history of water-clocks merits attention. One of the two oldest of all time-instruments, water-clocks played a founding and fundamental role in the making of the modern mastery of time.

APPENDIX I

Abbreviated Description of Guillaume Amontons' Water-clock (from *Procès verbale de l'Académie Royale des Sciences*, 7 January 1693, XIII, fols. 123–124)

Clepsidre d'une nouvelle Construction

AB est un Canon de verre d'environ 3 piés ½ de Longueur, sur quatre ou cinq Lignes de grosseure interieure, le plus egcl et uniforme en toute sa Longueur qu'il est possible: son extremité, A, est exactement fermée par un bouchon de même matiere, son extremité, B G H, est soufflée en forme urinal: ce Canon de verre est plain d'eau seconde a l'exception de l'espace, E D, qui renferme de l'air; D C, est un tuiau Capillaire d'environ deux pouces de Longueur et de la grosseur d'un crin de cheval, plus ou moins suivant qu'il sera dit ci après, Ce tuiau capillaire passe a travers et par le milieu de deux paillettes de Cuivre fort mince, l'une circulaire, E, l'autre triangulaire C, toutes deux arretées fixement sur led[it] tuiau avec un peu de cire et de telle grandeur qu'elles puissent librement couler d'unbout à l'autre du Canon de verre AB, ce Canon est appliqué sur une monture de bois sur la quelle est une graduation qui marque les heures depuis midi jusqu'à minuit, et depuis minuit jusqu'à midi, la grosseur du tuiau Capillaire est tellement proportionnée a celle du Canon AB, que l'eau seconde, degoutant goute a goute par l'extremité C, la paillette E, s'eleve et parcoure en 24 heures les 24 divisions de la graduation; après quoi pour renouveler le mouvement de cette Clepsidre on l'incline a l'horizon en la maniere representée, fig. 3e. pendant quoi l'air qui soutenoit le tuiau Capillaire s'échappe et s'engage dans l'infleure en GH, puis redresant la Clepsidre on donne lieu au tuiau Capilaire de descendre, voiés fig. 2nd. et lors qu'il est parvenu sur l'heure qu'on desire on l'i arreste en couchant la Clepsidre sur le Côté ; enfin on i fait rentrer l'air en la tenant en la scituation representée fig. 4e. alors la Clepsidre remise en sa scituation ordinaire continuera a marquer l'heure et ainsi du reste.

On avoit doné d'abord a cette Clepsidre la disposition de la 5e figure, mais plusieurs raisons ont fait qu'on s'est determiné uniquement a celle de la premiere. Pour peu de reflexion qu'on fasse sur cette Cleepsidre je m'assure qu'on la trouvera exemte, je ne veus pas dire, de tous les deffauts des autres, mais au moins très disposée a etre perfectionnée; il s'[...] des quantités egaux de liqueur dans des tems egaus, elle est sur et peu embarrassnte, d'un transport facile et qui même selon apparences, n'en doit pas sensiblement alterer son mouvement, puis les surfaces des liqueurs qui dans les autres Clepsidres ont beaucoup d'etenduë et s'elevent pour cette raison, lorque ces liqueurs [. . .] par vagues – qui en precipitent ou retardent le mouvement ne doit pas produire les mêmes effets en celle-ci ou les surfaces n'en ont point : on l'appreçeoit facilement si la liqueur contient quelques saletés qui puissent dans la suite alterer ou meme arreter son mouvement, ce qui n'est pas, selon moi, un petit avantage: En fin cette Clepsidre dont lieu de faire plusieurs experiences considerables sur l'ecoulement des liqueurs et peut-etre qu'etant bien perfectionné elle ne seroit pas inutile a la Navigation, a l'occasion de quoi l'assemblée juge que cette Clepsidre puisse etre de quelque precision sur Mer et qu'ille veuille bien me commetre le soin des experiences neccesaires a la perfectionner pour cet usage a la joignant aux idées passées les idées presents, je ne desspere pas qu'il n'en resulte quelque chose d'avantageus, sinon je serai toute aissé savoit d'avoir contribué auytant qu'il a eté en moi a l'avancement d'une si louable dessein. AMONTONS

APPENDIX II

A Water-clock by William Winde[38] (from 'Gulielmi Windaei Bergannsis ad Zoman Cursus Mathematicus Tomus Secundus Continens Doctrinum Tranguloru[m] Sphericum, Usum globorum Geographium, Astronomium, Horologiographiam, Navigationem, Opticam, Perspectiam, Staticam, Mechanicam, Mondini M.DC. LXXXII'. British Library, London, Kings MS 267 fol. 331^{v} [p. 605].)

Another Engine showing the hours.
Fig 483 AB is a pipe reaching to the bottome of this vessell C, w[hi]ch is to be filled with water, and about B it is divided into two branches BD and BE, and afterwards is continued by IH. The cocks let be wher [*sic*] the figure sheweth them about B and I; IH is a vessel holding as much as C; the Glasse pip[e]s DF, EG, should have a little hole at the top to lett the air in and out, and be distinguished into hours. Opening the cock at B, the water runneth out of the vessell C into the branches BD and BE, and riseth in the pipes DF and EG, and maketh in DF the forenoone hours, till the glasspipes be filled up. Then opening the cock at I, the water sinketh againe out off the glass-pipes DF and EG into the vessell H, and by its decrease sheweth the afternoone houres.

38 (Pre-1647–1722), soldier and architect, brought up in Bergen-op-Zoom (Brabant), active in England from 1660, member of the Royal Society 1662–85. See H. Colvin, *A Biographical Dictionary of British Architects 1600–1840* (London, 1978), pp. 902–905.

APPENDIX III

Known Makers of Compartmented Cylinder Water-clocks

Boyer	pewter-worker, Chartres (Tardy 82)
Duflos	rue de la Vieille-Monnaie, Paris, *ca.* 1693, prospectus of Thimothée Langlois (Planchon 9)
Arnold Finchett	Cheapside London, 2/4 eighteenth century, British Museum, London (Britten 15).
Finchett jr	London, 2/2 eighteenth century, British Museum, London
Hunat	Sens (Planchon 9)
Lacorde	Laisné, 1754 (Planchon 10)
Regnard	Sens, late seventeenth- /early eighteenth-century (Dubois 60).

Epilogue

STEPHEN CLUCAS

If this book tells us anything, it is that Oronce Fine was not a great mathematician. As Henrique Leitão's chapter shows, Fine's work was found seriously wanting by a series of the most able mathematicians of the sixteenth century: Pedro Nunes, Jean Borrel, Niccolò Tartaglia, Francesco Barozzi and Adrian van Roomen all took him to task in print. Nunes, in his *De erratis Orontii Finaei,* a work dedicated to what he saw as the manifold errors of Fine's *magnum opus* the *Protomathesis* (1532), criticised Fine for his ignorance of the art of mathematical demonstration,[1] his ignorance of the first rudiments of astrology,[2] and even of the first definitions of Euclid's *Elements*.[3] One cannot even say that Fine 'errs', Nunes adds,

> For one can only err if one dares to make mathematical demonstrations: but he rarely does so: unless, perhaps, you call those demonstrations Orontius's which are wholly and openly taken from Theon and Campanus: of which he makes no mention. In these things, therefore, he cannot err, unless Theon or Campanus erred before him.[4]

These damning criticisms – as Leitão shows – did little for Fine's international standing and a great deal for that of Nunes himself, who, as a newly-created professor at the University of Coimbra, had a great deal to gain from cutting the Royal Professor of Mathematics at Paris down to size. Fine's posthumous reputation declined and traces of it can be found in the few modern commentators on his

1 P. Nunes, *De erratis Orontii Finaei, regii mathematicarum Lutetiae professoris* (Coimbra, 1573 [first edition 1546]), p. 11. Fine errs, Nunes says, 'ob ignorantiam artis demonstrandi'.

2 'Oronce erred tremendously in the investigation of the longitude of places through ignorance of the first rudiments of astrology.' ['Orontium uehementer errasse in inuestigatione longitudinis locorum, ob ignorantiam primorum rudimentorum astrologiae.'] Nunes, *De erratis*, p. 35.

3 '[...] definitiones positas in initijs librorum Euclidis ignorare videatur.' ['(...) he seems to be ignorant of the definitions posited in the beginning of Euclid's books (i.e. the *Elements*).'] Nunes, *De erratis*, p. 28.

4 'Solum enim errat, cum mathematicas demonstrationes conficere audet: sed raro audet: nisi Orontij fortasse demonstrationes appelles, quas omnino palamque à Theone, & Campano mutuatus est: quorum tamen non meminit. In his enim errare non poterat, nisi prius aut Theon, aut Campanus errassent.' Nunes, *De erratis*, p. 1.

mathematical achievements. Emmanuel Poulle, as several of the contributors note, characterised Fine's works as 'encyclopaedic, elementary and unoriginal',[5] while Jean-Claude Margolin noted that despite the fact that Fine's contemporary reputation was 'immense' and his public lectures well-attended, 'he did not leave behind him a single work of the first rank.'[6] Nonetheless, as Margolin concedes, despite these setbacks to his reputation, Fine remained a significant figure:

> his reputation was, at the same time, not inconsiderable as a professor, mathematician, engraver and illustrator of scientific works – as a popularizer and [as a] maker of scientific instruments [...].[7]

Fine, Margolin, concludes, was not 'a great mathematician, but he was incontestably a great Professor and disseminator of mathematics.'[8]

Yet although Fine was unquestionably not a great mathematician, this volume shows us that the history of great mathematicians can obscure the value of other kinds of histories of mathematics. The 'revolutionary' ideas of 'great' mathematicians can overshadow the slower and equally important developments of workaday mathematics; they can obscure the outlines of a mathematical *culture*. In recent years more attention has been paid to mathematical cultures: the work of Antonella Romano and Ugo Baldini, for example, on Jesuit mathematical culture, or Stephen Johnston and Eric Ash's work on the culture of mathematical practitioners in Elizabethan England.[9] As Adam Mosley so aptly puts it in this volume, Fine offers us an excellent opportunity to study

5 E. Poulle, 'Fine, Oronce', in C. C. Gillispie (ed.), *Dictionary of Scientific Biography*, 18 vols. (New York, 1970–1990), vol. 15 (Supp.), pp. 153–157, at p. 156.

6 'Si la reputation de Fine fut immense, et si ses cours publique étaient très fréquentés, il n'a pas laissé une oeuvre de premier plan.' J.-C. Margolin, 'L'Enseignement des mathématiques en France (1540–70): Charles de Bovelles, Fine, Peletier, Ramus', in P. Sharatt (ed.), *French Renaissance Studies 1540–70: Humanism and the Encylopaedia* (Edinburgh, 1976), pp. 109–155, at p. 122.

7 '[...] mais sa réputation n'en était pas moins considérable à la fois comme professeur, mathématicien, graveur et illustrateur d'ouvrages scientifiques – [et comme] vulgarisateur et constructeur d'instruments scientifiques [...].' Margolin, 'L'Enseignement des mathématiques en France', p. 23.

8 'Oronce Fine n'a pas été sans doute un grand mathématicien: il fut incontestablement un grand professeur et un bon diffuseur des mathématiques.' Margolin, 'L'Enseignement des mathématiques en France', p. 128.

9 A. Romano, *La contre-réforme mathématique: constitution et diffusion d'une culture mathématique jésuite à la Renaissance* (Rome, 1999); U. Baldini, 'The Academy of Mathematics of the Collegio Romano from 1553 to 1612', in M. Feingold (ed.), *Jesuit Science and the Republic of Letters* (Cambridge, MA, 2003), pp. 47–98; S. Johnston, 'Making Mathematical Practice: Gentlemen, Practitioners and Artisans in Elizabethan England', unpublished PhD. diss. (University of Cambridge, 1994); E. Ash, *Power, Knowledge, and Expertise in Elizabethan England* (Baltimore, MD, 2004).

> [...] such apparently unrevolutionary topics as the nature of sixteenth-century mathematical education, the breadth and depth of the international mathematical community, and the extent to which particular tools and approaches were deployed and discussed within and without the early modern academy. Fine was much more representative of the mathematical culture of the period than the extraordinary figures who have tended to monopolise historians' attention.[10]

Fine was, in S. K. Heninger's words, 'a pedagogue with a flowing pen', publishing some thirty books over the course of his career, a number of which were reprinted more than once.[11] Writing both in Latin and French, Fine reached out to what Margolin calls the 'double clientèle' for mid-sixteenth-century French mathematics: the learned, Latinate audience of the colleges and universities, and the broader, public audience with various interests in practical mathematics.[12]

Several of the contributors to this volume note the academic orientation of Fine's work after his appointment as Royal Lecturer at the Collège Royal in 1531 and the importance of his mathematical textbook, the *Protomathesis*, for securing his pedagogical reputation. While this work has been criticised for its elementary nature or for its unoriginality (it could be seen as following in the footsteps of earlier encyclopaedic mathematical compendia such as those of Giorgio Valla, Jacques Lefèvre d'Etaples or Pedro Sanchez Cirvelo),[13] the close links between the *Protomathesis* and Fine's teaching duties at the Collège Royal gives it a different significance: it was one of the first attempts to formalize the teaching of mathematics as an autonomous academic discipline (rather than as a subordinate part of the quadrivium).[14] The two essays in this volume on Fine as Royal Lecturer (by Isabelle Pantin and Angela Axworthy) make clear the importance of the precedent which Fine's appointment created, and the contribution it made to the establishment of a thriving academic mathematical culture, of which later (and better known) figures such as Viète, Descartes and Pascal were the indirect beneficiaries.

Many of the contributors in the present volume stress that, in various ways, Fine was a participator in existing mathematical traditions rather than an innovator, but Eva Taylor, for one, viewed Fine's role as a synthesiser of existing mathematical knowledge in a very favourable light. '[Fine] did for France,' she wrote,

10 See above, p. 134.

11 S. K. Heninger Jr., 'Oronce Finé and English Textbooks for the Mathematical Sciences', in D. B. J. Randall and W. W. Williams (eds.), *Studies in the Continental Background of Renaissance English Literature. Essays presented to John L. Lievsay* (Durham, NC, 1977), pp. 171–85, at p. 173.

12 Margolin, 'L'Enseignement des mathématiques en France', p. 116.

13 Heninger, 'Oronce Finé and English Textbooks', p. 175.

14 On the importance of the *Protomathesis* and Fine's role as Royal Lecturer for the formalization of the discipline see Margolin, 'L'Enseignement des mathématiques en France', pp. 111, 122 and 141.

'what Gemma did for the Low Countries, and Mercator at a later date for Western Europe at large, he prepared maps and texts synthesizing [...a] vast mass of conflicting material.'[15] His textbooks might seem very modest achievements in retrospect, but as Adam Mosley points out in his chapter on Fine's cosmographical work, Fine was probably (if print-runs are an accurate guide) the second most popular cosmographical author in Europe after Peter Apian. S. K. Heninger has previously noted the popularity of Fine's *De mundi sphaera*, which he said was 'probably the best known text book of cosmography in England at the beginning of Elizabeth's reign', and Fine was treated as a great authority on the subject by Robert Recorde, who (according to Heninger) saw himself as the 'English Orontius'.[16] In his chapter on Fine's *De speculo ustorio*, Sven Dupré notes that Fine's works on burning mirrors was pillaged as a source of diagrams and ideas by Giovanni Battista della Porta in the late sixteenth century, while Pascal Brioist shows how similar unacknowledged use of Fine's work in practical geometry was made by the English mathematician Thomas Digges in his *Pantometria* (1571). In her account of Fine's *De solaribus horologiis* (Fig. 13.1) Catherine Eagleton stresses that Fine's work 'helped shape the genre of sundial books' in the sixteenth century, and notes that his work was utilised by a succession of authors on sundials, including Sebastian Münster (1533), Jean Bullant (1564) and Athanasius Kircher (1646).[17] While Fine's books may have been neither original, nor innovative, then, they were not without their influence, and made distinct contributions to particular areas of mathematical knowledge, such as cosmography, horology and practical geometry.

One of Fine's key contributions to learned culture of the Renaissance was as a mapmaker and illustrator of scientific books.[18] [Fig. 13.2] As Jean-Jacques Brioist emphasises in his chapter on Fine's planispherical geometry, Fine's engraved maps, and particularly his cordiform world map of 1534, made his name as a cartographer, and 'rank amongst his most original works', demonstrating 'a far more innovative use of geometry' than is to be found in his printed works.[19] Fine's world map was an 'instant success' and was 'reprinted for two decades in numerous books', and in all likelihood influenced the scheme of Mercator's *Orbis Imago* (1538).[20] As a cartographer and illustrator, then, Fine was justly renowned, and it was in this area that he can be best seen as putting his theoretical knowledge of mathematics to practical use in a way which was influential on his contemporaries.

15 E. G. R. Taylor, *Tudor Geography: 1485–1583* (London, 1930), p. 86.

16 Heninger, 'Oronce Finé and English Textbooks', p. 176.

17 See above, pp. 94–97.

18 For a bibliography see above, pp. 1–2.

19 See above, p. 154.

20 See above, p. 152. Sven Dupré notes that Fine's high-quality engravings of conic sections, which far exceeded those of his contemporary Antonio Gogava, enjoyed a similar popularity. See above, p. 65.

Fine's influence as a presenter of non-verbal information should also, perhaps, be considered in terms of the mathematical table. The *Protomathesis* is bristling with tables of various kinds, and Fine gives them a great deal of prominence, not only by describing in some detail how they are to be used by mathematicians, but also by dedicating a special section of the contents to an exhaustive list of them: 'An index of the tables, scattered here and there in this whole work, calculated by the author'.[21] Fine presents a table of proportionals, for example, 'adapted indifferently to all astronomical calculations',[22] a length set of sine tables,[23] and a 'Table of longitudes from the West, and latitudes from the equator, of the chief places, cities and towns [...] recently verified by the author.'[24] Fine's emphasis on his authorship of these tables is pronounced, and clearly must have been influential in the vogue for mathematical tabulation later in the sixteenth and in the seventeenth century.

Fine's problematic relationship to practice makes him an ideal test-case for historians who wish to interrogate the historical valency of the term 'mathematical practitioner'. Sven Dupré, Pascal Brioist, Catherine Eagleton, Adam Mosley and Anthony Turner all examine Fine's reliance on 'bookish' rather than artisanal traditions, and note his primary orientation towards an academic university culture which privileged knowing over doing. That said, Pascal Brioist notes that Fine had acquired first-hand practical knowledge as an astrologer, and that his Parisian lectures were open to a wider, non-academic audience.[25] Brioist praises Fine's vernacular treatise *Composition et usage du quarré geometrique* (1556) as 'an early example of the "manufacture and use" genre' of instrument books,[26] and Eagleton notes that in Fine's *De solaribus horologiis*, several of the instruments are printed 'with their composite pieces separated [...] so that an instrument could be made from them.'[27] Adam Mosley, however, remains sceptical about the 'practical' nature of Fine's engagements, and convincingly presents the French mathematician as a prime example of a 'theoretical practical' outlook, separating the 'actual application' of practical mathematics from 'the *promotion*, through teaching and authorship, of such an application'.[28] Fine, Mosley points out, was ultimately 'more

21 'Index tabularum, sparsim in hoc toto opere per authorem supputatum.' Fine, 1532*b*, sig. AA^{r-v}:

22 '[...] tabula proportionalis [...] omnibus astronomicis calculationibus indifferenter accomoda.' Fine, 1532*b*, fol. 31^{r}.

23 Fine, 1532*b*, fols. 59^{r}–63^{v}.

24 'Tabula Longitudinum ab occidente, atque Latitudinum ab Aequatore, insignorum locorum, ciuitatum & oppidorum [...] Ab Authore recenter verificata.' Fine, 1532*b*, fols. 147^{r}–148^{v}.

25 See above, p. 54.

26 See above, p. 57.

27 See above, p. 94.

28 See above, pp. 131 (author's own emphasis).

interested in expounding principles than in describing procedures [...] *doing* took second place to *knowing* and *understanding* in Fine's presentation'.[29] Fine's historical importance, for Mosley, is that his work shows us more clearly than that of some of his contemporaries how practical mathematics 'straddled the divide between the academy and other sites of mathematical expertise'.[30]

Fine's work, striving as it was to dignify mathematics as a discipline, made appeal both to the utility of mathematics and to its traditional propaedeutic role as a preparation for 'higher' studies such as metaphysics and theology. Thus we find emphasis on such practical issues as making 'the use of geometrical and celestial instruments easier',[31] together with claims of the usefulness of arithmetic when considering 'universal philosophy or divine things',[32] or of the intermediary status of mathematics 'between natural or physical and supernatural or metaphysical knowledge'.[33] Fine was deeply influenced by the peculiar blend of neoplatonism and practical mathematics to be found in an earlier generation of French humanists such as Jacques Lefèvre d'Etaples and Charles de Bovelles.[34] As Pantin's chapter amply demonstrates, Fine's work was heavily indebted to this scholarly milieu, and his work allows us to trace the influence of such humanistic trends on the development of mid-sixteenth century mathematics, and helps explain why the works of later sixteenth-century mathematicians, such as John Dee's 'Mathematicall Praeface' to Henry Billingsley's translation of Euclid's *Elements* (1570), share this seemingly paradoxical mixture of mystical speculation and concern for the 'mechanicien'.

Fine's work also gives us some insight into another historical trend, that of the influence of late fifteenth- and early sixteenth-century German mathematical thought on the rest of Europe. Fine's links with the German mathematical tradition emerge at various points in this volume: not only do we learn that he edited

29 See above, p. 133 (author's own emphasis).

30 See above, p. 136.

31 '[...] geometricorum & coelestium instrumentorum [...] redderemus faciliorem.' Fine, 1532*b*, fol. 64r.

32 '[...]uniuersa philosophia siue quae diuina'. Fine, 1532*b*, fol. 1r.

33 'Sunt enim Mathematicae mediae inter naturalem seu physicam ausculationem, & supernaturalem siue Metaphysicam.' Fine, 1532*b*, fol. 1r.

34 On the combination of practical and Platonic mathematical interests in the Parisian milieu and its influence on the Low Countries see M. Mulsow, 'Seelenwagen und Ähnlichkeitsmaschine – Zur Reichweite der praktischen Geometrie in der *Ars Cyclognomica* von Cornelius Gemma (1569)' in J. J. Berns and W. Neuber (eds.), *Seelenmaschinen: Gattungstraditionen, Funktionen und Leistungsgrenzen der Mnemotechniken vom späten Mittelalter bis zum Beginn der Moderne* (Vienna, 2000), pp. 249–277 and *idem*, '*Arcana naturae*: Verborgene Ursachen und universelle Methode von Fernel bis Gemma und Bodin' in T. Leinkauf and K. Hartbecke (eds.), *Der Naturbegriff in der Frühen Neuzeit: Semantische Perspektiven zwischen 1500 und 1700* (Tübingen, 2005), pp. 31–68.

Georg von Peurbach's *Theorica planetarum* and Gregor Reisch's *Margarita philosophica*, but, as Eagleton shows us, Fine's work on the ship-shaped sundial drew on a German manuscript tradition and both Jean-Jacques Brioist and Isabelle Pantin note Fine's unacknowledged reliance on Johannes Werner and Walter Lud for his work on planispherical projections.

Oronce Fine may not have been a 'great' mathematician – no new demonstrations or mathematical breakthroughs can be accredited to him, but his work as an illustrator and editor of scientific books, a composer of a host of mathematical textbooks in Latin and French, his work as a cartographer and as an exponent and promotor of mathematical instruments, his role as a founder of mathematics as a legitimate academic discipline, all entitle him to serious historical consideration. As this volume so amply shows, Oronce Fine was in many ways one of the most typical and characteristic figures of this phase of the development of Renaissance mathematical culture in Europe.

INDEX

1.1: Portrait of Oronce Fine, from A. Thevet, *Les vrais pourtraits et vies des hommes illustres* (Paris, 1584). © The British Library Board. Shelfmark C.22.f. 4, p. 564.

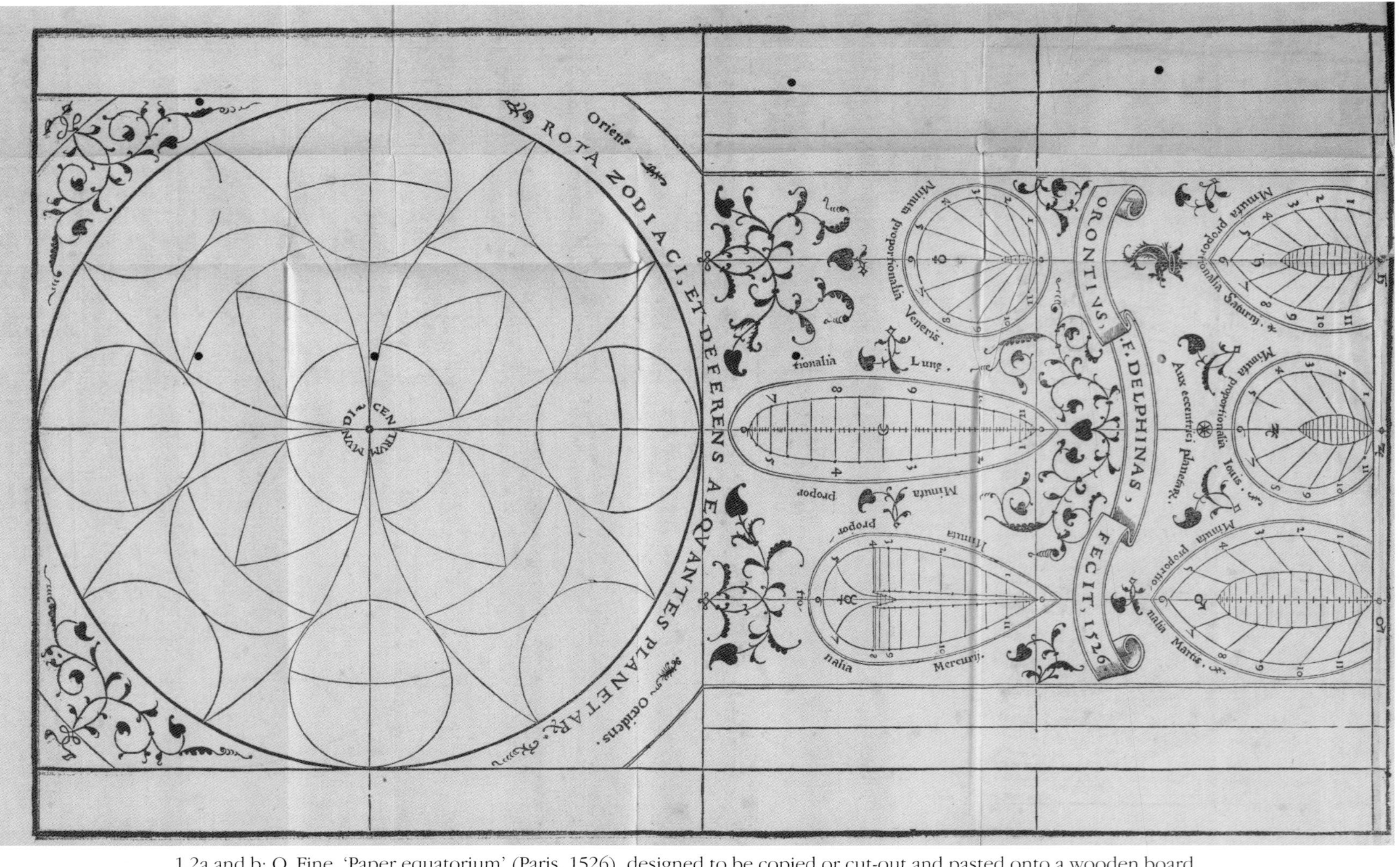

1.2a and b: O. Fine, ‘Paper equatorium’ (Paris, 1526), designed to be copied or cut-out and pasted onto a wooden board.
Reproduced courtesy of The Houghton Library, Harvard.

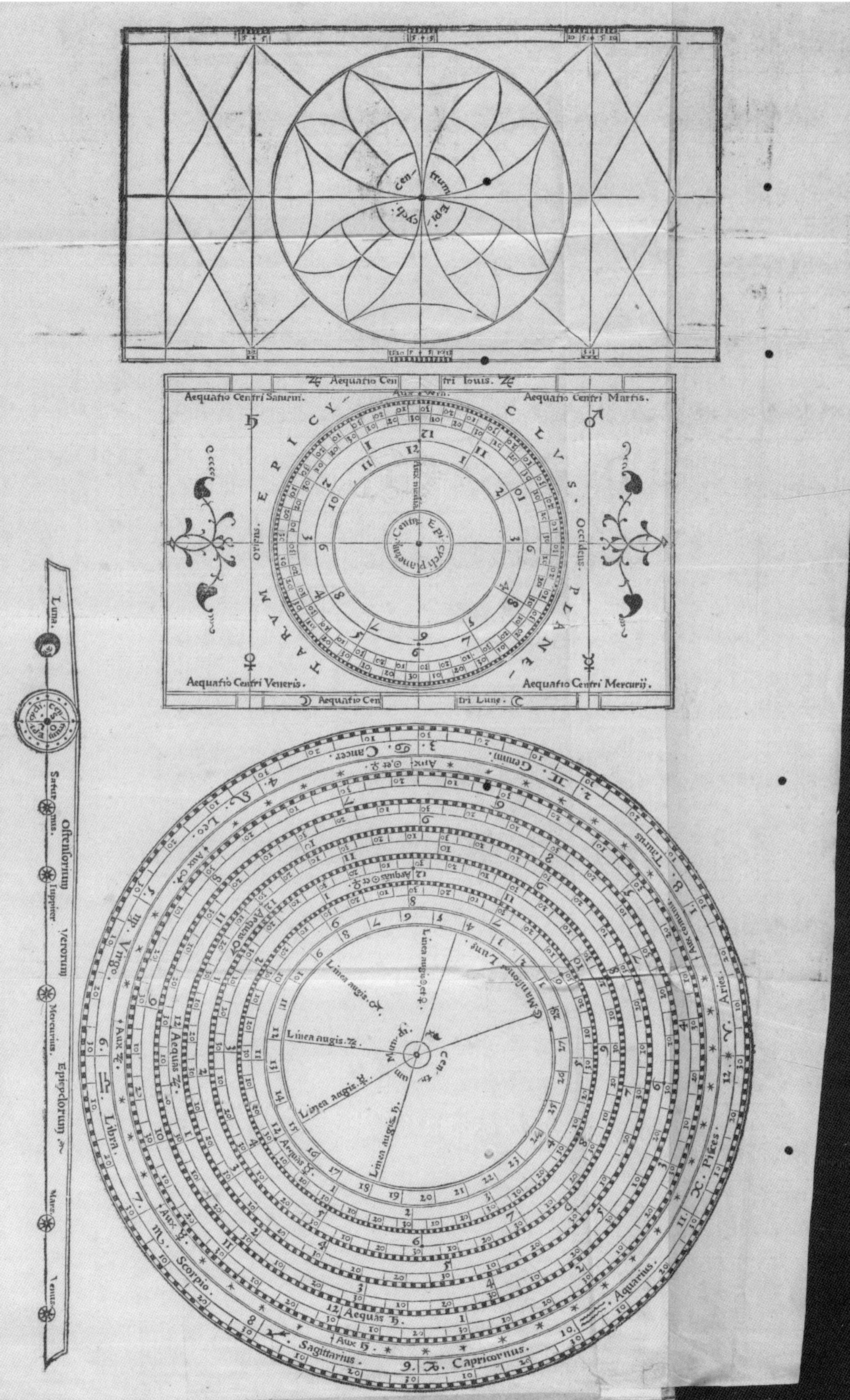

Centrum Epicycli
Aequatio Centri Iouis.
Aequatio Centri Saturni.
Aequatio Centri Martis.
Aux vera.
EPICYCLUS. PLANETARUM
Aux media
Centrum Epicycli planetae
Oriens.
Occidens.
Aequatio Centri Veneris.
Aequatio Centri Mercurij.
Aequatio Centri Lunae.
Ostensorium Verorum Epicyclorum
Luna.
Saturnus.
Iuppiter.
Mercurius.
Mars.
Venus.
Aries.
Taurus.
Gemini.
Cancer.
Leo.
Virgo.
Libra.
Scorpio.
Sagittarius.
Capricornus.
Aquarius.
Pisces.
Aux communis
Mansiones Lunae.
Centrum Mundi.
Linea augis.
Aequans.
Aux

1.3. 'The author and Urania'. Woodcut illustration to O. Fine, *De mundi sphaera* (Paris, 1542). Reproduced courtesy of St Andrews University Library.

1.4: 'A mathematician's study', illustration from J. Van der Straet, *Nova Reperta* (Antwerp, 1600). The British Library, reproduced by permission.

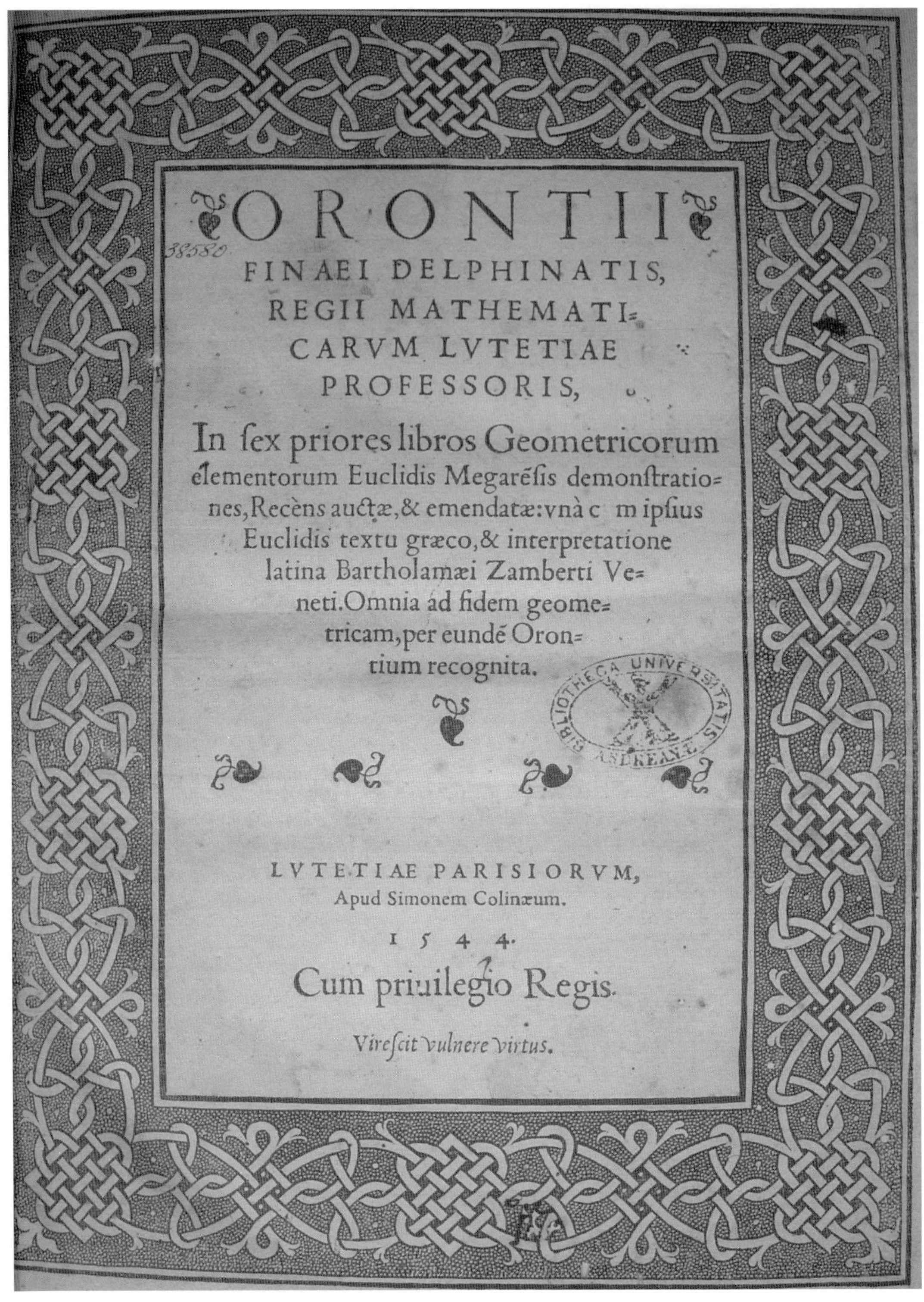

ORONTII
FINAEI DELPHINATIS,
REGII MATHEMATI-
CARVM LVTETIAE
PROFESSORIS,

In ſex priores libros Geometricorum
elementorum Euclidis Megarēſis demonſtratio-
nes, Recēns auctæ, & emendatæ: vnà c m ipſius
Euclidis textu græco, & interpretatione
latina Bartholamæi Zamberti Ve-
neti. Omnia ad fidem geometricam, per eundē Oron-
tium recognita.

LVTETIAE PARISIORVM,
Apud Simonem Colinæum.

1 5 4 4.

Cum priuilegio Regis.

Vireſcit vulnere virtus.

2.1: O. Fine, *In sex priores libros geometricorum elementorum Euclidis...* (Paris, 1544 [first edition 1536]), title-page. Reproduced courtesy of St Andrews University Library.

3.1 (opposite): O. Fine, *Protomathesis...* (Paris, 1532), title-page. Reproduced by permission of the National Library of Scotland.

ORONTII FINEI DELPHINATIS, LIBERALIVM DISCIPLINARVM PROFESSORIS REGII,

PROTOMATHESIS:

Opus uarium, ac scitu non minus utile quàm iucundum, nunc primùm in lucem fœliciter emissum. Cuius index uniuersalis, in uersa pagina continetur.

PARISIIS ANNO 1532.

Cum gratia & priuilegio Christianissimi Francorum Regis, ad Decennium.

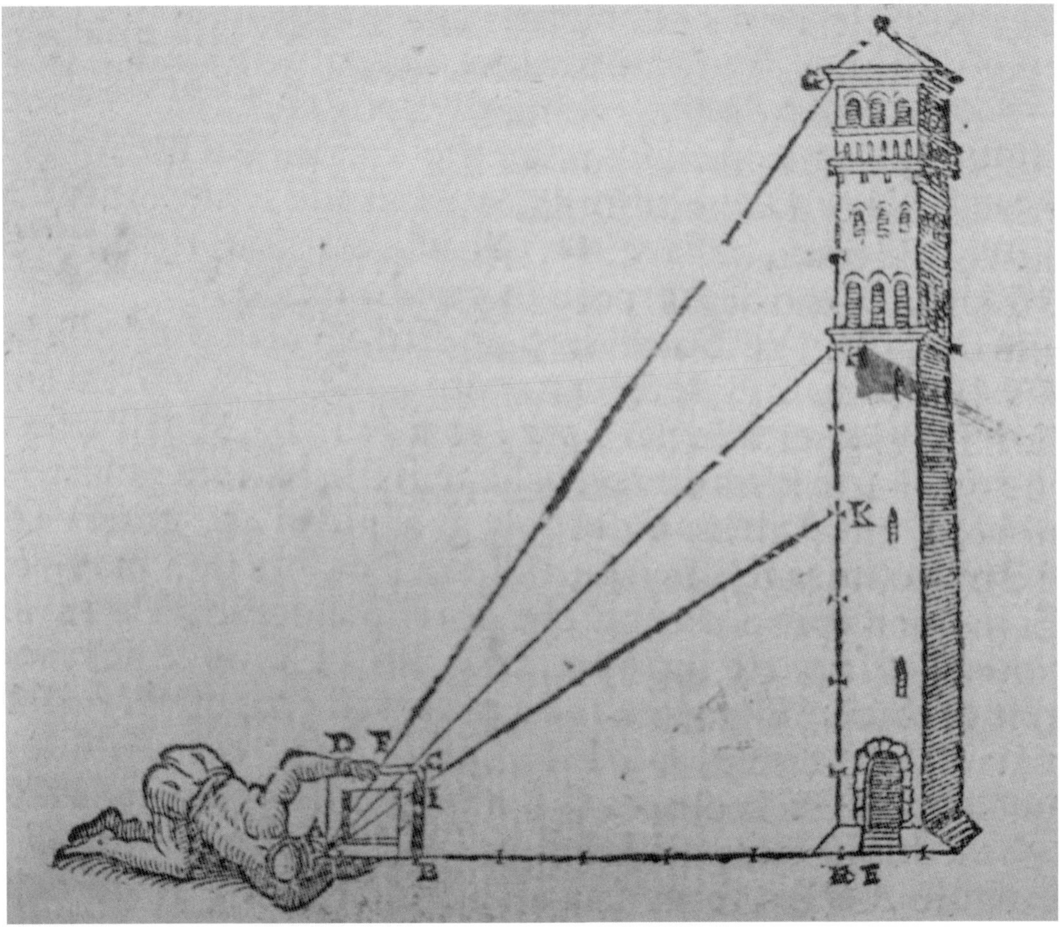

4.1 (above): Measuring using the geometrical square, from O. Fine, *Opere*, trans. C. Bartoli (Venice, 1587), p. 169. Reproduced courtesy of the editor.

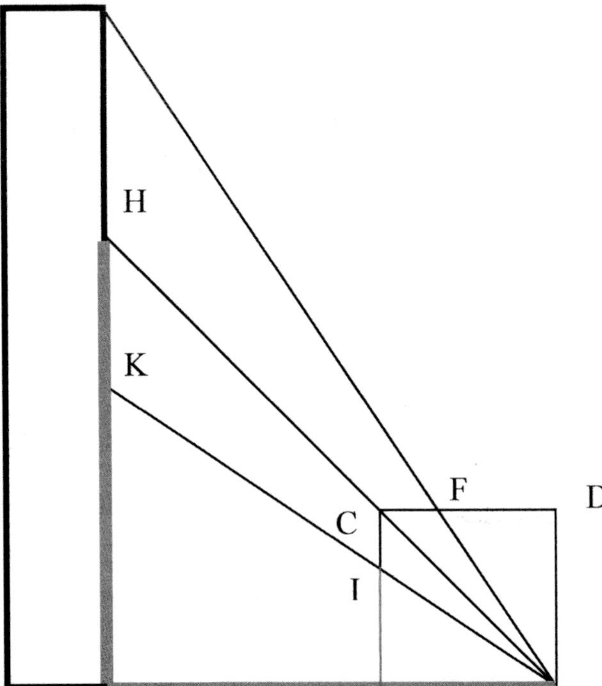

4.2 (left): Diagrammatic representation of Fig. 4.1.

4.3 (opposite, above): 'Geometrical square', from O. Fine, *Opere*, trans. C. Bartoli (Venice, 1587), p. 233. Reproduced courtesy of the editor.

4.4 (opposite, below): Bodily contortions using the geometrical square, from O. Fine, *Opere*, trans. C. Bartoli (Venice, 1587), p. 183. Reproduced courtesy of the editor.

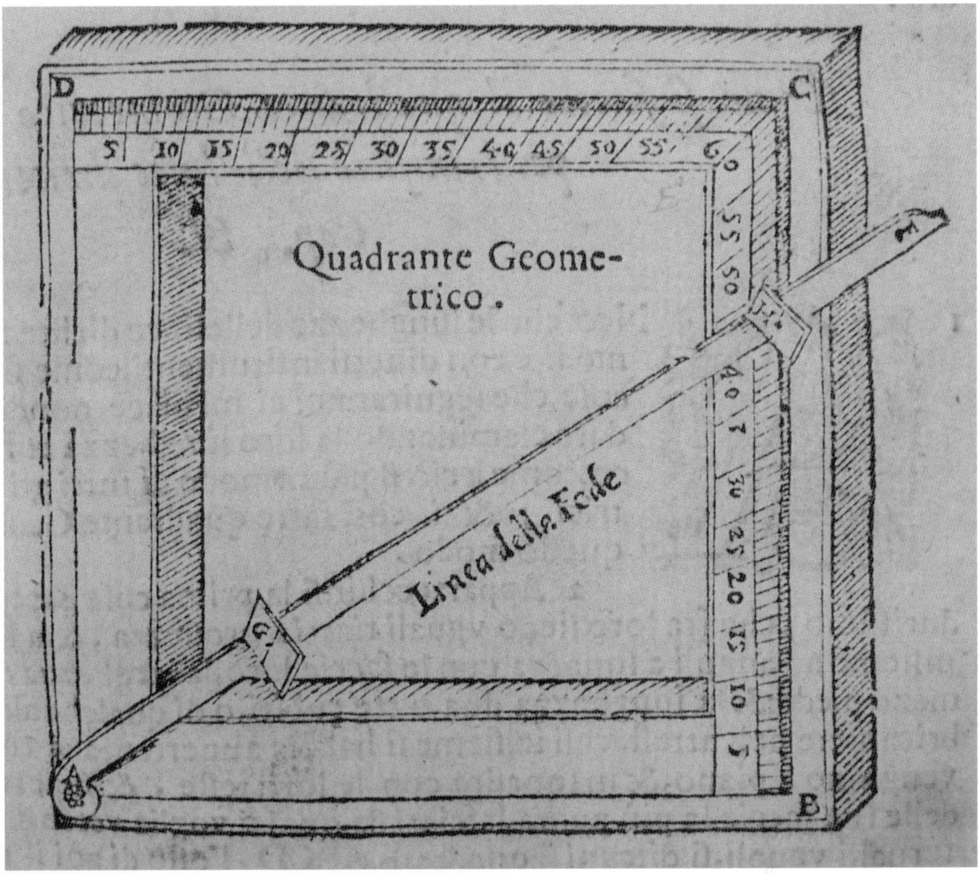

D C

5 10 15 20 25 30 35 40 45 50 55 60

Quadrante Geome-
trico.

55 50 40 35 30 25 20 15 10 5

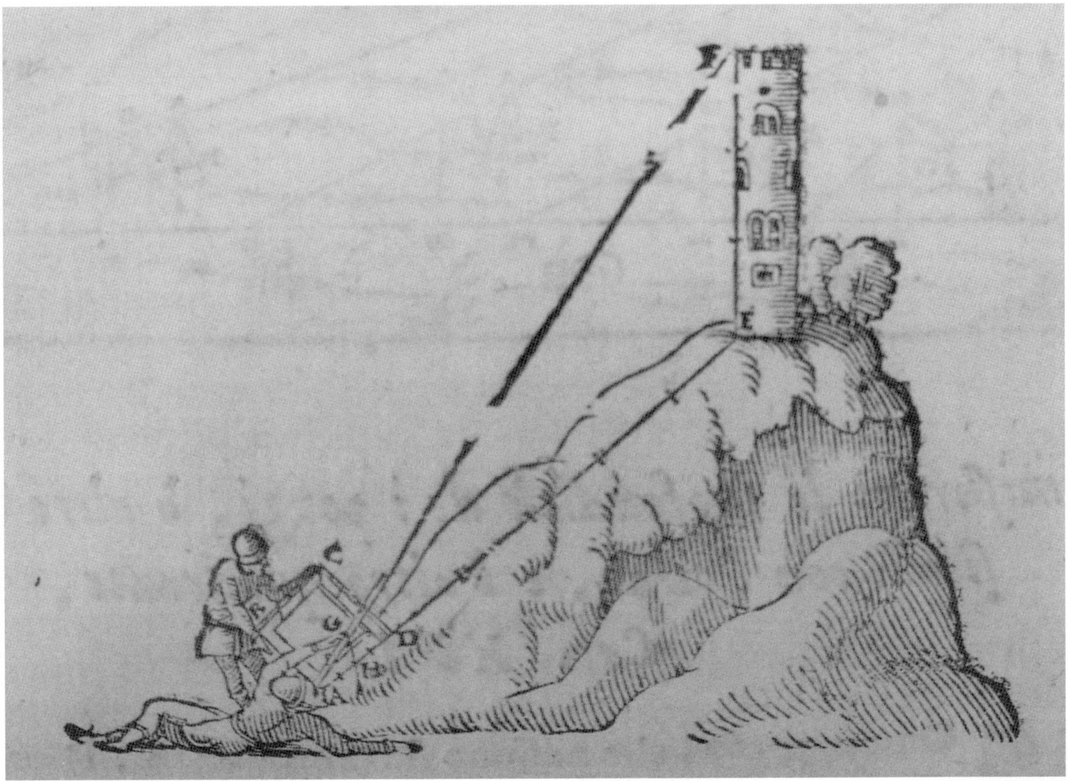

Linea della Fede

B

L
B
E
A
M
D

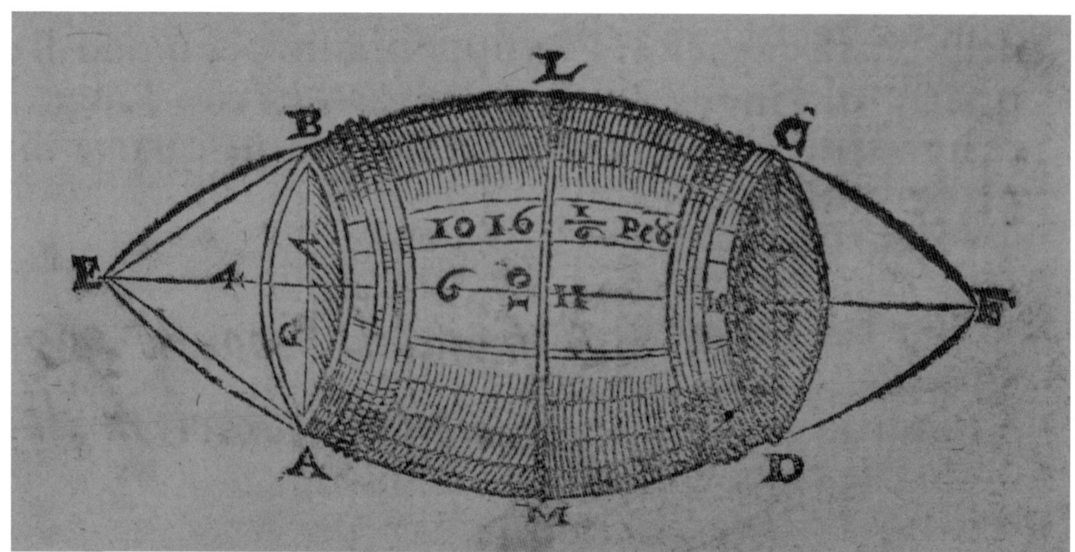

4.5: A measured barrel, from O. Fine, *Opere*, trans. C. Bartoli (Venice, 1587), p. 233. Reproduced courtesy of the editor.

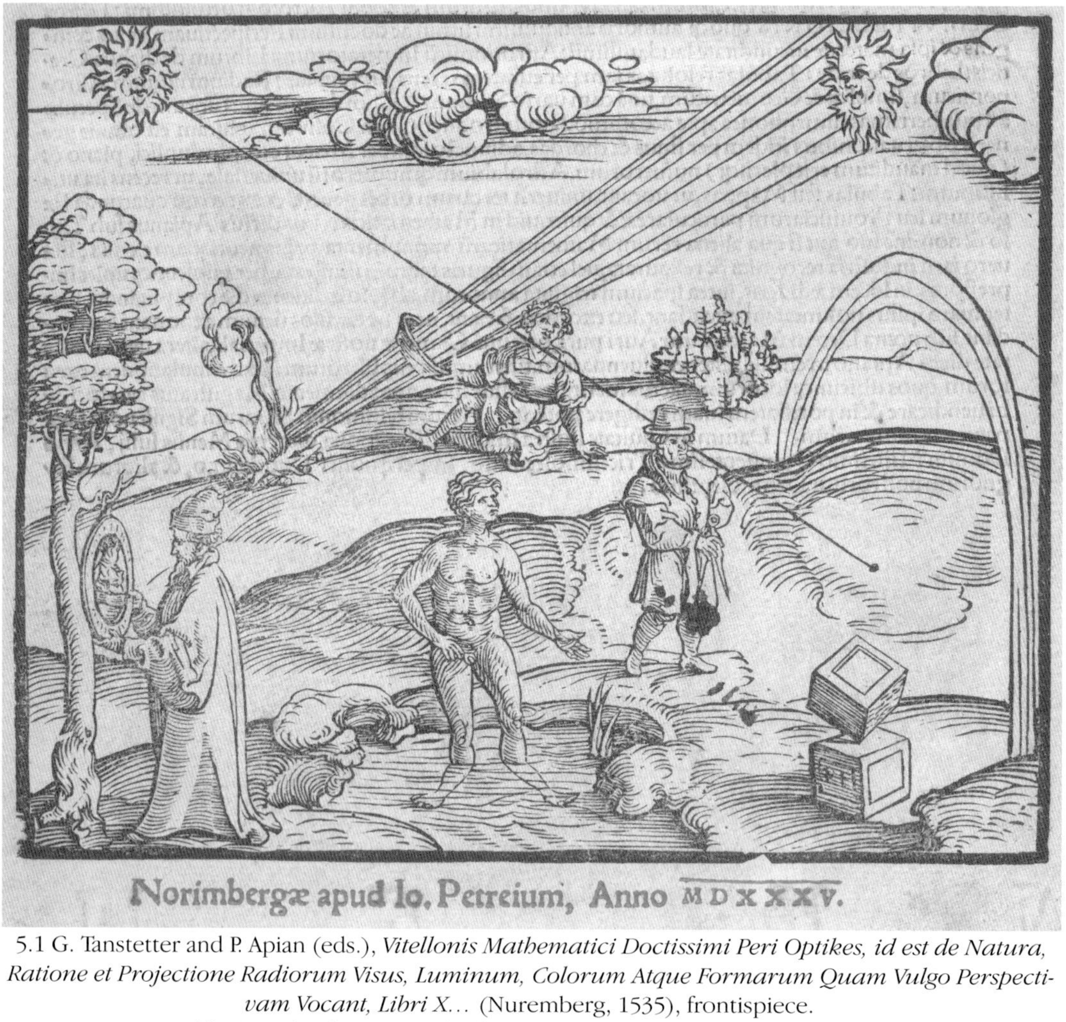

Norimbergæ apud Io. Petreium, Anno MDXXXV.

5.1 G. Tanstetter and P. Apian (eds.), *Vitellonis Mathematici Doctissimi Peri Optikes, id est de Natura, Ratione et Projectione Radiorum Visus, Luminum, Colorum Atque Formarum Quam Vulgo Perspectivam Vocant, Libri X…* (Nuremberg, 1535), frontispiece.

VITELLONIS THVRINGOPOLONI OPTICAE LIBRI DECEM.

Instaurati, figuris nouis illustrati atque aucti: infinitisq; erroribus, quibus antea scatebant, expurgati.

A'

FEDERICO RISNERO.

BASILEAE.

5.2: F. Risner (ed.), *Opticae Thesaurus Alhazeni Arabis Libri Septem, Nuncprimùm Editi. Eiusdem Liber de Crepusculis et Nubium Ascensionibus. Item Vitellionis Thuringopoloni Libri X* (Basel, 1572), illustrated title-page.
Photographic Archive: Institute and Museum of the History of Science, Florence.

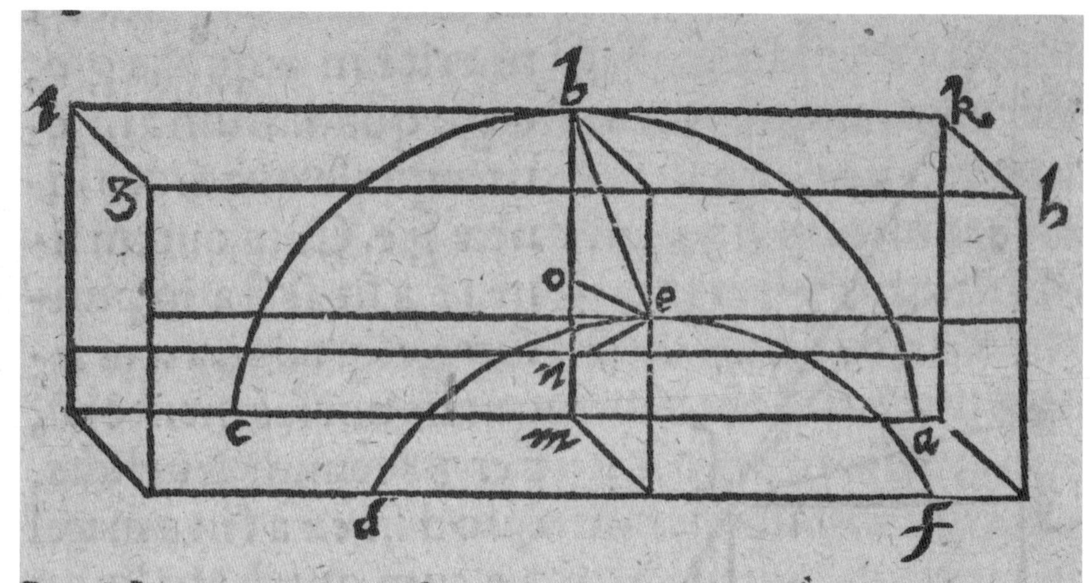

5.3: Detail from: *Cl. Ptolemaei Pelusiensis Mathematici operis quadriparti … De sectione conica, orthogona, quae parabola dicitur: De Speculo ustorio, libelli duo, hactenus desiderati: restituti ab Antonio Gogava Graviensi* (Louvain, 1548). The British Library Board. All Rights Reserved. Shelfmark 531.g.2(4).

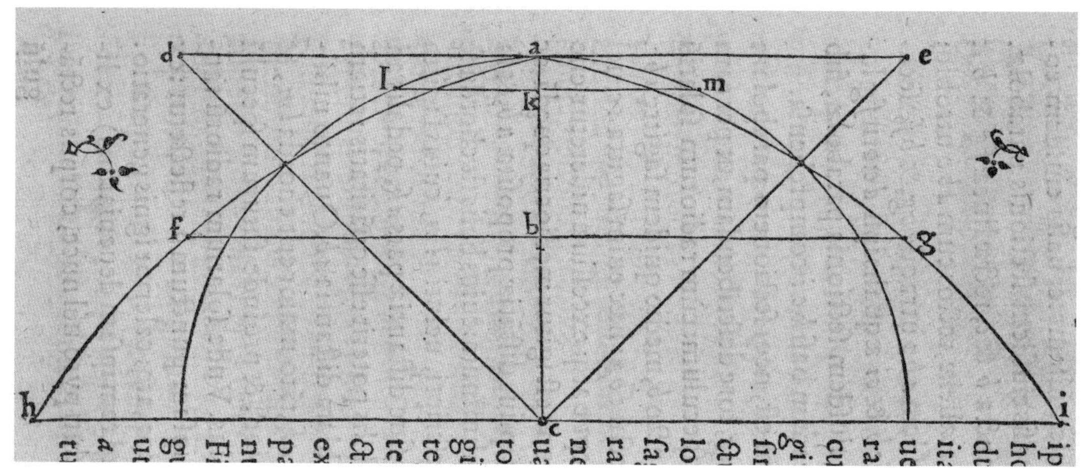

5.5: O. Fine, *De speculo ustorio…* (Paris, 1551), p. 18v.
The British Library Board. All Rights Reserved. Shelfmark 529.g.7(1).

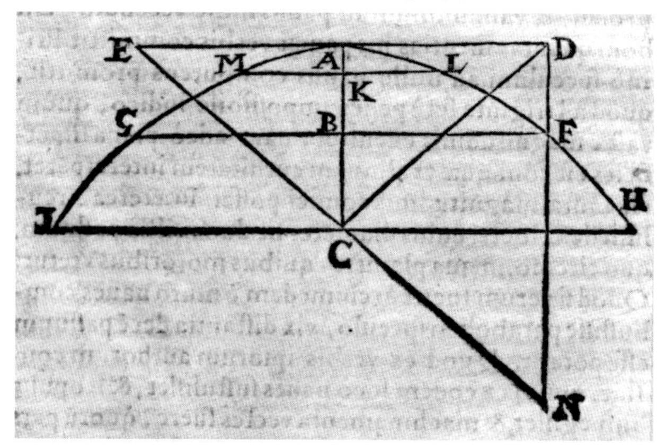

5.6: Detail from G. B. Della Porta, *Magiae naturalis libri viginti* (Frankfurt, 1591). Ghent University Library, Ma 1232a., p. 602.

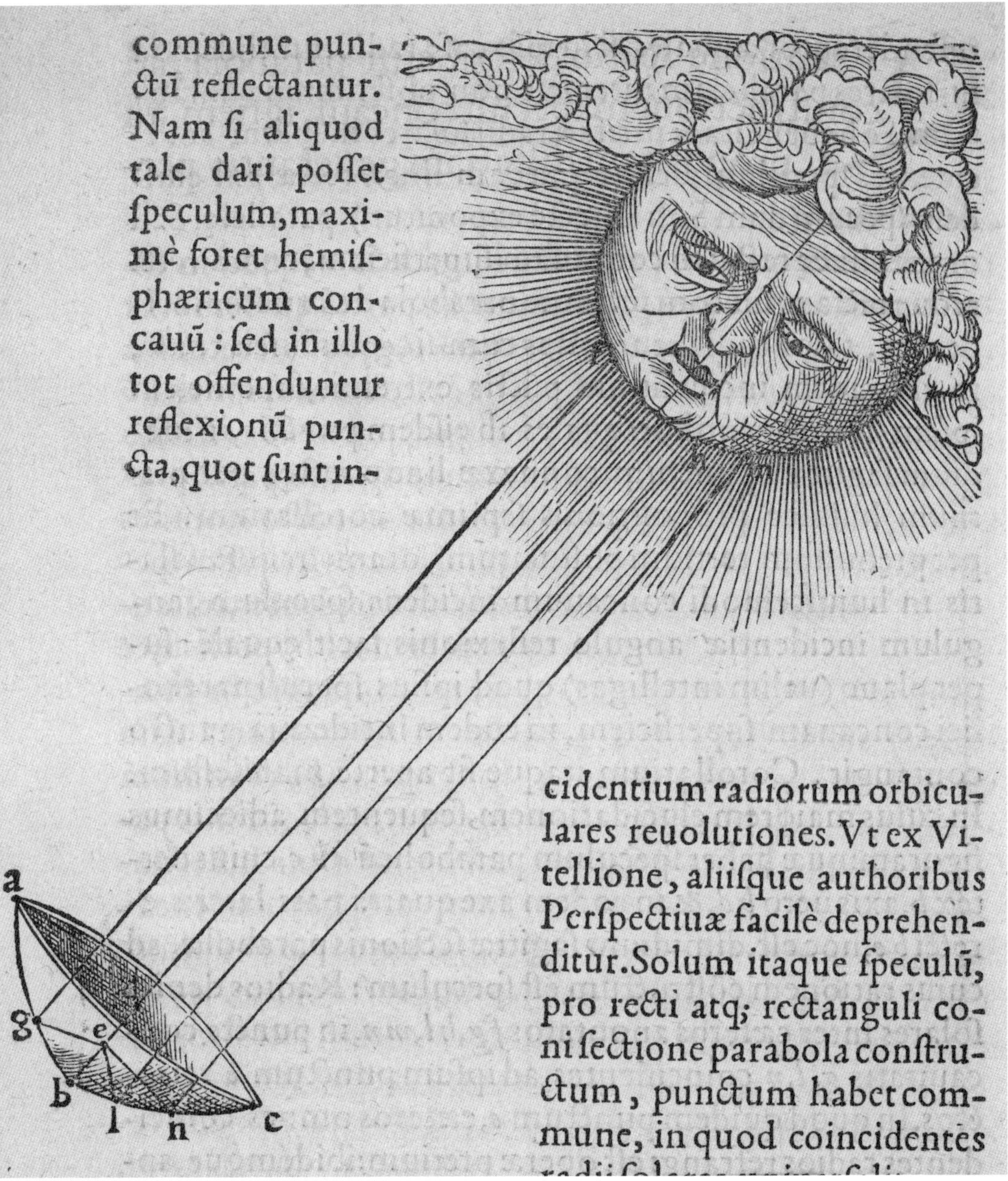

commune punctũ reflectantur. Nam ſi aliquod tale dari poſſet ſpeculum, maximè foret hemiſphæricum concauũ : ſed in illo tot offenduntur reflexionũ puncta, quot ſunt incidentium radiorum orbiculares reuolutiones. Vt ex Vitellione, aliiſque authoribus Perſpectiuæ facilè deprehenditur. Solum itaque ſpeculũ, pro recti atq; rectanguli coni ſectione parabola conſtructum, punctum habet commune, in quod coincidentes

5.4: O. Fine, *De speculo ustorio…* (Paris, 1551), p. 17^{v}.

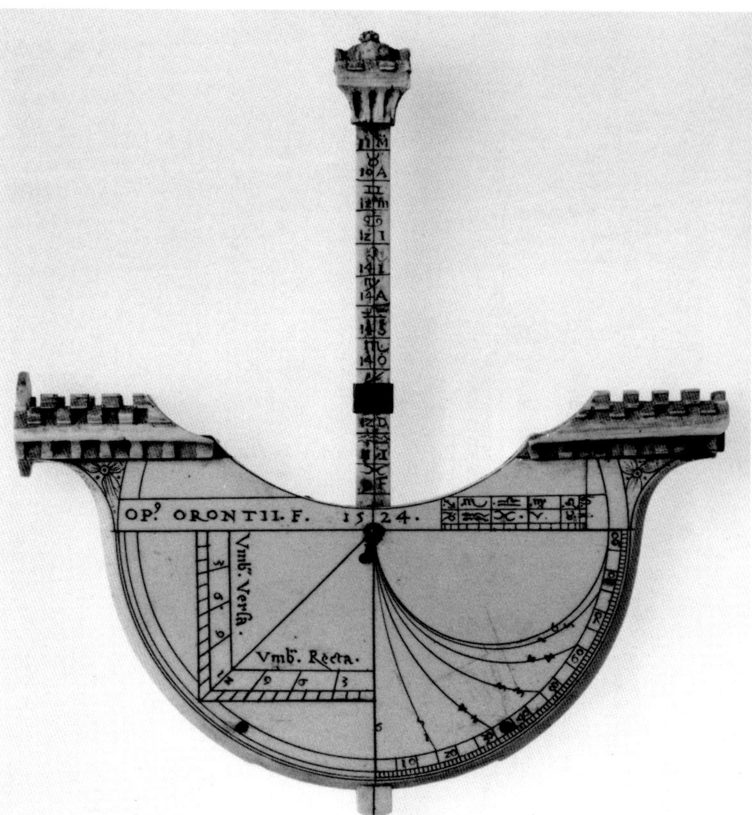

6.1a and b: Ship-shaped dial in the Museo Poldi Pezzoli, Milan (inv. no. 4277). Reproduced by kind permission of the Museo Poldi Pezzoli.

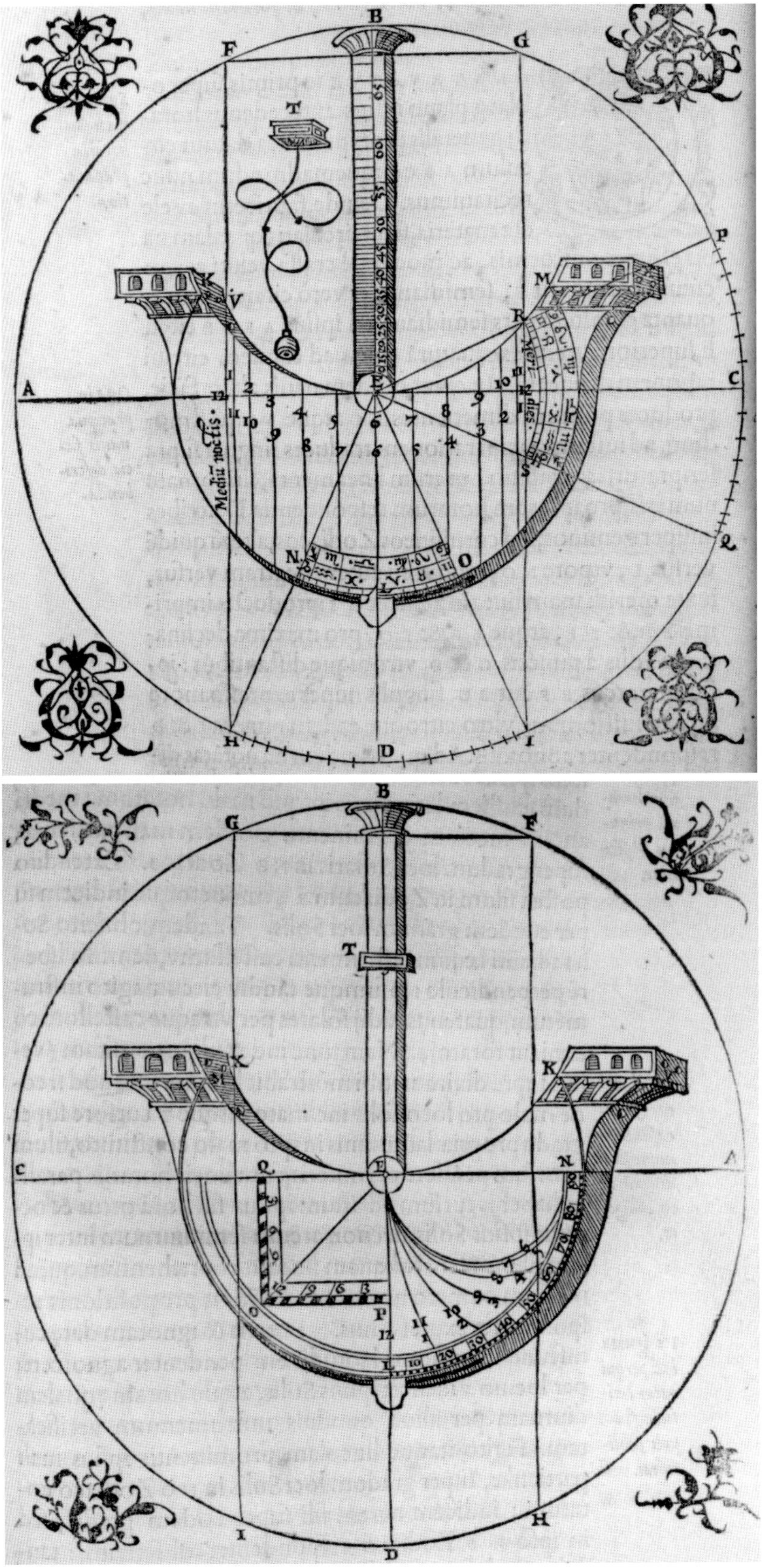

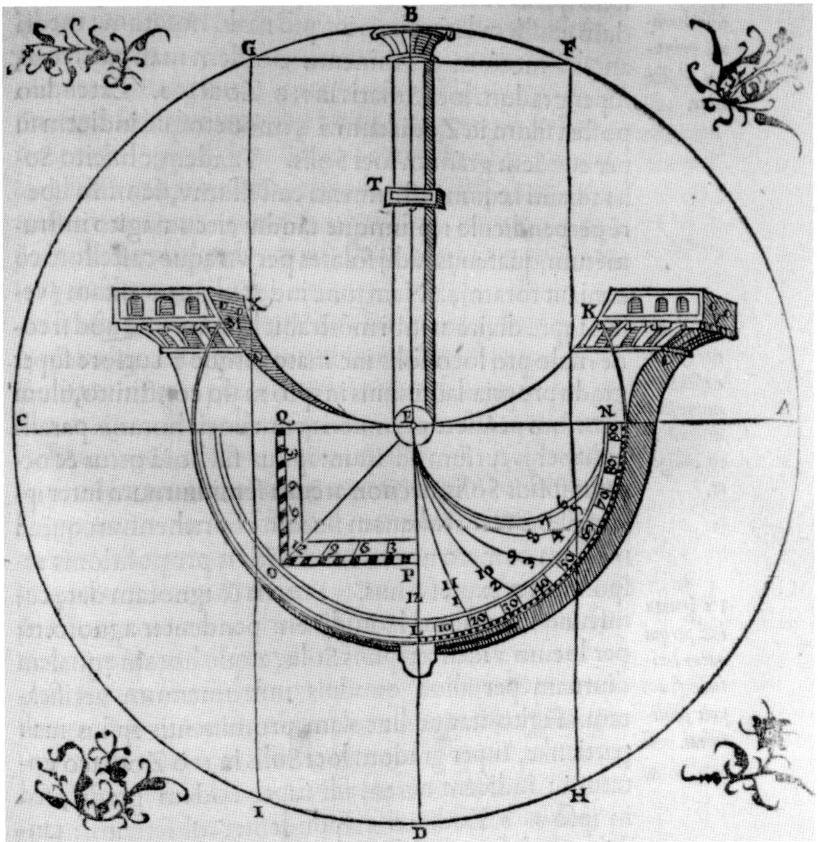

6.2a and b: Fine's printed version of the ship-shaped dial, from O. Fine, *De solaribus horologiis…* (Paris, 1560), pp. 184 and 187. Reproduced by kind permission of the Whipple Library, University of Cambridge.

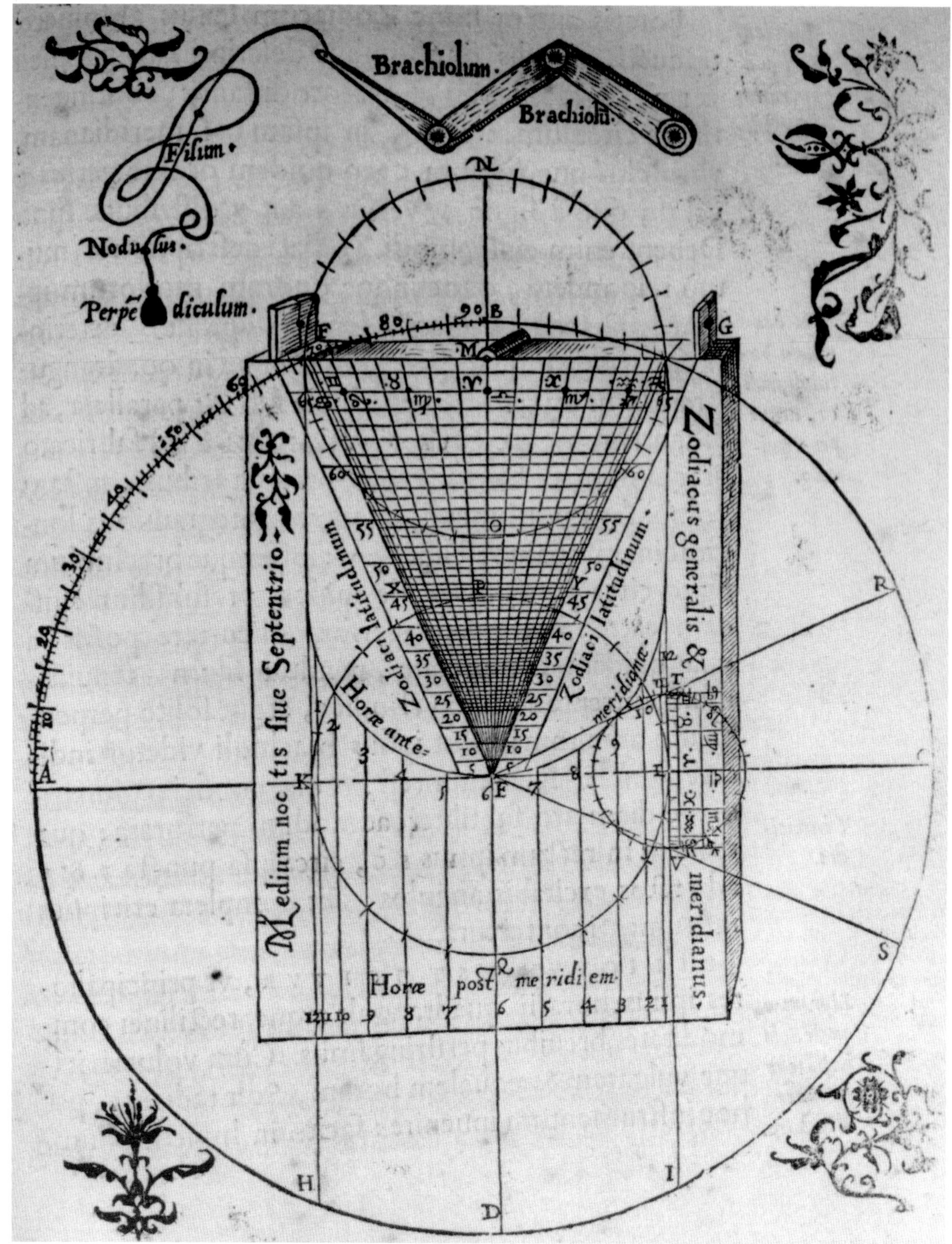

6.3: Fine's printed version of a universal rectilinear dial, now often known as the Regiomontanus dial. From O. Fine, *De solaribus horologiis...* (Paris, 1560), p. 179. Reproduced by kind permission of the Whipple Library, University of Cambridge.

6.4 (opposite): Clavius's diagram of the universal rectilinear dial: Clavius (1581), p. 628. Reproduced by kind permission of the Whipple Library, University of Cambridge.

ſcriptus circulus in 24. horas ęquales diuidatur, quæ rurſus, ſi placet, in ſemihoras, & in quadrãtes horarum ſubdiuidantur. Rectæ enim ductæ per bina quęuis puncta æqualiter à punctis L, O,

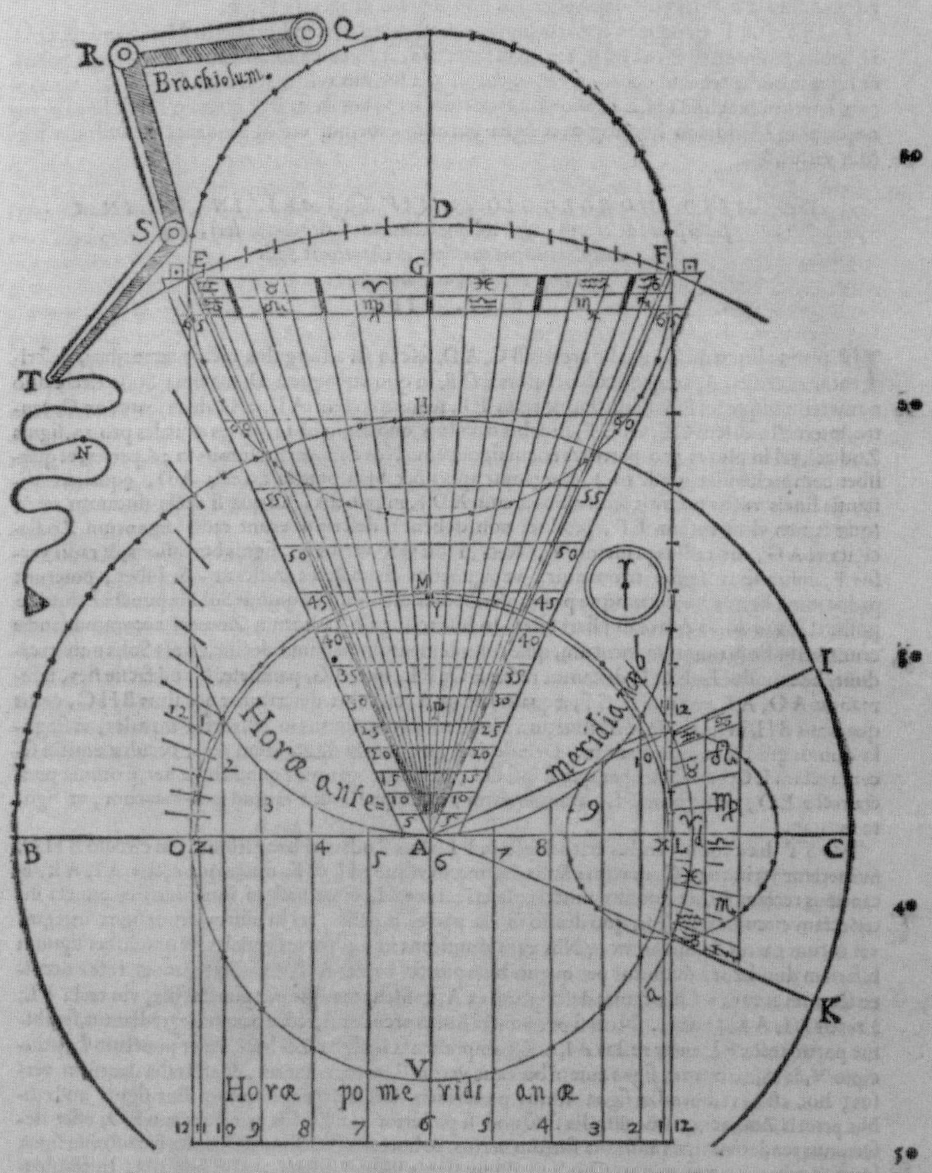

remota dabunt 12. horas à meridie, & 12. à med. noc. quarum M A, ſextam horam dabit, antemeridianę vero horæ à recta E O, verſus F L, & pomeridianæ à recta F L, verſus E O, numerantur. Terminabuntur autem hæ lineæ horarum, in ſuperiore quidem parte, in circunferentia circuli ex M, ad interuallum M A, deſcripti, in inferiori autem in linea ipſi L O, parallela tanto ſaltem interuallo diſtante à recta L O, quanta eſt portio rectæ A I, vel A K, inter A, & rectam F L.

RVRSVS fabricandum erit brachiolum ex dura aliqua, & ſolida materia, conſtans tribus volubilibus ſegmentis Q R, R S, S T, quæ in longum extenſa longitudinem efficiant rectæ G A, æqualem. Huius brachioli extremum punctum Q, figendum eſt in puncto G, ita vt alterum extremum T, liberæ per omnia loca trianguli A E F, diſcurrere poſſit, & in quocunque loco firmari, ne volu-

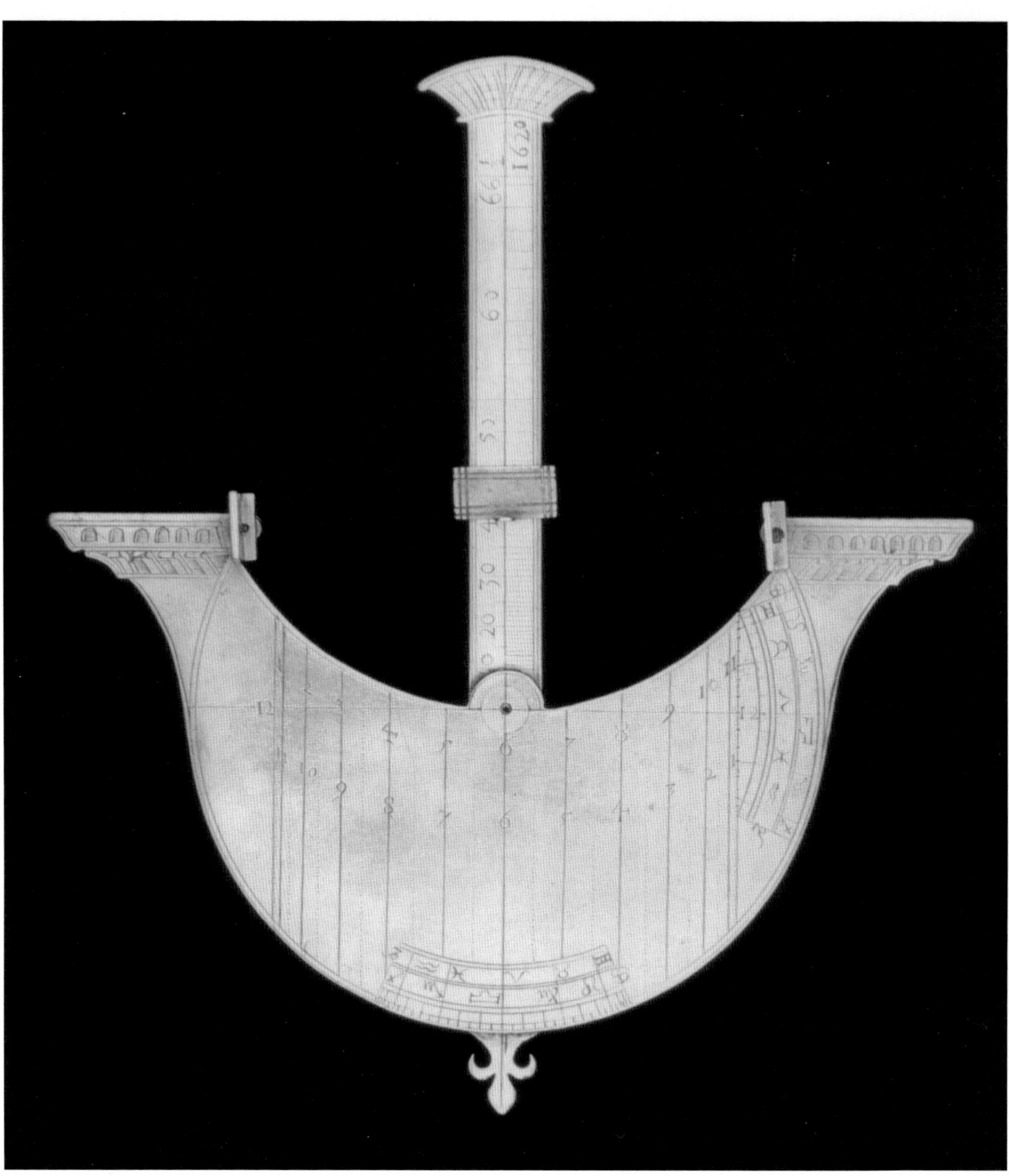

6.5, a and b: The Whipple ship-shaped dial (inv. no. 731).
Reproduced by kind permission of the Whipple Museum of the History of Science, University of Cambridge.

SF

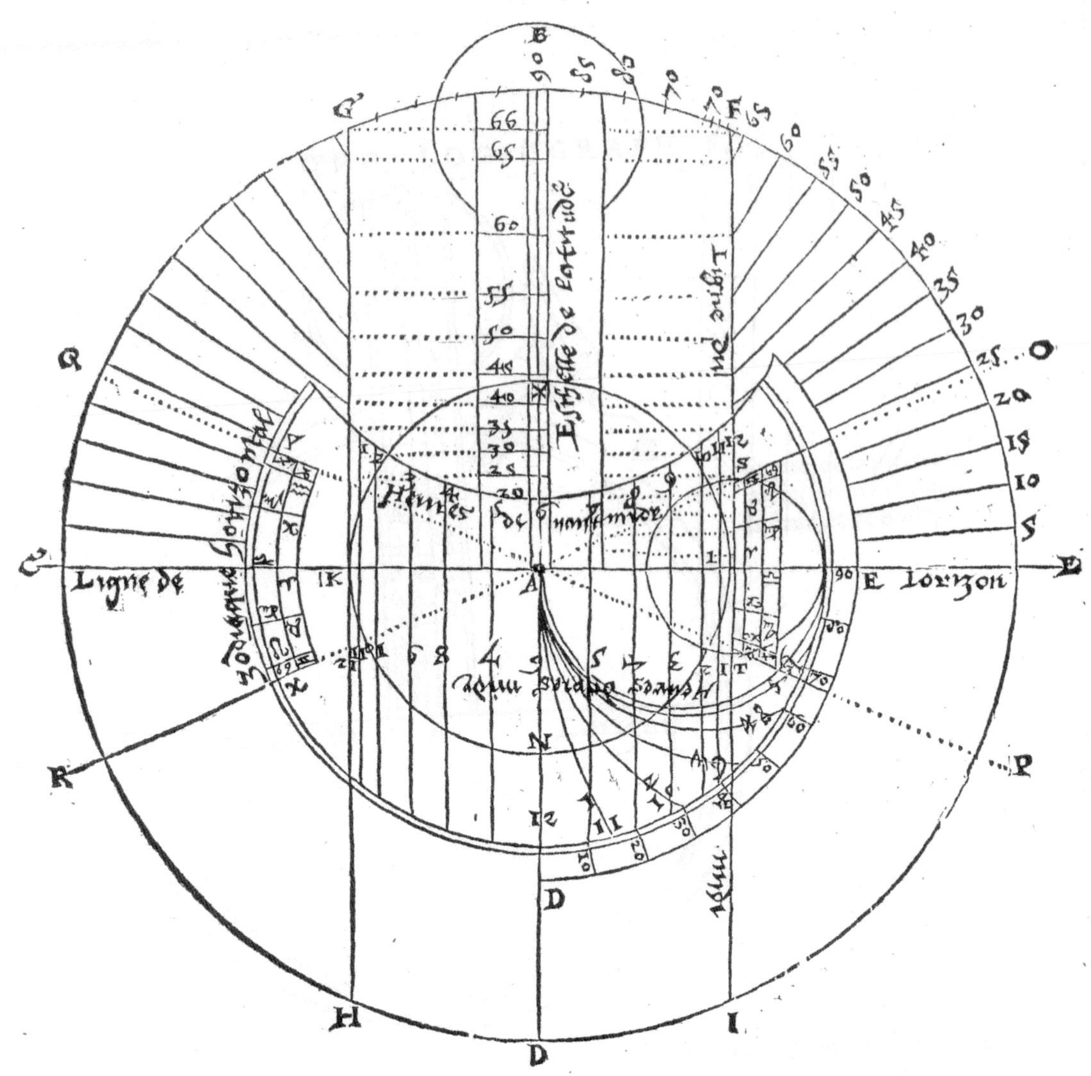

6.6 a, b and c: Diagrams showing (a) the construction of the ship-shaped dial, and (b and c) showing the two parts from which the dial can be made, by fixing them together at the centre and adding a jointed arm. From J. Bullant, *Petit traicte de geometrie et horologiographie* (Paris, 1564), pp. 106 and 107. Reproduced by kind permission of the Museum of the History of Science, Oxford.

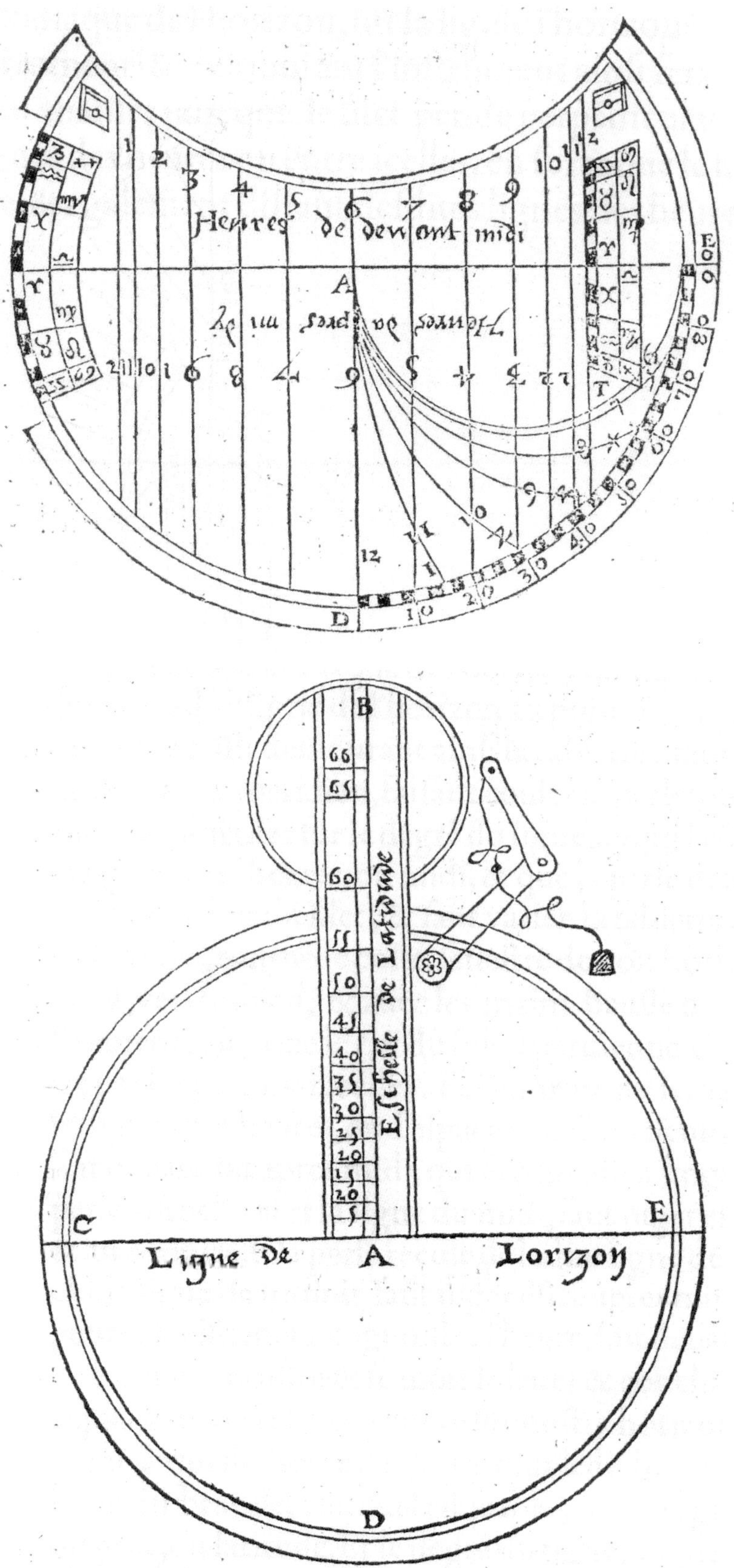
Heures de devant midi
Heures da pres midy
A
D
E
T
B
Eschelle de Latitude
C
E
Ligne de
A
Lorizon
D

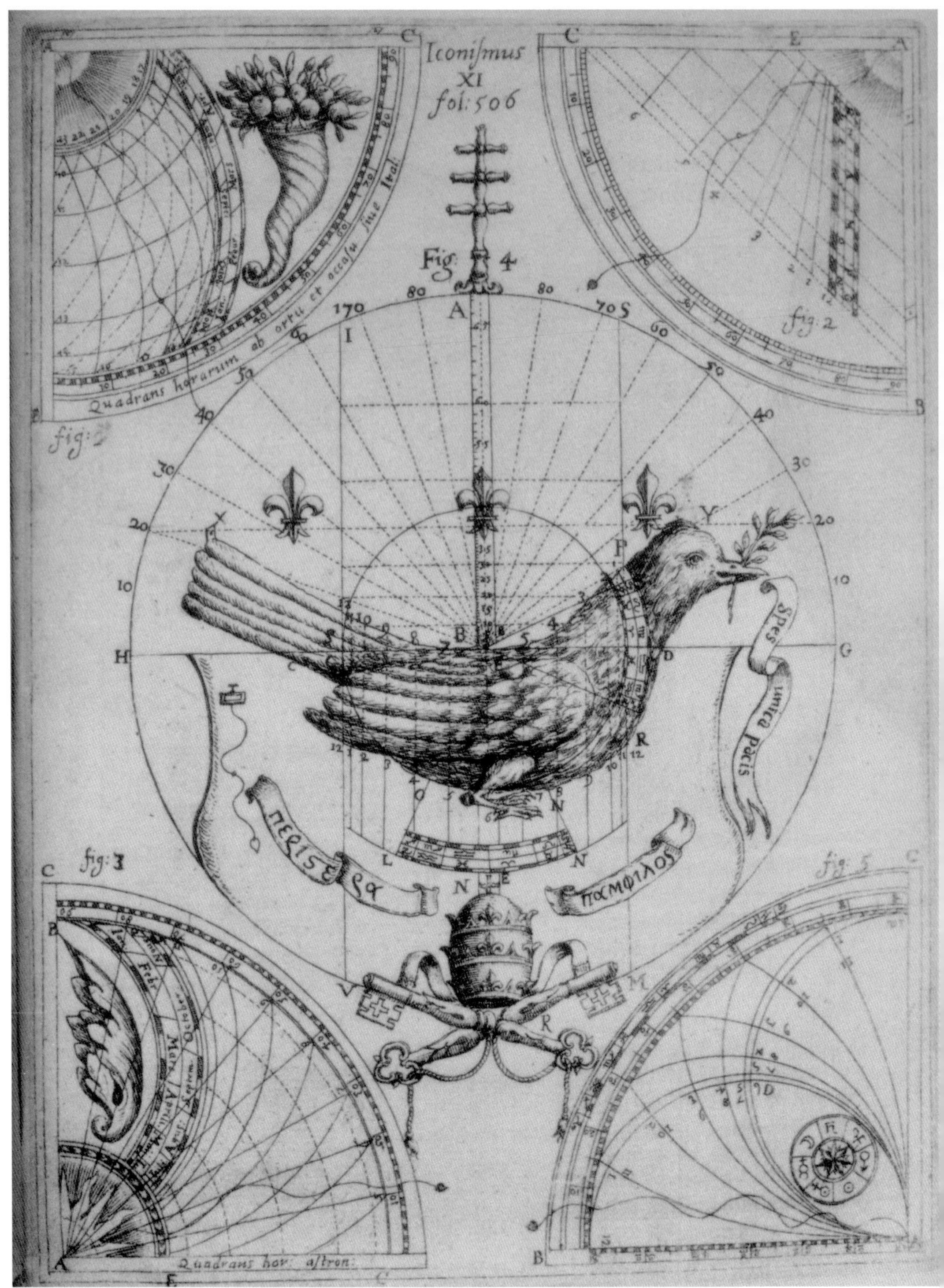

6.7: Kircher's *columba* dial. From Kircher, *Ars magna lucis et umbrae* (Rome, 1646), facing p. 506 in most copies of the book, but wrongly placed at p. 560 in this copy.
Reproduced by kind permission of the Whipple Library, University of Cambridge.

TEXTVS DE SPHÆ-
RA IOANNIS DE SACROBOSCO: INTRODVCTORIA AD-
ditione (quantum necessarium est) cõmentarioq; ad vtilitatem studentium
philosophiæ Parisiensis Academiæ illustratus. Cum compositione Annuli astronomici Boneti Latensis: Et Geometria Euclidis Megarensis.

PARISIIS.
¶ Vænit apud Simonem Colinæum, e regione scholæ Decretorum.
1521

7.1: J. de Sacrobosco, *De sphaera* (Paris, 1521), title-page.
Reproduced by kind permission of Gloucester Cathedral Library.

DISCE PRIVS NVMEROS TELLVRISQVE ORDINE MENSVS: NAM FACILEM CVRA HAEC STERNET AD ALTA VIAM.

TYPVS VNIVERSI ORBIS

VRANIA

Authoris Tetrastichon.

Cùm natura sagax numero mensuq́; crearit
Singula, ponderibus clauserit inde suis:
Non poteris rerum proprias discernere causas,
Ni teneas numeros, & geometra simul.

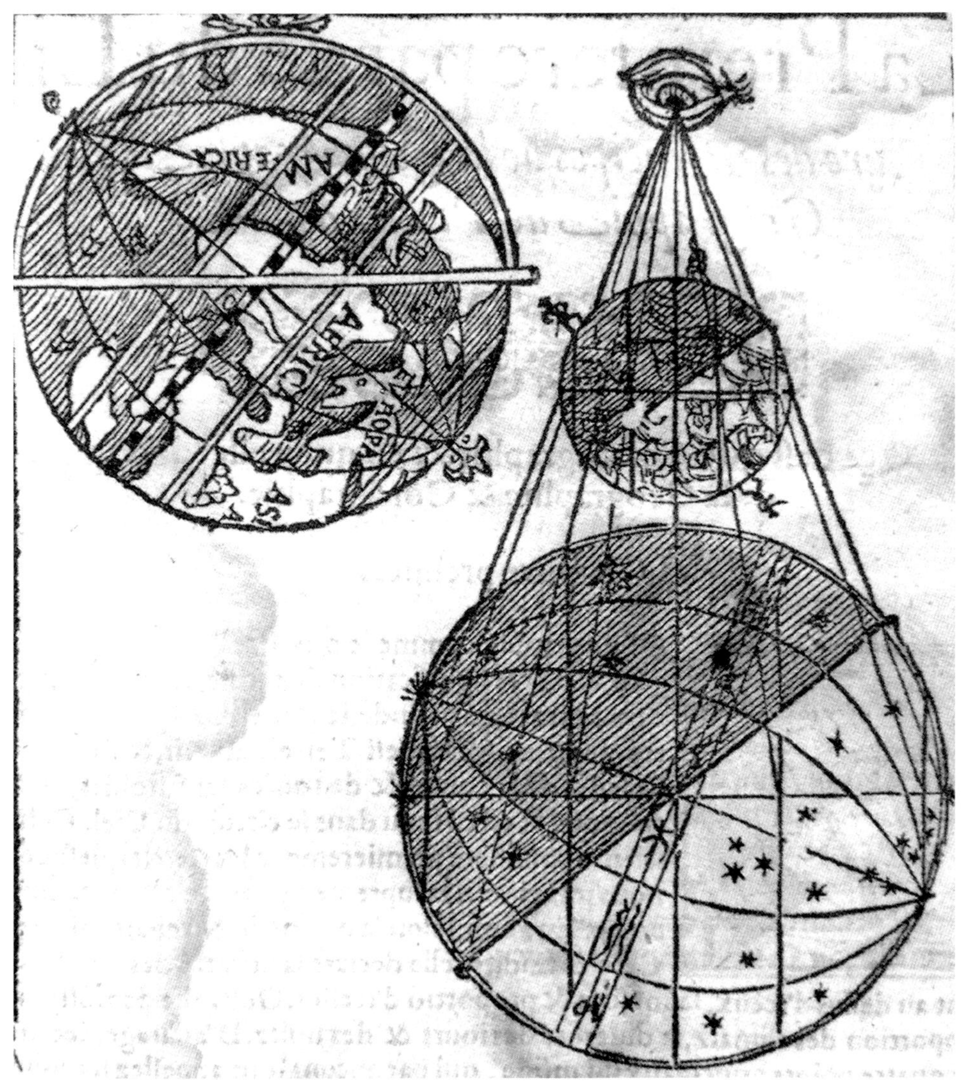

7.3: Detail from P. Apian, *Cosmographie*, ed. G. Frisius (Antwerp, 1581), p. 6.
By permission of the Bibliothèque Nationale de France.

7.2 (opposite): 'The author and Urania', in O. Fine, *Protomathesis*… (Paris, 1532).
Reproduced by permission of the National Library of Scotland.

7.4: Ptolemy measuring the heavens, from G. Reisch, *Margarita Philosophica* (n.p., 1504). Reproduced permission of the Bibliothèque municipale de Lille.

7.5 (opposite): Ptolemy (ed. G. Mercator), *Tabulae geographicae* (Cologne, 1578), title-page. The British Library, reproduced by permission.

TABVLAE GE
OGRAPHICAE
Cl: Ptolemęi ad
mentem autoris
restitutæ &
emendatę
Per Gerardum Mercatorem
Illustriss: Ducis Cliuię &c:
Cosmographum.
EVROPA
ASIA
APHRICA
Arabia
India
Oceanus
PTOLEMÆVS
MARINVS

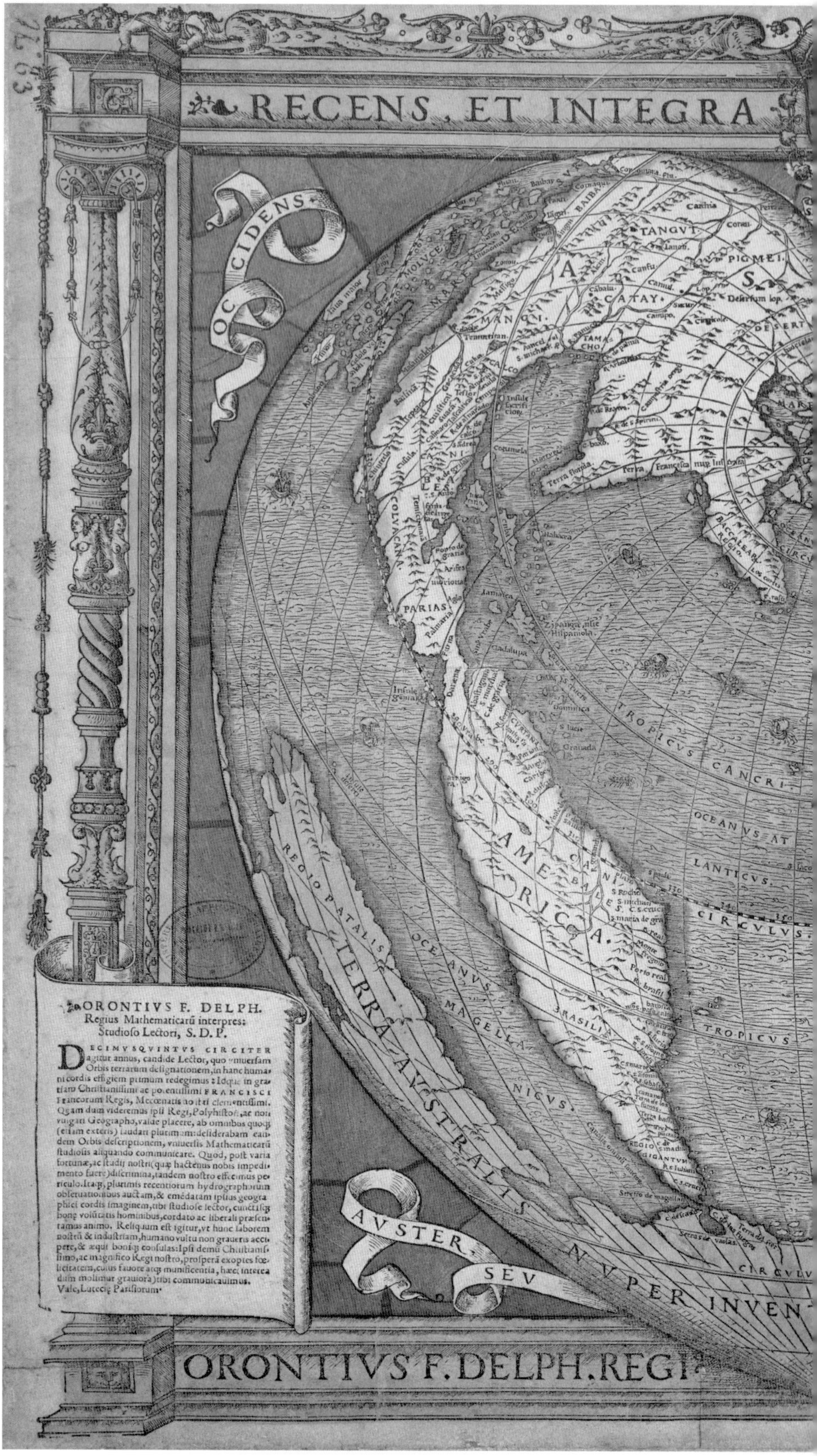

9.1: O. Fine, *Recens et integra orbis descriptio* (Paris, 1534).
By permission of the Bibliothèque Nationale de France.

ORBIS DESCRIPTIO
ORIENS
MERIDIES
ED NONDVM PLENE EXAMINATA
OCEANVS INDICVS
AEQVINOCTIALIS
SCYTHIA
TARTARIA MAGNA
INDIA
AFRICA
ANNOTATIO.
Ex hac plana terrarum orbis descriptione, duorum quorumcunq; locorum, datarum longitudinum atq; latitudinum, directũ itineris interuallum (modò illud nonaginta non superet gradus) prope verum supputare licebit. Numeratis itaq; eorundem locorum lõgitudinibus atq; latitudinibus, eisdẽve locis in Charta coassumptis: imposito vnũ circini pedẽ super altero locorũ, aliũ vero extendito in reliquũ. Dein traducito circinũ inuariatũ in eã rectã, quæ figurã bifariã diuidit, & in suos gradus distributa est: & animaduertito, quot gradus capiat ipse circinus. Hos enim si per 62 miliaria, aut gallicas leucas 31, seu 20 cõmunes, quidecimve maiores multiplicaueris: viatoriã eorundẽ locorũ distãtiã obtinebis.
REGALI PORRO CAVTVM est sanctione, ne quispiã hãc geographici cordis effigiem, hinc ad decennium, absq; manifesto opificis consensu imprimat, seu venditet, aut quouis modo distrahat, sub graui mulcta, concesso apud Lugdunũ diplomate luculenter expressa.
Parisiis,
MATHEMATIC? FACIEBAT.

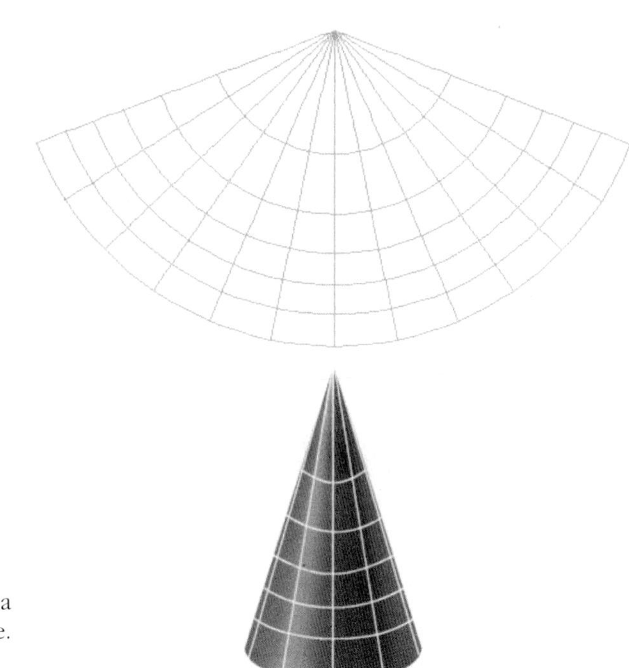

9.2: Unfolding of a circular cone.

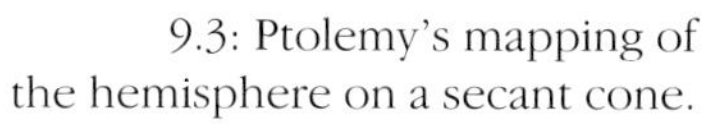

9.3: Ptolemy's mapping of the hemisphere on a secant cone.

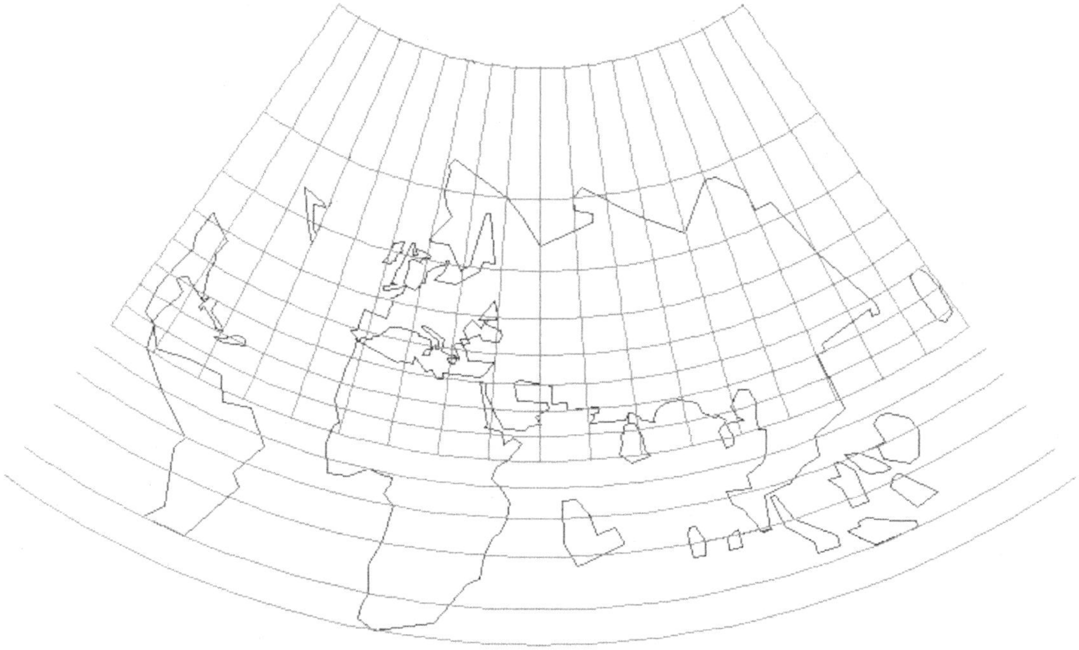

9.4: Waldseemüller's map using Ptolemy's first system.

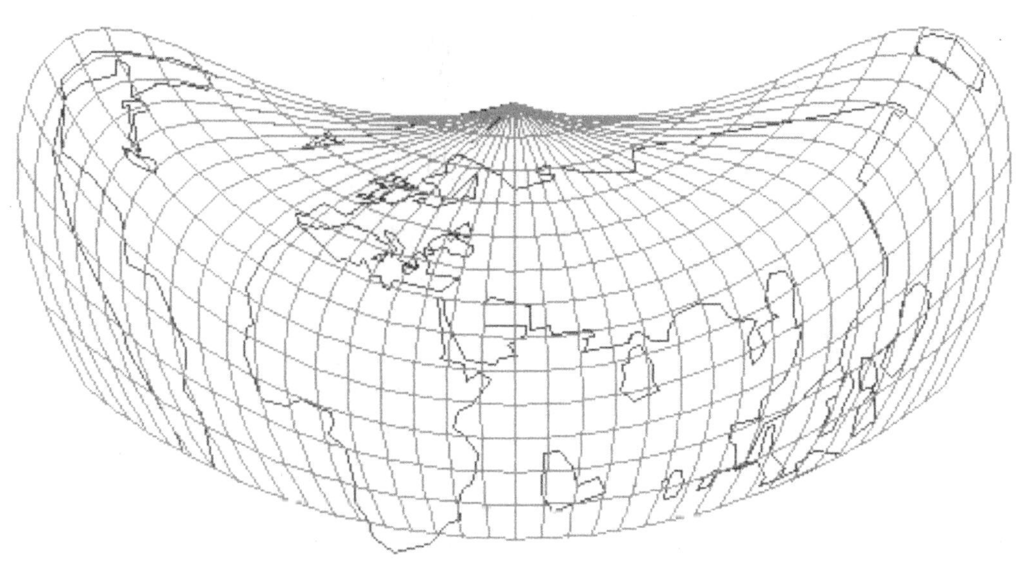

9.5: Waldseemüller's map using Ptolemy's second system (computed).

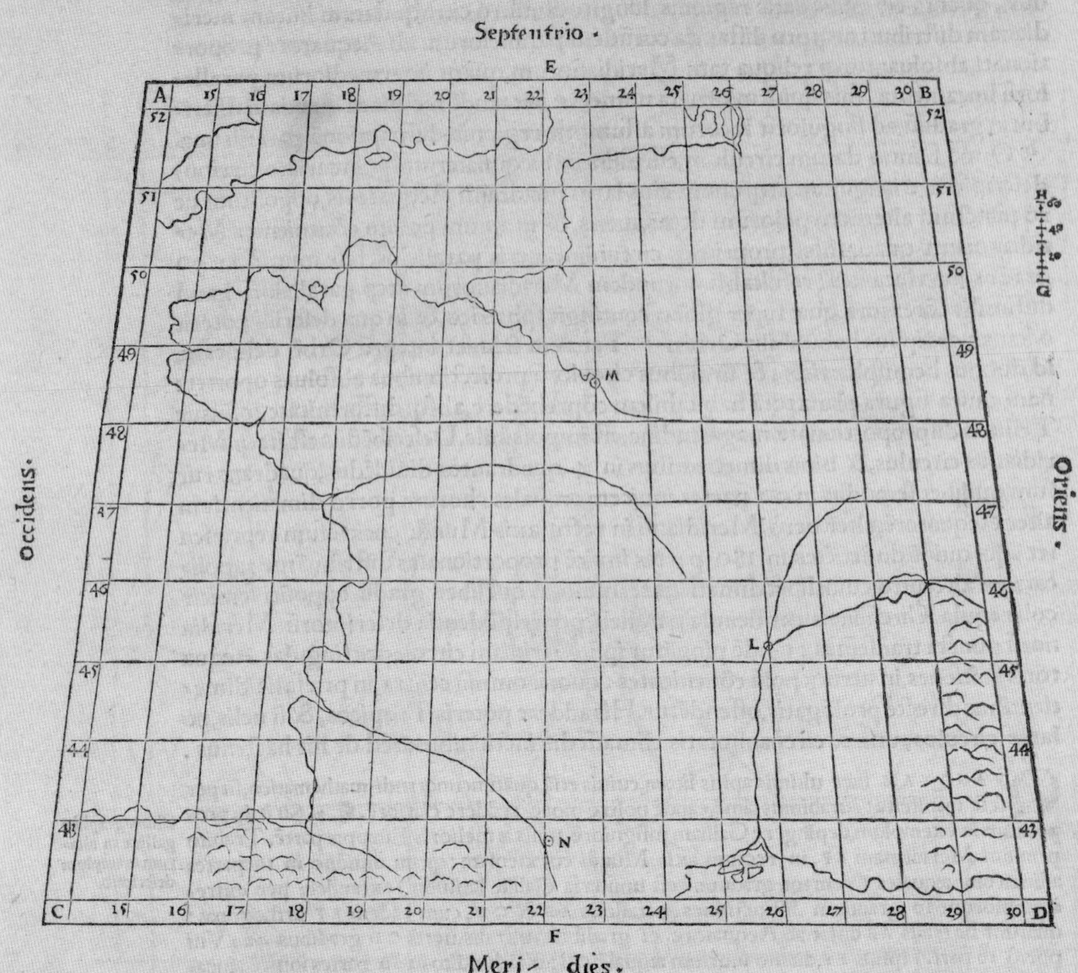

Chorographiæ ex curuis lineis cõtextæ figura tionis exẽplũ.

¶ b Apperiamus cõſequenter, qualiter faciẽda ſit Meridianorũ atq; parallelorũ cõtextura, quæ b ſimilis exiſtat octauæ parti ſphæricæ cõuexitatis. Sit igitur circulus liberæ quãtitatis A B C : cu ius circũferentia in tres partes æquas diuidatur, in ipſis quidẽ ſignis A, B, C . Impoſito deinde cir cini pede in ſigno A, extende reliquum in B, uel in C : & ducito arcũ B C. Rurſum inuariato cir circino

9.6: Fine's sketch for the use of Ptolemy's mesh for regional maps, from *Protomathesis* (Paris, 1532). Reproduced by permission of the National Library of Scotland.

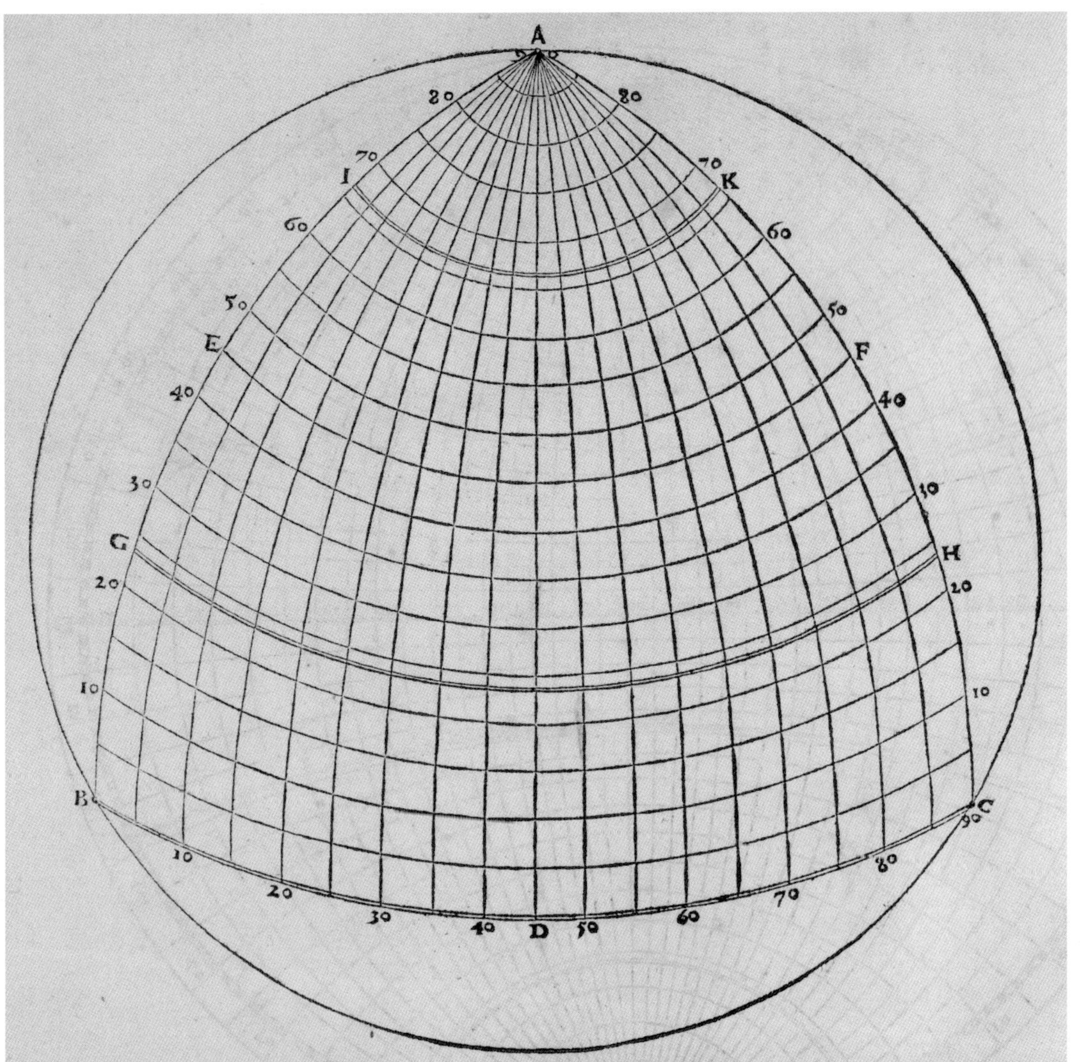

9.7: Fine's octant mesh for mapping an eighth of the globe, from *Protomathesis* (Paris, 1532). Reproduced by permission of the National Library of Scotland.

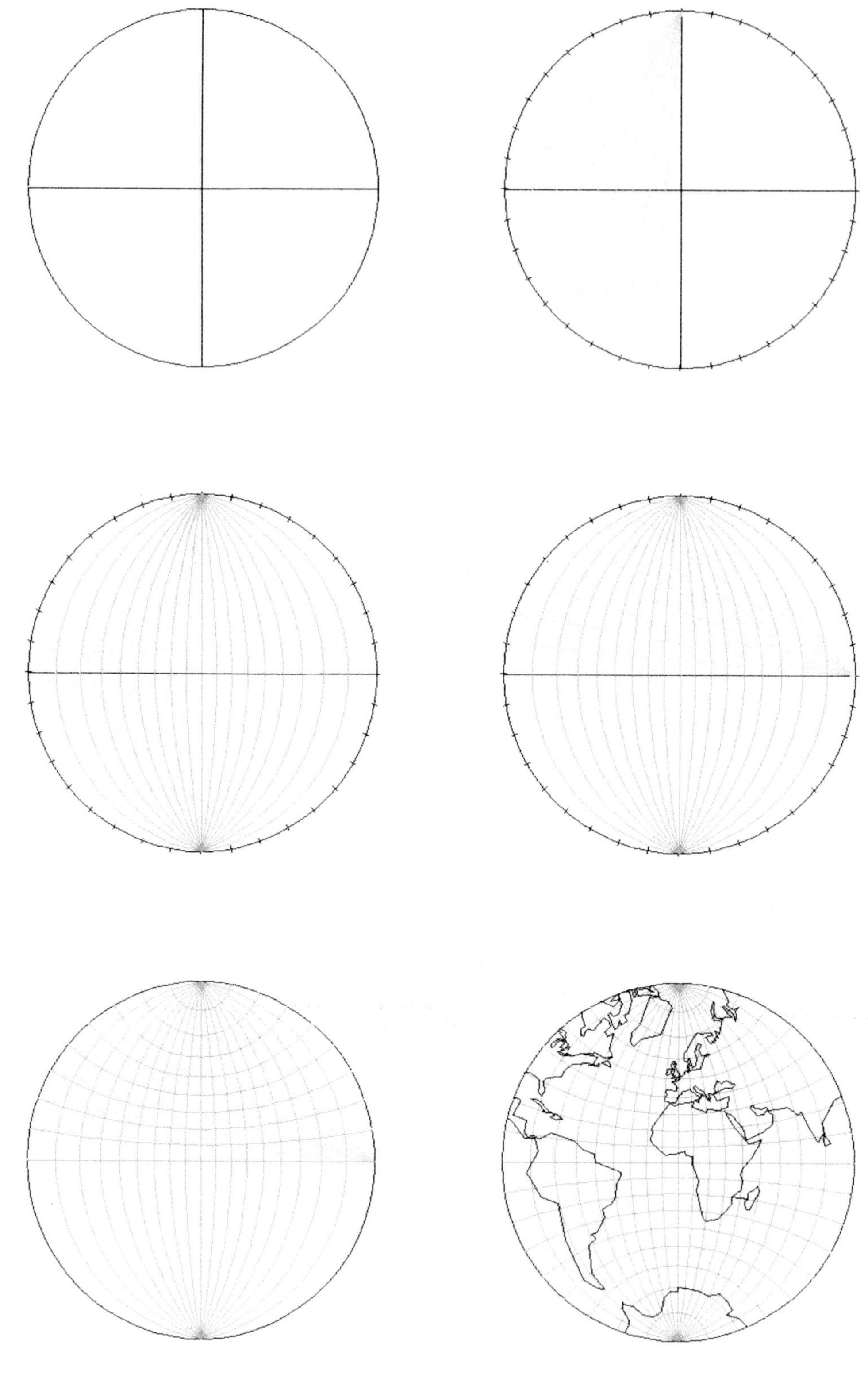

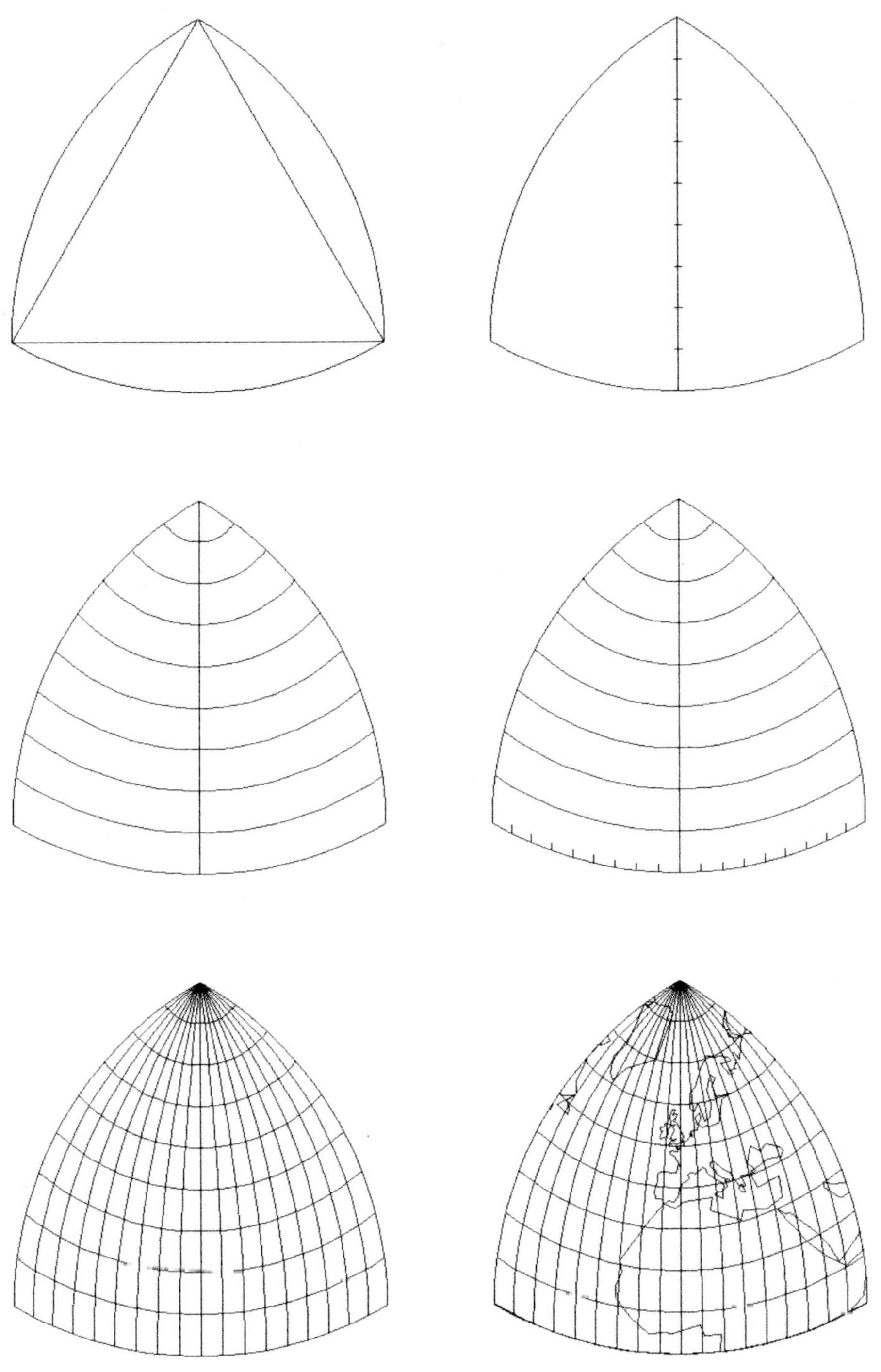

9.9: Fine's method for producing an octant map.

9.8 (opposite): Oronce Fine's method of equatorial stereography.

9.10: Mapping points on a sphere.

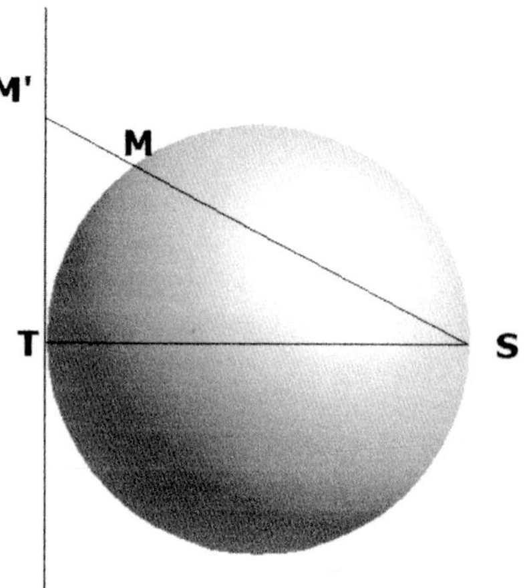

9.11 (below): Equatorial stereography: a form of mapping which keeps meridians converging to the poles.

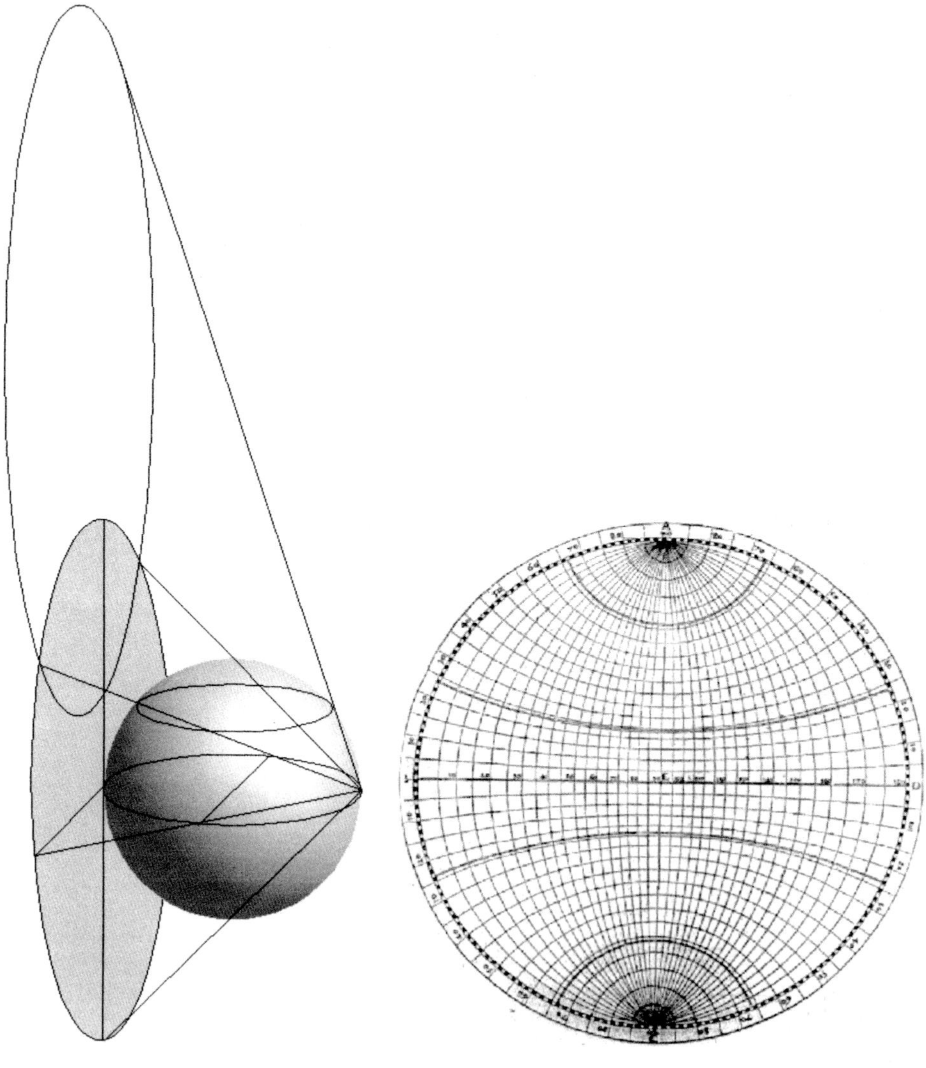

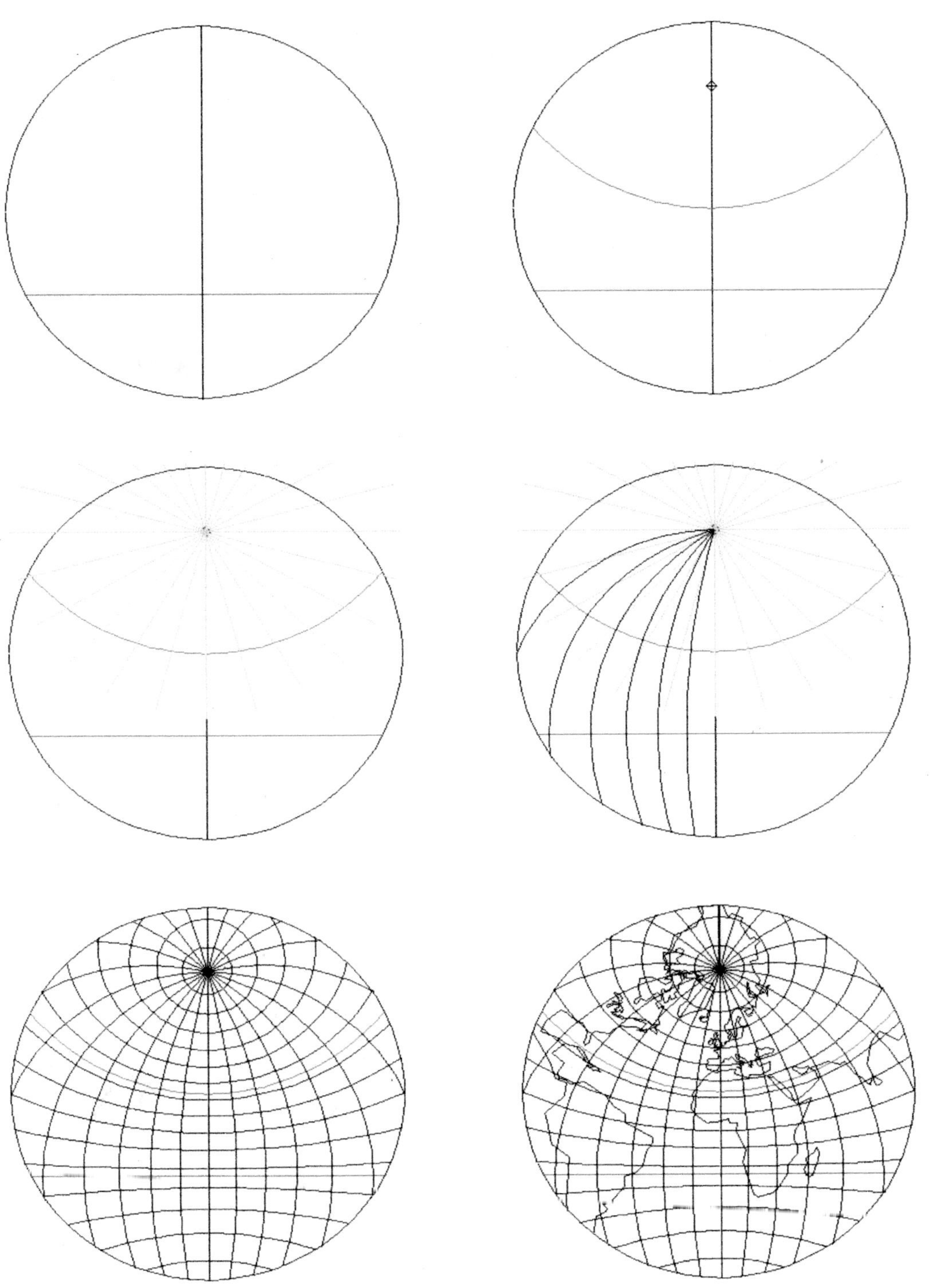

9.12: Oronce Fine's method of oblique stereography.

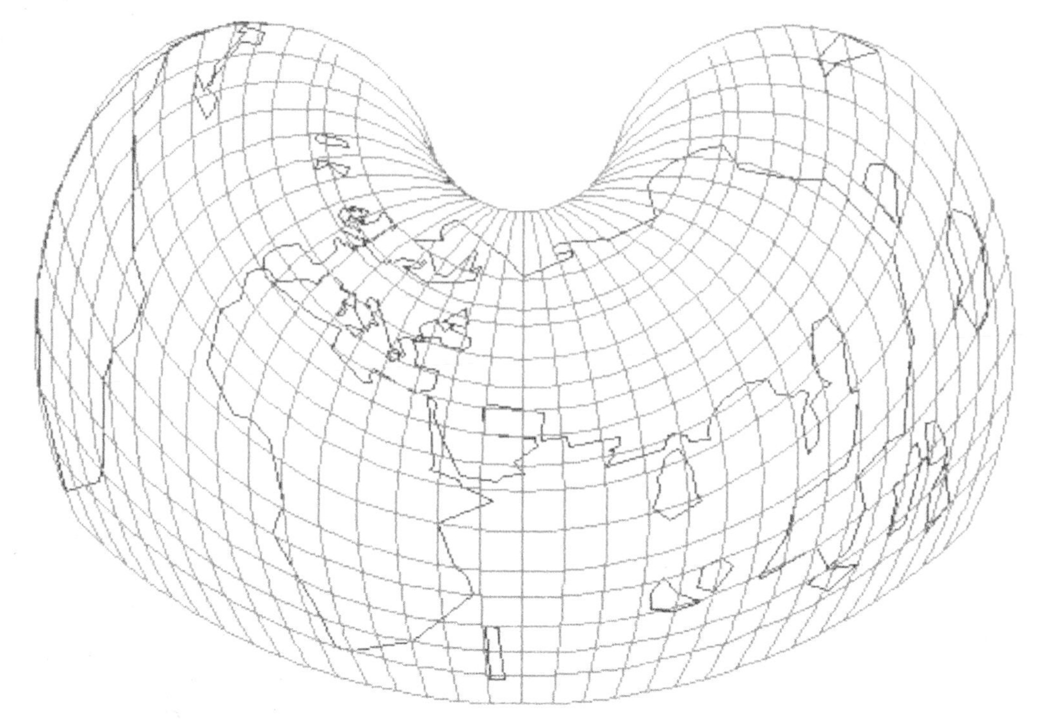

9.13: Sylvanus's 1511 heart-shaped map (reconstructed).

9.14 (opposite): From the octant map to the double heart-shape in five steps.

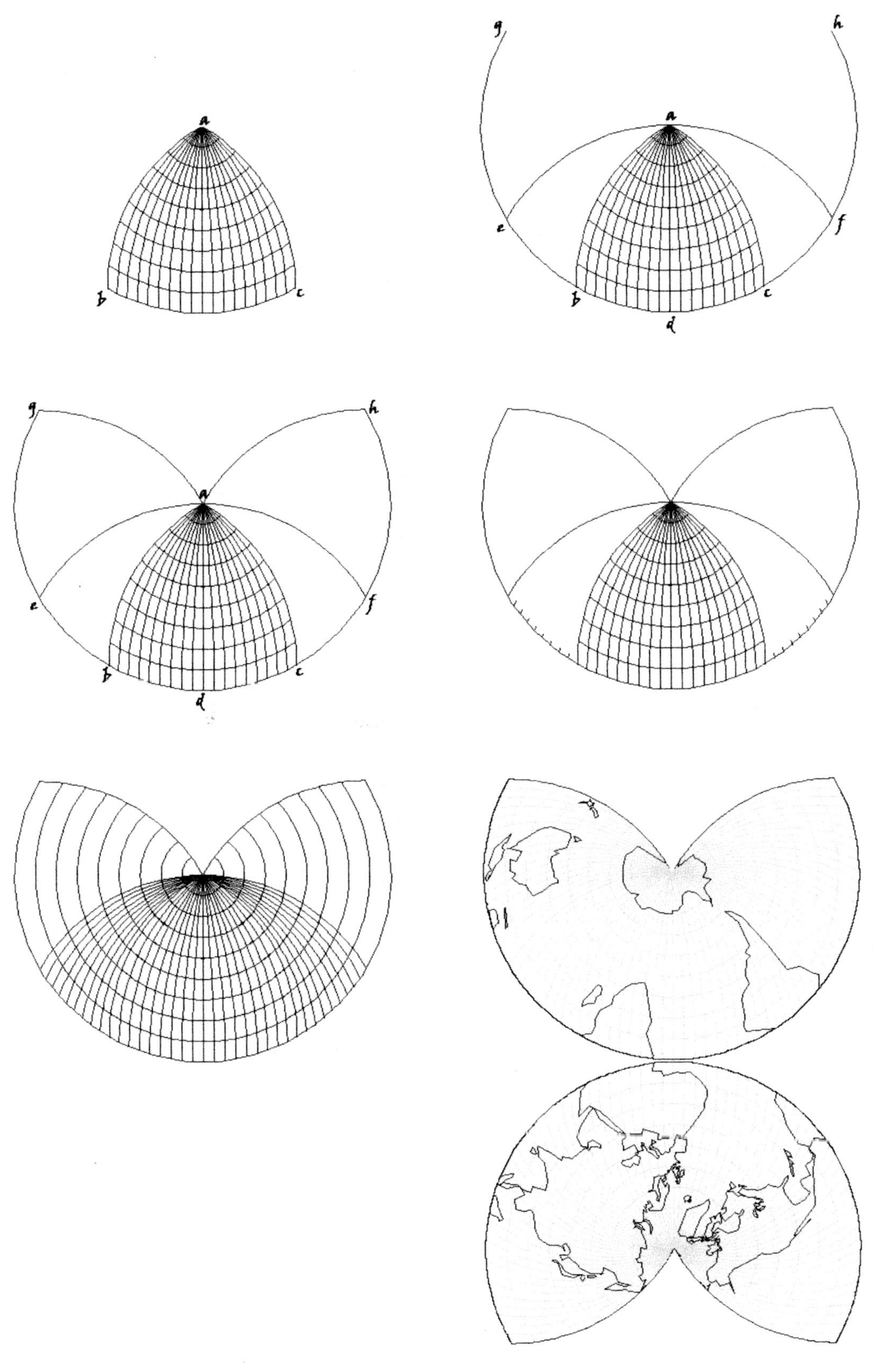
a
b
c
g
h
a
e
f
b
c
d
g
h
a
e
f
b
c
d

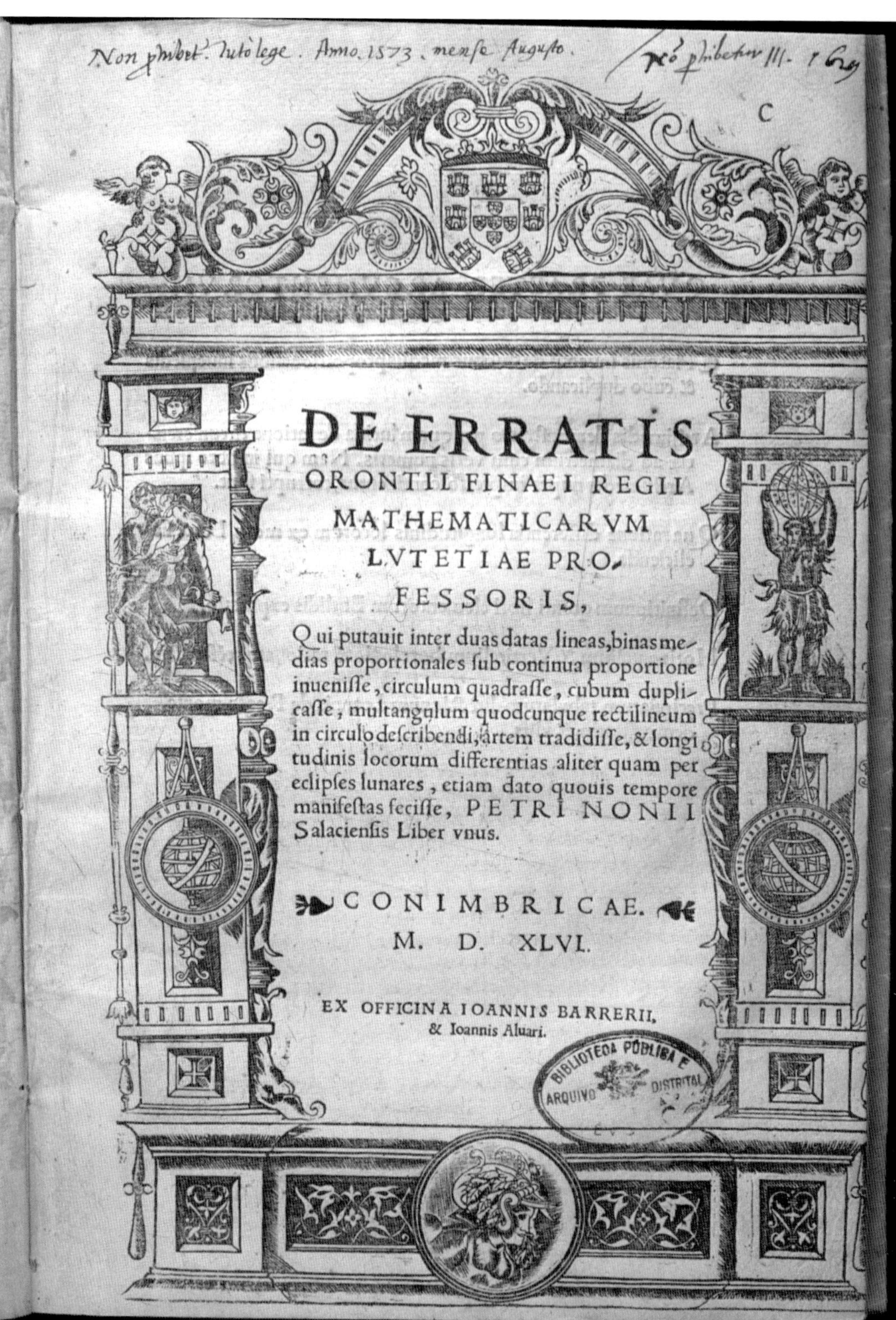

C

DE ERRATIS ORONTII FINAEI REGII MATHEMATICARVM LVTETIAE PROFESSORIS,

Qui putauit inter duas datas lineas, binas medias proportionales sub continua proportione inuenisse, circulum quadrasse, cubum duplicasse, multangulum quodcunque rectilineum in circulo describendi, artem tradidisse, & longitudinis locorum differentias aliter quam per eclipses lunares, etiam dato quouis tempore manifestas fecisse, PETRI NONII Salaciensis Liber vnus.

CONIMBRICAE.
M. D. XLVI.

EX OFFICINA IOANNIS BARRERII, & Ioannis Aluari.

10.1: P. Nunes, *De erratis Orontii Finaei* (Coimbra, 1546), title-page.
By permission of the University of Lisbon.

DE ERRATIS ORONTII FINAEI, REGII MATHEMATICARVM LVTETIAE PROFESSORIS.

Qui putauit inter duas datas lineas, binas medias proportionales ſub continua proportione inueniſſe, circulum quadraſſe, cubum duplicaſſe, multangulum quodcunque rectilineum in circulo deſcribendi, artem tradidiſſe, & longitudinis locorum differentias aliter quàm per eclipſes lunares, etiã dato quouis tempore manifeſtas feciſſe.

PETRI NONII Salacienſis Liber vnus.

SECVNDA EDITIO.

CONIMBRICAE
Excudebat Antonius à Mariis.
Anno. 1571.

10.2: P. Nunes, *De erratis Orontii Finaei* (Coimbra, 1571), title-page. By permission of the University of Lisbon.

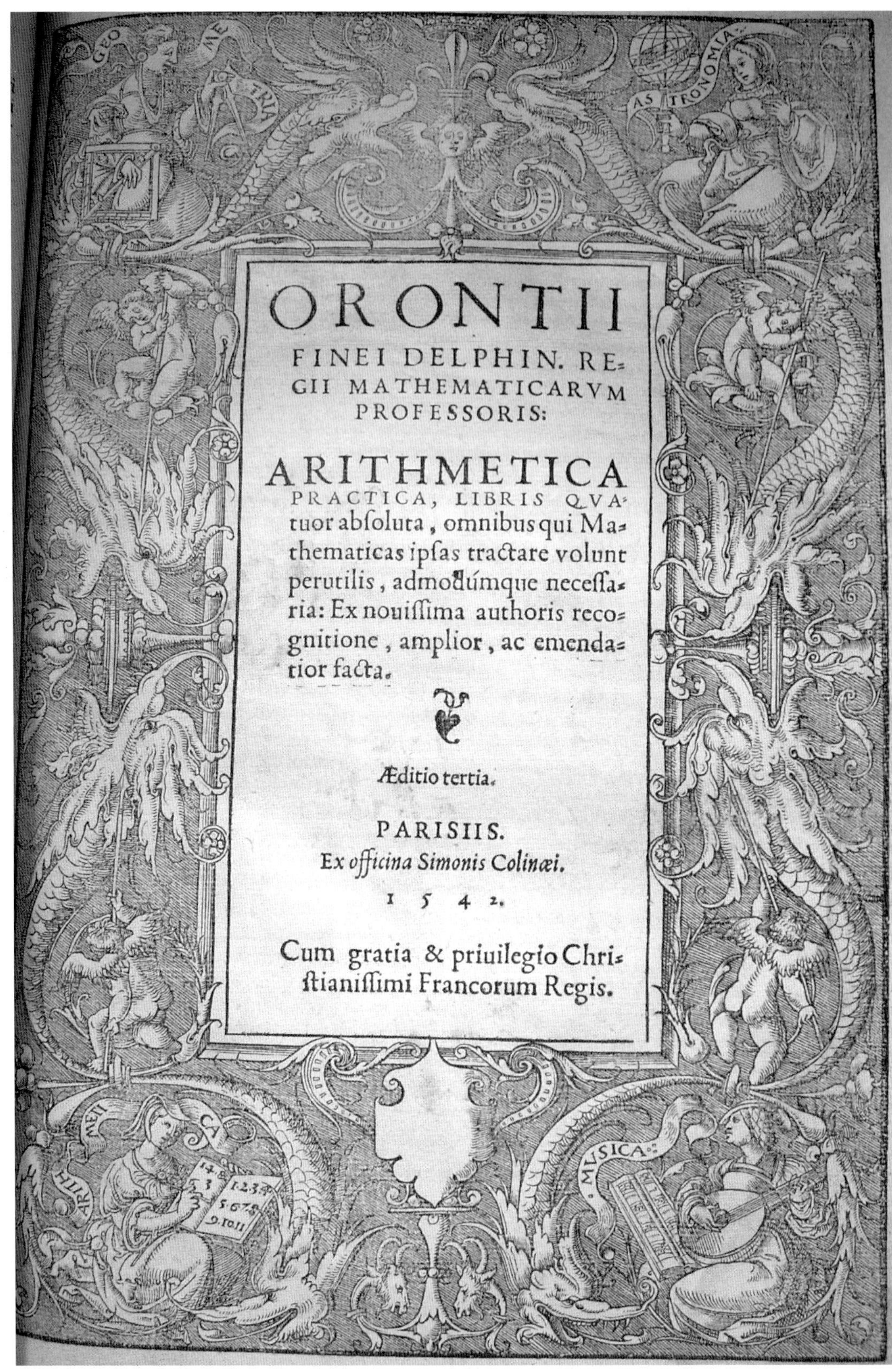

ORONTII
FINEI DELPHIN. RE-
GII MATHEMATICARVM
PROFESSORIS:

ARITHMETICA
PRACTICA, LIBRIS QVA-
tuor absoluta, omnibus qui Mathematicas ipsas tractare volunt
perutilis, admodúmque necessaria: Ex nouissima authoris reco-
gnitione, amplior, ac emenda-
tior facta.

Æditio tertia.

PARISIIS.
Ex officina Simonis Colinæi.
1 5 4 2.

Cum gratia & priuilegio Christianissimi Francorum Regis.

11.1: O. Fine, *Arithmetica practica...* (Paris, 1544), title-page.
Reproduced courtesy of St Andrews University Library.

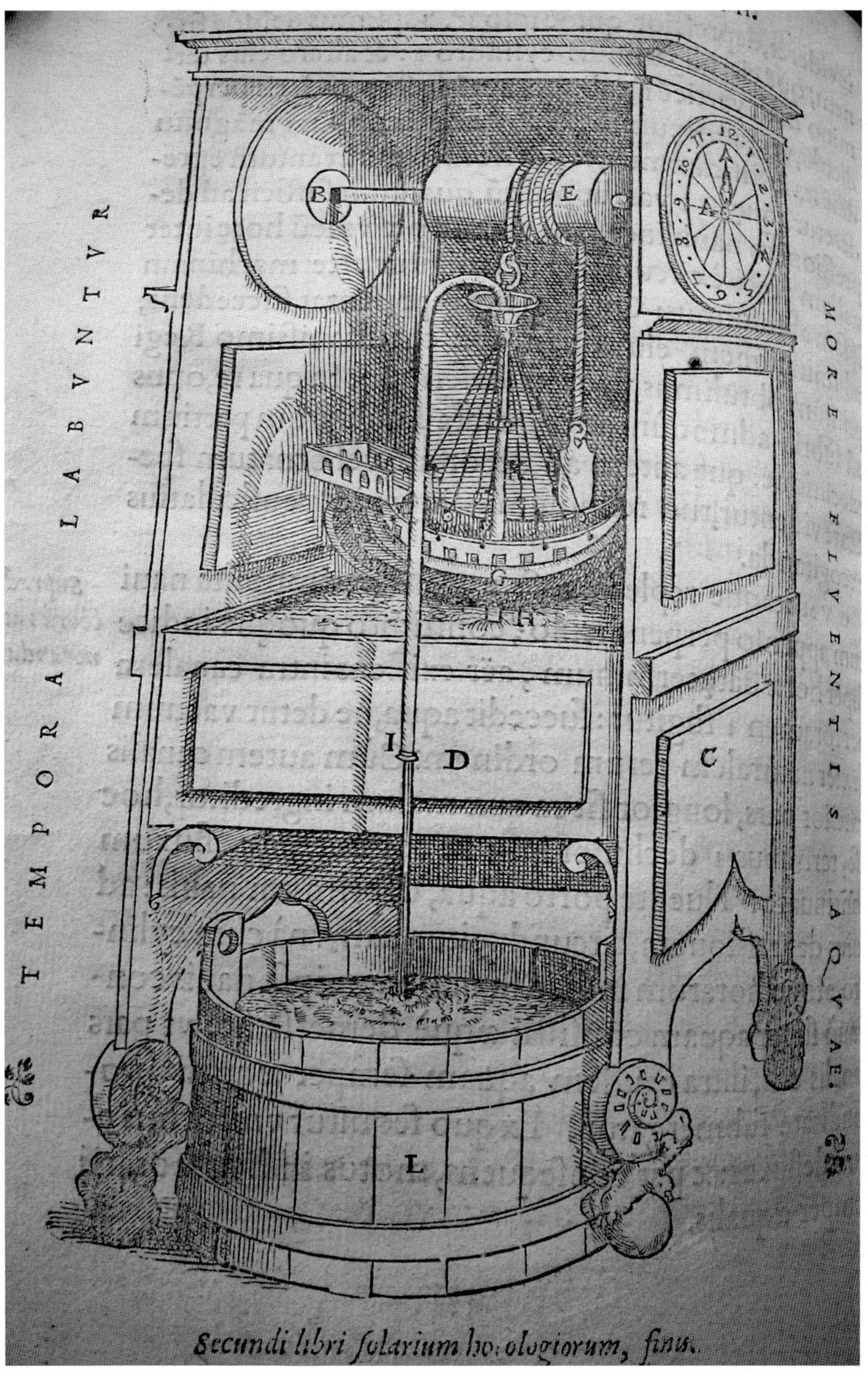

12.1: Oronce Fine's water-clock, from O. Fine, *De solaribus horologiis*… (Paris, 1560). Reproduced courtesy of St Andrews University Library.

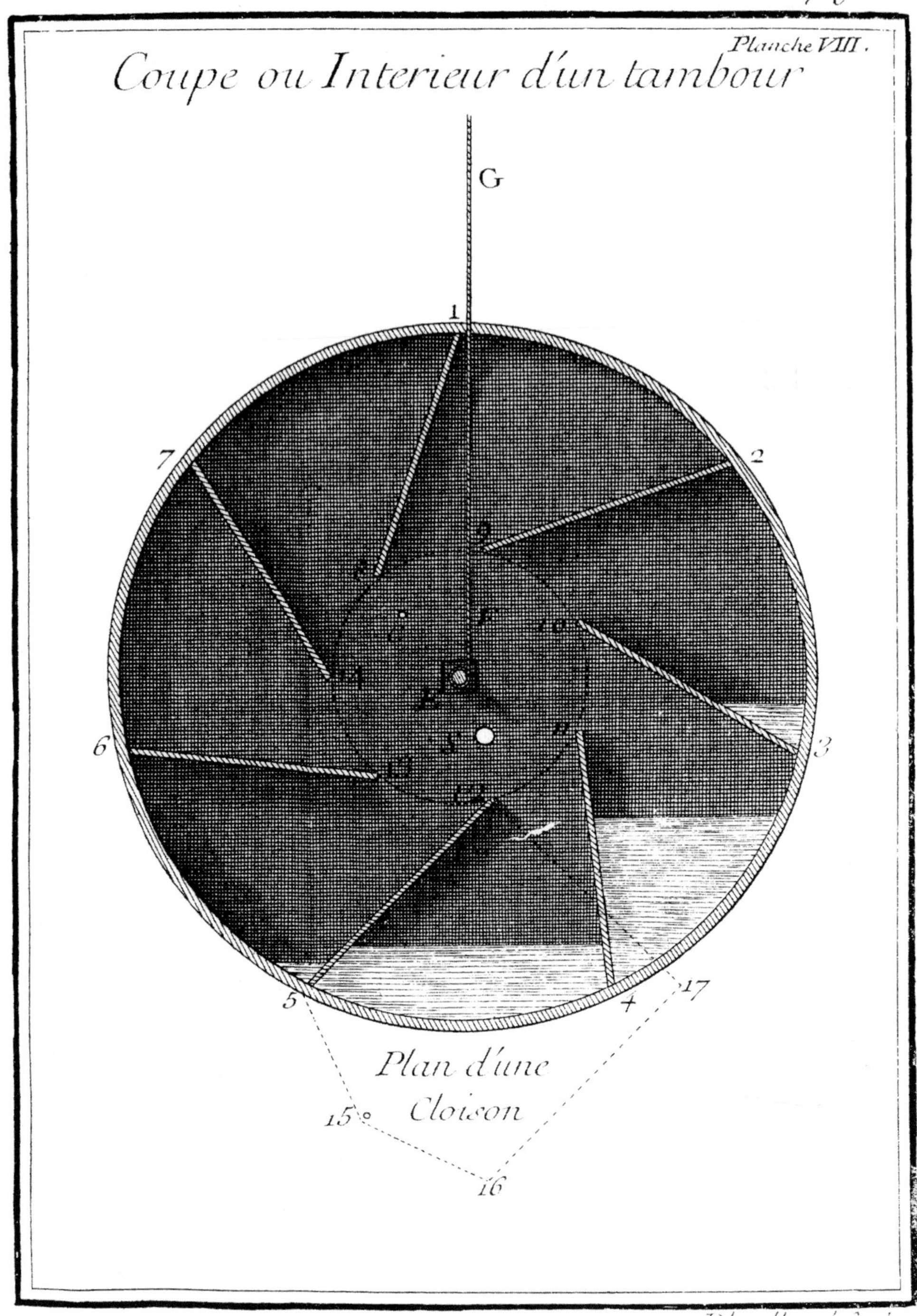

12.2 a and b: Section and arrangement of the drum for a compartmented cylindrical clepsydra from 'Traité des horloges Elementaires, ou de la manière de faire des Horloges avec l'Eau, la Terre, l'Air & Feu', translation of D. Martinelli, *Horologi elementari divisi in quattro parti...* (Venice, 1669) in J. Ozanam, *Récreations mathématiques et physiques* (Paris, 1694). Reproduced courtesy of the editor.

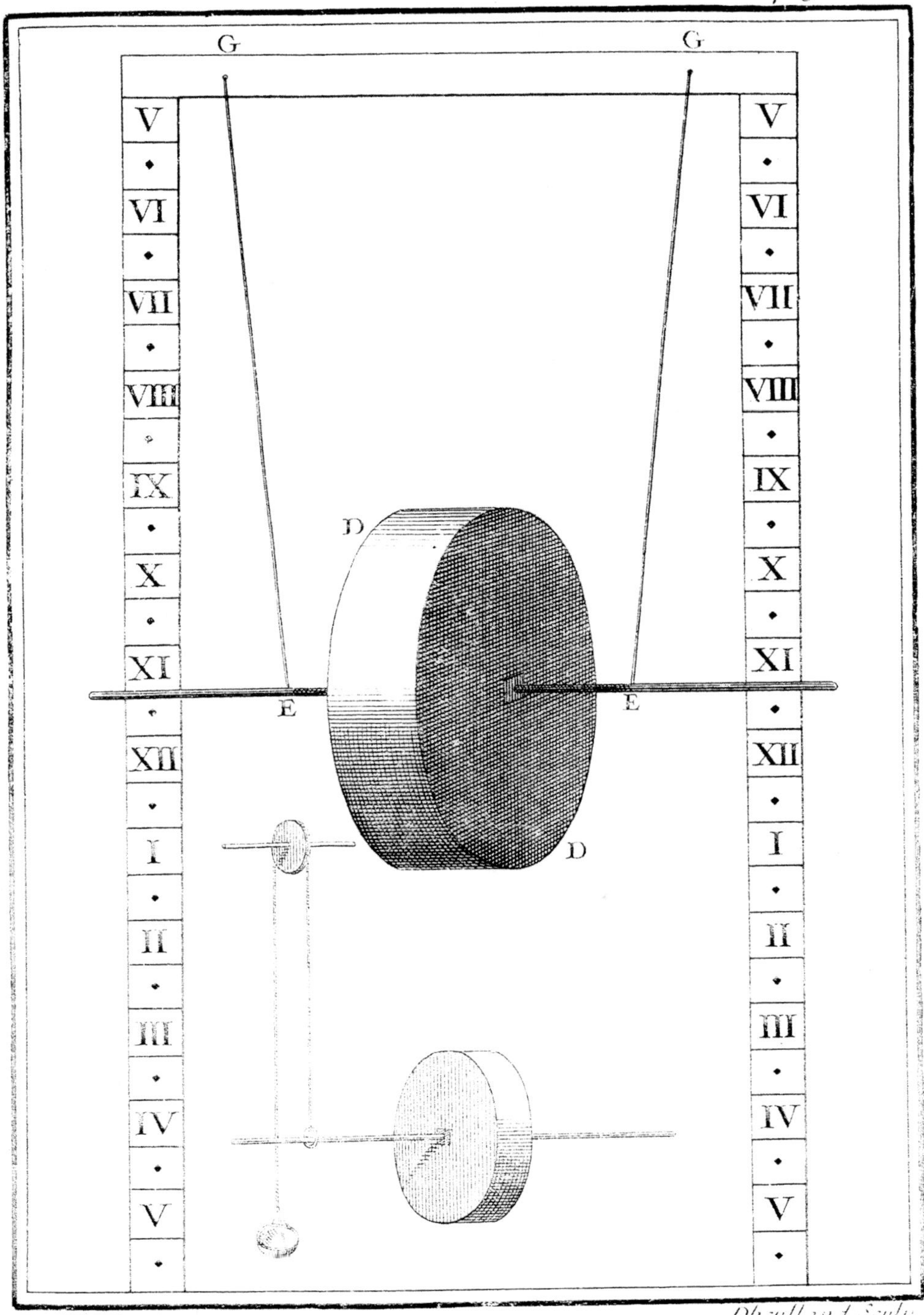

Dheulland Sculp.

12.3: Water-clock by William Winde (1582), from *Gulielmi Windaei Bergannsis ad Zoman Cursus Mathematicus Tomus Secundus*. The British Library, Kings MS 267.

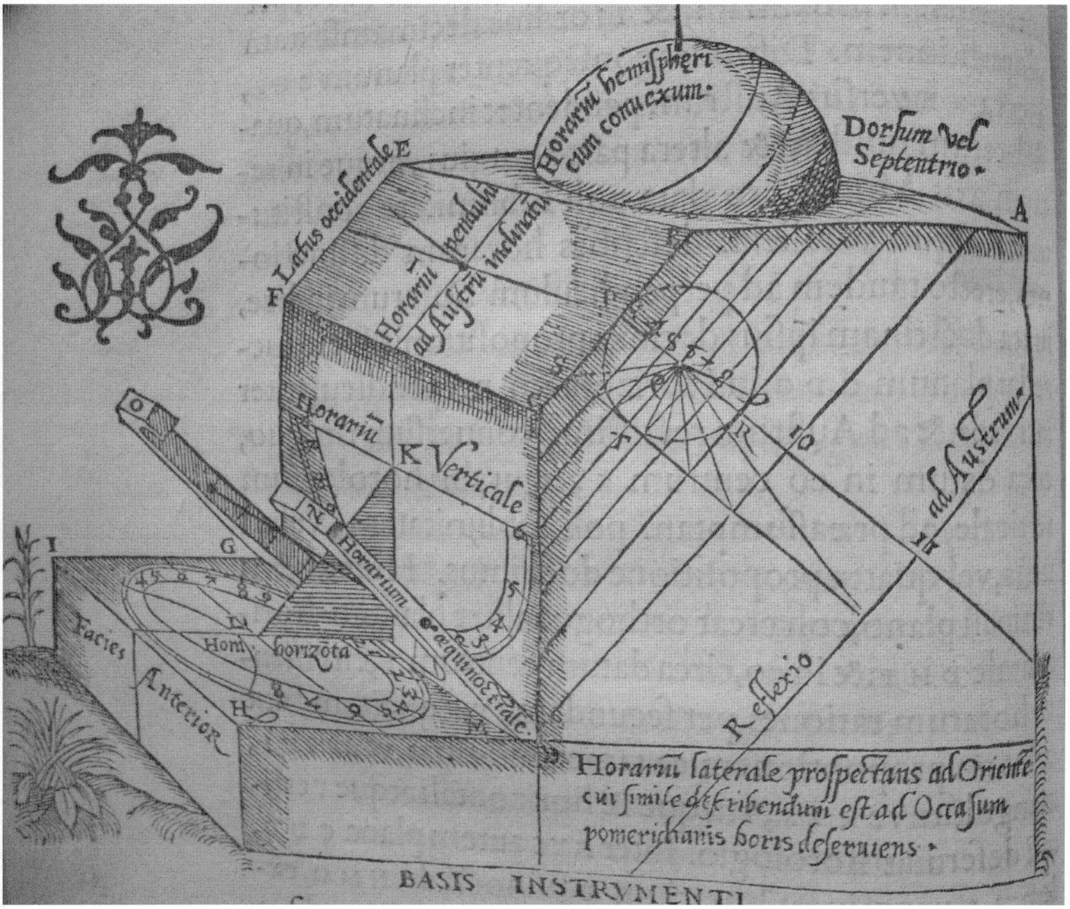

13.1 (above): Polyhedral dial, from O. Fine, *De solaribus horologiis…* (Paris, 1560). Reproduced courtesy of St Andrews University Library.

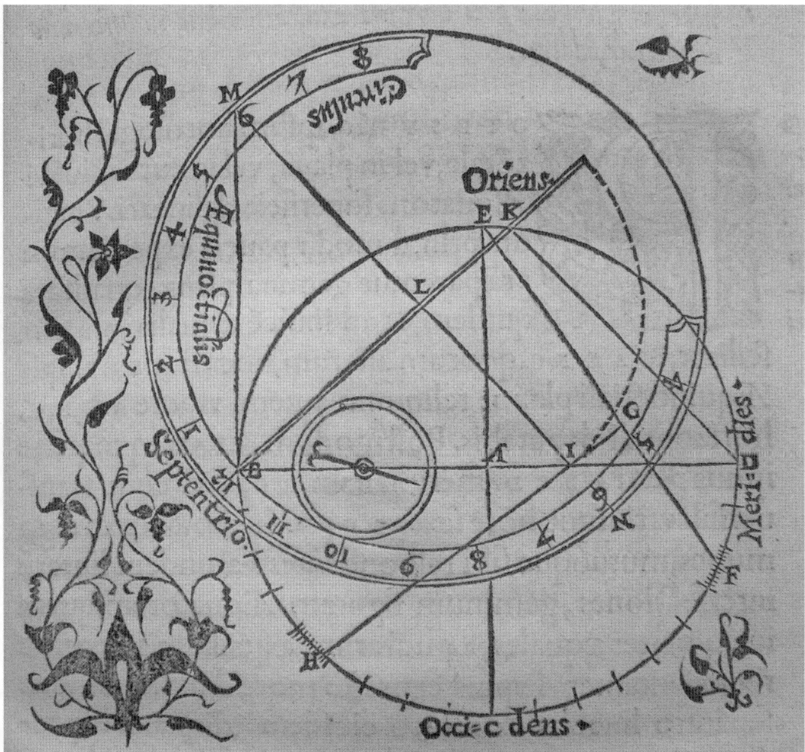

13.2 (left): Woodcut diagram with decorative flourish, from O. Fine, *De solaribus horologiis…* (Paris, 1560). Reproduced courtesy of St Andrews University Library.

De

OMNES
graphia, G
dierum &
ræ ſitu con
partem, fru
um erit ita
& horarur
rere diſcrir
Sphæræ ui
ctè declara
alius artificialis dicitur. Naturalem ſol
trum corporis ſolaris, ad regulatum V

Woodblock decorations from O. Fine, *Protomathesis...* (Paris, 1532).
Reproduced courtesy of the National Library of Scotland.

LIBRI SECVNDI ET VLTI-
MI, TOTIVSQVE GEO-
METRIAE PRA-